Sustainable textiles

The Textile Institute and Woodhead Publishing

The Textile Institute is a unique organisation in textiles, clothing and footwear. Incorporated in England by a Royal Charter granted in 1925, the Institute has individual and corporate members in over 90 countries. The aim of the Institute is to facilitate learning, recognise achievement, reward excellence and disseminate information within the global textiles, clothing and footwear industries.

Historically, The Textile Institute has published books of interest to its members and the textile industry. To maintain this policy, the Institute has entered into partnership with Woodhead Publishing Limited to ensure that Institute members and the textile industry continue to have access to high calibre titles on textile science and technology.

Most Woodhead titles on textiles are now published in collaboration with The Textile Institute. Through this arrangement, the Institute provides an Editorial Board which advises Woodhead on appropriate titles for future publication and suggests possible editors and authors for these books. Each book published under this arrangement carries the Institute's logo.

Woodhead books published in collaboration with The Textile Institute are offered to Textile Institute members at a substantial discount. These books, together with those published by The Textile Institute that are still in print, are offered on the Woodhead web site at: www.woodheadpublishing.com. Textile Institute books still in print are also available directly from the Institute's website at: www.textileinstitutebooks.com.

A list of Woodhead books on textile science and technology, most of which have been published in collaboration with The Textile Institute, can be found towards the end of the contents pages.

Woodhead Publishing in Textiles: Number 98

Sustainable textiles

Life cycle and environmental impact

Edited by
R. S. Blackburn

The Textile Institute

CRC Press
Boca Raton Boston New York Washington, DC

WOODHEAD PUBLISHING LIMITED
Oxford   Cambridge   New Delhi

Published by Woodhead Publishing Limited in association with The Textile Institute
Woodhead Publishing Limited, Abington Hall, Granta Park, Great Abington
Cambridge CB21 6AH, UK
www.woodheadpublishing.com

Woodhead Publishing India Private Limited, G-2, Vardaan House, 7/28 Ansari
Road, Daryaganj, New Delhi – 110002, India
www.woodheadpublishingindia.com

Published in North America by CRC Press LLC, 6000 Broken Sound Parkway,
NW, Suite 300, Boca Raton, FL 33487, USA

First published 2009, Woodhead Publishing Limited and CRC Press LLC
© Woodhead Publishing Limited, 2009
The authors have asserted their moral rights.

British Library Cataloguing in Publication Data
A catalogue record for this book is available from the British Library.

Library of Congress Cataloging in Publication Data
A catalog record for this book is available from the Library of Congress.

Woodhead Publishing ISBN 978-1-84569-453-1 (book)
Woodhead Publishing ISBN 978-1-84569-694-8 (e-book)
CRC Press ISBN 978-1-4398-0128-4
CRC Press order number N10023

The publishers' policy is to use permanent paper from mills that operate a
sustainable forestry policy, and which has been manufactured from pulp which is
processed using acid-free and elemental chlorine-free practices. Furthermore, the
publishers ensure that the text paper and cover board used have met acceptable
environmental accreditation standards.

Typeset by SNP Best-set Typesetter Ltd., Hong Kong
Printed by TJ International Limited, Padstow, Cornwall, UK

Contents

Contributor contact details

(* = main contact)

Editor

Dr R. S. Blackburn
Green Chemistry Group
Centre for Technical Textiles
University of Leeds
Leeds L52 9JT
UK

E-mail: r.s.blackburn@leeds.ac.uk

Chapter 1

A. Sherburne, MA, FRSA
Kingston University
Kingston upon Thames
Surrey KT1 1LQ
UK

E-mail: sherburne@btopenworld.
  com

Chapter 2

L. Grose
Associate Professor
California College of the Arts San
  Francisco
190 Sunset Way
Muir Beach
CA 94965
USA

E-mail: lyndagr@aol.com

Chapter 3

I. M. Russell
CSIRO Division of Materials
  Science and Engineering
PO Box 21
Belmont
Victoria 3216
Australia

E-mail: ian.russell@csiro.au

Chapter 4

Dr Ivan Chodák
Polymer Institute of the Slovak
  Academy of Sciences
CEDEBIPO
872 36 Bratislava
Slovakia

E-mail: upolchiv@savba.sk

Dr R. S. Blackburn*
Green Chemistry Group
Centre for Technical Textiles
University of Leeds
Leeds LS2 9JT
UK

E-mail: r.s.blackburn@leeds.ac.uk

Chapter 5

P. H. Nielsen,* H. Kuilderd,
  W. Zhou and X. Lu
Novozymes A/S
Sustainability Development
Krogshøjvej 36
DK-2880 Bagsværd
Denmark

E-mail: phgn@novozymes.com

Chapter 6

Dr John R. Easton
Ecology Solutions Manager
DyStar Textile Services
DyStar UK Ltd
Akcros Site
Lankro Way
Eccles
Manchester M30 0BH
UK

E-mail: Easton.John@DyStar.com

Chapter 7

Dr T. Stegmaier, M. Linke,*
  A. Dinkelmann, Dr V. von
  Arnim and Prof. Dr H. Planck
Institute for Textile Technology
  and Process Engineering (ITV)
Denkendorf
Koerschtalstr. 26
D-73770 Denkendorf
Germany

E-mail: Thomas.Stegmaier@
  itv-denkendorf.de

Chapter 8

Dr Jana M. Hawley
Professor and Department Head
Department of Apparel, Textiles
  and Interior Design
Kansas State University
225 Justin Hall
Manhattan
KS 66506-1405
USA

E-mail: hawleyj@ksu.edu

Chapter 9

Professor Margaret Rucker
231 Everson Hall
Department of Textiles and
  Clothing
University of California
Davis
CA 95616
USA

E-mail: mhrucker@ucdavis.edu

Chapter 10

Sam B. Moore*
Hohenstein Institute America, Inc.
4003 Birkdale Court
Elon
NC 27244
USA

E-mail: s.moore@hohenstein.de

Dr Manfred Wentz
Oeko-Tex Certification Body
9016 Oak Branch Drive
Apex
NC 27539
USA

E-mail: mcwentz@gmail.com

Chapter 11

Dr P. J. Wakelyn*
National Cotton Council of
  America (retired)
1521 New Hampshire Avenue
Washington, DC 20036
USA

E-mail: pwakelyn@cotton.org

Dr M. R. Chaudhry
International Cotton Advisory
  Committee
1629 K St, NW
Suite 702
Washington, DC 20006
USA

Chapter 12

Professor S. Black
Director of Centre for Fashion
  Science
London College of Fashion
University of the Arts London
20 John Princes Street
London W1G 0BJ
UK

E-mail: s.black@fashion.arts.ac.uk

Chapter 13

B. R. George,* B. A. Haines and
  E. Murphy
School of Engineering & Textiles
Philadelphia University
4201 Henry Avenue
Philadelphia
PA 19144
USA

E-mail: georgeb@PhilaU.edu

Chapter 14

Dr S. Nazaré
Centre for Materials Research and
  Innovation
University of Bolton
Bolton BL3 5AB
UK

E-mail: S.Nazare@bolton.ac.uk

Chapter 15

Dr K. Fletcher
Reader in Sustainable Fashion
London College of Fashion
20 John Princes Street
Oxford Circus
London W1G 0BJ
UK

E-mail: kate@katefletcher.com

Woodhead Publishing in Textiles

Preface

Developing economies together with the increasing global population are having a direct impact upon production and consumption of both fossil and natural resources. This has the potential to lead to further significant environmental degradation if appropriate measures are not put in place to control, manage or mitigate the impact of such developments. There is a need wherever possible to dissociate growth from increased consumption of fossil fuels, water, energy, and reduce the use and resulting impact upon natural resources. Output to the environment from production and waste must also be reduced in volume and impact.

'Sustainability' has become a ubiquitous word, but what level of understanding exists in terms of what constitutes a sustainable product or process? Most people would associate sustainability with environment, but many would not consider that sustainability also embraces economy: the comment that products 'should not cost the earth' genuinely has a double meaning – both meanings being equally important. The third aspect of sustainability is impact on society: sustainable textiles should not cause detriment to those societies where products are grown or disposed of, and workers in textiles supply chains should not be exposed to hazardous working conditions.

Misunderstanding of what is 'good' and 'bad' prevails, not just from the consumer, but within the textiles industry. Terms like 'organic', 'natural' and 'biodegradable' are generally perceived to be 'good', whereas 'GM', 'synthetic' and 'chemical' are perceived to be 'bad'; an interesting perspective considering that water is a chemical (dihydrogenoxide) and arsenic is natural. Many of these terms are used cleverly to market claimed sustainability benefits, but a product can only be sustainable when the whole 'cradle-to-grave' life cycle is sustainable.

A previous book in this series entitled *Biodegradable and sustainable fibres* concentrated largely on fibres that were sustainable in terms of the raw materials from which they were produced and also their potential for disposal through degradation processes. However, this perspective examines only the 'cradle' and 'grave' of textiles and does not look at everything else that goes on in the textiles supply chain, one of the longest and most complicated supply chains of any industrial product. While it is positive to utilise renewable and/or biodegradable (the two terms are mutually exclusive) resources as the starting materials for textile products, arguably the most significant contributions to product life cycle sustainability are energy and water savings in processing, as these savings usually realise an economic benefit also.

In this book, sustainability through the textiles supply chain will be examined. The first chapter considers the designer's perspective in achieving sustainable textiles, how the designer must take into account the technologies available and how to achieve desirable fashion at the same time as reducing the use of resources and the generation of waste. Two subsequent chapters consider two natural fibres in existence for millennia – cotton and wool – and how the processing of these fibres can be sustainable into the future, before the fourth chapter discusses new sustainable synthetic fibres. Sustainability in the wet processing stages of preparation, dyeing and finishing are considered in three separate chapters on enzyme biotechnology, key sustainability issues in dyeing, and environmentally friendly plasma technologies, respectively. Supply chain sustainability is completed with a chapter examining systems in textile recycling.

The second half of this book presents applications and case studies in sustainable textiles, with chapters focusing on recycled fibres, eco-labelling, organic cotton, nanotechnology, nonwovens, flame-retardants, and a final chapter examining systems change for sustainability in textiles.

R. S. Blackburn

Part I

Sustainability through the supply chain

Achieving sustainable textiles: a designer's perspective

A. SHERBURNE, Kingston University, UK

Abstract: This chapter is based on research to establish what constitutes an environmentally friendly textile, from the perspective of a practising designer. It identifies the environmental issues, the new goals and objectives, and the role of the designer in manifesting the necessary changes across the whole industry. The chapter identifies key environmental design strategies, and gives practical advice in the form of questions and exercises so that designers may engage their creativity to develop interpretations and alternatives based on their own methods of life cycle analyses. Because designers are in such key positions, the chapter also endeavours to explain the unique qualities that comprise the nature of design practice to others working within the field. There are also websites and suggestions for further enquiry.

Key words: textile and fashion design, environmental design, life cycle analysis, eco-design, biomimetics, biomimicry, renewable, recycling and reuse, sustainability.

1.1 Introduction: key issues affecting textile and fashion design

Designers have privileged access to the production process since they are responsible for specifying up to 70% of the subsequent material and production processes in any given project (McAlpine, undated). However, they do not have the power to implement environmental changes in the production of products because much of their work is confined by a brief that is set by the employer. (Unless of course they run their own business, where arguably they may have the freedom and flexibility to make changes.)

A brief is the means by which an employer purchases the appearance and often the practical instructions for production from a designer. The overall objective of all manufacturing businesses is to profit by fulfilling the demands of the marketplace. The price point for the product is the single most important aspect. The product has to sell, it has to be affordable, fit for purpose and desirable, or else the company can not survive. The number of colours in a print, the quality of cloth, finishes, accessories such as buttons and zips, ease and speed of production, hand finishes and embellishments – all are crucial to the bottom line. Is it possible to change these priorities? This is a challenge, not only for the eco-designer, but for everyone from producer to consumer. How can we convert the destructive consequences of the way the profit imperative is systematically allowed to undervalue material and human resources, into a constructive and new method of use and recognition of the true values that conserve these resources?

The whole field of environmental design is very alive, constantly changing and evolving, and is as much about changing systems and mind sets as with immediate problem solving. It has to be lively, because putting these ideas into practice is challenging. Resources and materials, processes, supply chains, new ways to sell, use and return goods, reuse and recycle, along with the underlying understanding about whys and wherefores, all are being defined and refined as we go along. On a personal level it is also about having enough courage to choose what we do with our time, this might mean making less money, but needing less, doing more for ourselves and our immediate locale, inventing systems of change, and being aware of how we are enjoying all our activities of living. Enjoyment is one of many examples of a greater value. How much have we tried to buy enjoyment that is mediated through products, but devalued by having to work too hard doing something we disliked? Ghandi said, 'Be the change you want to see in the world.'

However, a thriving economy does not demand that we have to make less money. The success of the service industries has proved this. Because environmental design is still in this state of flux, it can not be implemented without a self-conscious and informed effort. It soon becomes clear that whole new chains of production, relationships, dependence and interdependence need to be invented and developed. This author believes that informed and regular referral to every actual activity as well as the requirement to make profit in new ways, will enable us to rebalance the product lifecycles with the true environmental and human values. If we are courageous enough to accept what elements constitute a future environmentally friendly series of ideals, and are honest enough not to lie to ourselves when we do things that are still cleaving to the old, environmentally abusive patterns, we can align our steps. We may not get there, but having a courageous and ambitious aim, helps us decide what to do, and how to do it. This is a creative process.

1.1.1 Starting where we are

Most businesses are already locked in economically competitive chains of supply and demand, and are not easily able to make unilateral changes to their methods or products. In current practice, a designer might be given a cost–benefit analysis between the available resources on the one hand

and on the other, the required product and the price point where it enters the marketplace. There are many scenarios: cheaper materials; cutting costs and processes; abandoning whole factories, manufacturing capacity and workforces in search of cheaper options in order to avoid legislation that protects the environment and workforce. All this can be accomplished by moving production to countries where the abuse is tolerated. These factors all lead to an ability to undercut competitors in the marketplace, and although there might be a low profit margin on each item, high sales volumes yield a larger overall profit. This system does nothing to ensure that the skills, the machinery and the ability to manufacture are conserved, if the profits tumble, everything can be lost. The baby gets thrown out with the bath water. Again, we need to realign our values.

Such practices have made clothes and textiles available to the masses, but these products are not of good quality, and there are knock-on effects. Loss of employment, loss of the ability of a given local population to provide for its own basic needs, loss of skills, the creation of monopolies and general destabilisation. For example, when the world bank wrote off Zambia's debt in the 1980s, it demanded the implementation of a free market, which destroyed the country's small indigenous manufacturing capacity because of cheap Chinese imports (Woolridge, 2006).

In the UK these cheap Chinese imports are also adversely affecting the charity trade in clothes. People buy them instead of quality items, do not wear them often and throw them away. The Salvation Army finds that the poor quality of these items placed in the recycling system, cannot be reused or sold on so easily, and there are no longer enough good-quality clothes for its charity shops. The second-hand cheap Chinese imports are also sent in mixed batches to Zambia where the Saluala industry, which means 'to select from a pile in the manner of rummaging' (Tranberg Hansen, 2000), employs two million workers in re-selection and reuse. White collar workers in Lusaka choose the designer labels, then wholesale lots work their way down through the social and economic orders and clothe the whole country. In the villages it is possible for a family to survive by specialising in one particular type of garment. Care is taken to clean and present the items beautifully. Nothing is wasted. However, when the quality of those clothes diminishes, their potential for reuse becomes increasingly compromised. No country is immune. The UK has lost much of its textile manufacturing capacity in this way, and India is currently concerned about losing out to China.

1.1.2 What is wrong? – the facts

Theodore Roszak in his book '*The Voice of the Earth*' (Roszak, 1992), thinks that if psychosis is an attempt to live a lie, then our psychosis is believing we have no ethical obligation to our planetary home. Are designers, architects and engineers responsible personally, and legally liable, for creating tools, objects, appliances and buildings that bring about environmental deterioration? Probably not, unless one is looking for a scapegoat rather than a solution. Designers should be able to understand the environmental issues within their own practice, even if they are not required to do so by their employers. This could make them indispensable and key to the survival of a company, not least because of increasingly stringent environmental legislation. The following five headings describe the issues that make the life cycles of textiles and clothing unsustainable. If we include issues of fair-trade and ethical treatment of workers the list is even longer.

1 *Water.* Misappropriation and inappropriate use. Excessive use. Contamination.
2 *Chemicals.* Profligate use of pesticides and herbicides in agriculture and of toxic chemicals in production.
3 *Asset stripping* of non-renewable resources, including energy sources. Undervaluation and non-regulation.
4 *Waste.* Too much is destroyed. Systems need to be developed to recycle all non-renewables or compost renewables.
5 *Transport.* Unnecessary demand caused by capitalist exploitation of cheap labour makes this also unethical as well as increasing the profligate use of non-renewable resources such as oil, and the attendant pollution.

Production

Many aspects of textiles and fashion get grouped together under the banner of unsustainability. The environmental issues are pollution, poisoning, excessive use of pesticides and insecticides, de-oxygenation of water supplies with subsequent loss of animal life, salination of water and reduction in soil fertility, and loss of natural biodiversity.

Ethics

Environmental abuse combines with ethical issues when there is excessive use of water and when land is misappropriated away from food production. Further ethical issues involve atrocious treatment of workers, abuses of human rights, forced labour and deaths from poisoning, all of which are unacceptable sacrifices to commerce. There are now issues surrounding the appropriation of agricultural land for biofuel production too. How can we enable carbon reduction without compromising the even more basic need for water and food.

Market forces

The toxins are not just confined to pesticides and insecticides, but are also present in fabric treatments, dyes and printing, and are used in the creation of synthetic fibres derived from petroleum-based feedstocks. Fossil fuels are used to run agricultural machinery and to transport fibres, fabrics and finished goods backwards and forwards across the globe to find the cheapest processors and to get the cheapest deals in unregulated, developing countries. There appears to be no moral obligation where commerce is concerned. In an interview on Radio 4 (12 January 2006) John Snow, the US treasury secretary, acknowledged that the internal economy in the United States, was growing because of the consumer boom fired by cheap imports from China. This is at a terrible environmental cost to China, where there are also no core labour rights such as collective bargaining (*Ethical consumer*, 2002) These imports also threaten the economic viability of manufacturing within the United States, creating unemployment which has contributed to the recent credit crunch, repossessions of homes and the repercussions across the international money markets.

The paradox is that the workforce relies for its survival on a system that appears to be destroying the ability of the globe to support that workforce. But any new solution has to build on the existing system to make changes. This is because money is the best internationally recognised method with which to share resources (Porritt, 2007). To destroy the economic balance could also lead to anarchy and war.

Use

The largest environmental impact of textiles occurs when they are being used by the consumer (estimated at 75–95% of the total environmental impact) and is accounted for mainly in the use of electricity to heat water and run laundry and drying processes (www.informationinspiration.org.uk and sevenethgen.com). This contributes to greenhouse gases and global warming.

Disposal

Much of this painfully achieved produce then gets thrown away, to be buried, or burned, releasing ozone-depleting methane gases from landfill sites where the fibres cannot decompose properly, or through burning, asthma releasing causing airborne particulates.

It is therefore sensible to look honestly at where and why the systems and processes are destructive, and find ways to change production, use and disposal.

1.1.3 Where can designers start to make a difference?

To start with, designers should not stop expressing the creativity that makes them tick! Genuinely sustainable textile and fashion design should prioritise the creative thought processes (even though so many practical issues still have to be solved). Creative thought processes include the vision that is more like reverie, and is more intuitive, i.e. knowing just what is right. Designers also need to be informed, so that they can choose the right production systems and materials, and in many cases be proactive innovators of those systems themselves. To become informed, designers have to become familiar with making life cycle assessments (LCAs) of the constituent parts of the textiles they are using and the garments they are designing. This will be explored later in the chapter.

In many cases the production system gives birth to the creative potential too. Starting with an environmentally friendly system and set of criteria with which to work can enable the development of its own desirable aesthetic. For example, the London-based company Junky Styling restyles discarded clothes (Fig. 1.1). Their outstanding and innovative designs add value. They went on to invent 'the wardrobe surgery', restyling other people's clothes to order. Why this is such an important innovation in environmental design will be explained later. Junky did not start out to be eco-friendly, yet they have become a model for financially and aesthetically successful design that utilises a local waste stream. In fact, they had no money and needed cutting edge clothes to get into clubs. Necessity was the mother of invention.

We cannot go back to the dark ages, forget space age technology or un-invent the wheel, and no-one wants to wear sackcloth. Every technology has to be recognised and incorporated in acceptable ways, and in combination with the necessity of sustainability. Many elements are not mutually exclusive. It seems counterintuitive, and even though it is derived from oil, polyester is currently more environmentally friendly than organic cotton, because it uses less energy during laundering. Everyone on the planet needs to align to environmental and sustainable ways of living. Textiles need joined up thinking and action from scientists, farmers, factories, shops, the public and governments. Designers have a unique knack of interpreting the zeitgeist, they are often inspirational, and crucially they are the ones who will be responsible for what environmental design looks like and how it's made. Can designers also be the key people who, by understanding the specific environmental issues of their specialism, are able to advise everyone else in the production chain to change?

Given the obvious complexity of the issues, and that we are all in bondage to the system, it is essential to unpick points where interventions can be effective enough to make real changes.

1.1 Junky Styling, Cuffs Dress (published with kind permission from Junky Styling).

1   Individual governments, if given accurate information, can legislate, but diplomacy, trade agreements, international alliances and global economics keep whole countries in bondage too. Governments can put standards in place and help to establish certification systems.

2   Consumer demand is one of the most powerful points for change, but this can not happen without educating the consumer and then providing actual alternatives. Coherent labelling is essential. Things also have to be affordable and fit for purpose.

3   Designers divine and define what we want and need. This vision is the unique definitive talent of designers; it is the key, because it is actually what makes something saleable or not.

1.1.4 How to understand how designers understand

Something can be certified, labelled, affordable and fit for purpose, but designers add the seen and unseen beauty. Although LCA is the best and most accurate way to solve issues one by one, it is the way in which a designer is able to think out of the box that is powerful and revolutionary. The industry needs to understand that designers work and think differently, and must learn to support them in this key role, and feed them with new ideas, technology, access to materials and processes, and information in a way that they can understand and work with. This means making designer-friendly interfaces that are free at the point of access, regularly updated and easy to navigate. These should be written in the simplest ways, avoiding jargonese, and where possible give access to samples, opportunities for hands-on experience of new processes and technologies, and effective contact details. Information given to students should be quick, basic, concise, precise, up to date and *free*.

Designers are inspired by everything and everyone in the world around them. We can tell this by looking at the international fashion shows which are often an instant barometer of the most pressing global issues, reflected in a seemingly fickle mirror, which, like laughter, is often the only way to face unpalatable truths (starving models in the abundant West, blackened eyes, bleached white cotton that pretends purity, bright and polluting chemical colours that reflect the 'natural' feathers of exotic rain forest birds, etc.). The heart of the problem is in the life games in which we are all players; of power, wealth, belief and imagination.

As an artist, maker, designer and craftsperson, the current author has always used imagination to understand, engage with and transform personal experiences of the big three (power, wealth and belief). Artists, designers and craftspeople have always made these elements at first tangible and then palatable, by interpreting them, and making them real, useful and attractive. Great artists like Michelangelo made The Divine visible, Leonardo Da Vinci dreamt of machines to control and navigate the physical world, medieval stone masons who had understanding of mathematics designed the great cathedrals, and craftsmen built them. Similarly, John Galliano's imagination might alight on Chinese hill tribes, and mix the visual evidence of their culture with romantic French revolutionary heroes, to refresh our own notions of who we could be. But why are these two groups of people so inspirational? The hill tribes live outside a money-based economy. This in itself does not mean that great visual expression would necessarily emerge in such an economy, but the existence of these very beautiful, individual, complex and fine masterful textiles and costumes could only emanate from communities of such a high order of peaceful cohabitation and harmony with their environment, as to enable this high form of self actualisation. We also love the French revolutionary hero because at the height of his own personal power he is ready to sacrifice everything to create a fairer society for others.

Do we believe we are powerless to do anything more than look like our heroes? Galliano is certainly a visionary, who can show the best of ourselves to ourselves. Both sources of inspiration look great, and both come from two recognisably important wellsprings of the human psyche. Do we recognise what we would really like to be, can we dare to change, rather than use clothes to disguise, hide, protect or forget? It is to do with choices. Can new designers give us the other options, not just in how we look, but in new systems with which to use our textiles and clothes; how we purchase, share, swop, collect, change, sell, exchange, use, reuse our clothes and textiles. Can we be the revolutionaries who then live in the sustainable societies and lands of our own creation?

Imagination fires all the creative industries, and also reflects the destructive and frightening by-products of power, wealth and belief. Do we believe that we have no influence over these realities, and how far do the cultural reinterpretations that bring us hoodies, combat prints and the like, also perpetuate and encourage violent and negative imagery, behaviour and role models? How far are we really in bondage to the system and how much are we inadvertently supporting and perpetuating it? How much do we recognise our own differing levels of debt bondage? How far can designers intervene, and can they inspire new alternatives? Do we have to compromise our environmental wisdom just to afford the rent? Can we afford to live slowly? What does the highest, most creative ecological manifestation of our imaginations actually look like? Can we make beauty that we deserve?

Designers can have a major role in changing manufacturing systems, but this ignores the particular (and bankable) talent of the designer; namely the power of alchemy, of transformation by imagination. Can we pay our own way and still be able to change the ways that money buys goods and services? Can we balance the real environmental value of a garment, which takes into account the real embedded energy and the real raw materials, with the amount of money that it cost? The Factor 10 Institute (www. factor10_institute.org) says that we have to get ten times more out of every product than we currently do, just to maintain the planet in its current unsustainable state. We also need two earth-planets to keep producing at the current rate.

Can we find a way to relate money to the real value of the resources we use? Can we find a way to use money to prevent the abuse of resources?

Can we think freely outside the destructive box that gives us our daily bread and shelter? Can we stop the invisible power of money from overriding and undervaluing the real cost of materials and processes? Can we accept and live with the answers to these questions, particularly if it means new limitations? Can the new designers make us choose the new options because that is what we really really want.

There are very real practical strategies that will enable designers to specify new environmentally friendly life cycles. There are inspiring new business opportunities that will evolve where the design of systems of use can be realised by new entrepreneurs. But the greatest contribution that any new designer can make is to manifest the subconscious desire of the whole of humanity for health and to live freely and harmoniously. What will these textiles look like? Will they be restrictive and dictatorial, or will they fulfil the promise of individuality, diversity, personal choice, abundance and freedom that could be offered by globalisation, universal access to millions of differing and specific needs and desires, made possible by the unifying power of the World Wide Web? Will it be enjoyable: envisioning, making, selling, using and reusing these textiles?

Apart from the limitations placed on them by their terms of employment, other immediate problems facing the proactive and aspiring designer of sustainable textiles are associated with the intransigence of the existing system. Most businesses are already locked in economically competitive chains of supply and demand, and are not easily able to make unilateral changes to their methods or products. Changes in supply chains or production methods could also affect diplomatic and trade agreements with other sovereign states, or reduce exports, create unemployment in some industries, or affect the national well-being in other ways. Intervening in some businesses is even dangerous. There are power structures, personal interests, hidden agendas and many cultural and social interrelationships that are simply not visible, particularly when we start working with third-world countries. This is why my own particular focus is on clarifying the environmental issues where designers could make a contribution. Sometimes these interventions can actually save money because they are more efficient and less wasteful. The issues need to be solved because they make sense on many levels. Many businesses, charities, agencies and government initiatives are already aware of and engaged with ethics and fair-trade, such as 'Labour behind the label', 'Fashioning an ethical industry', PAN UK (Pesticide Actor Network UK), Katherine Hamnett, RED at Gap, Oxfam, etc. However, it is important to note that environmentally friendly, ecological design goes *even deeper*, because without it human beings might not survive at all.

Most of the current discussion is likely to have less immediate use for designers who are already in employment (at least until more legislation comes into force), and will be of more benefit for their personal development, freelance activities, for self-employed designers, proactive eco-design companies, and of course teachers and students. Creating change will rely on building new arrangements. The word 'arrangements' is preferred to the word 'business', because until new patterns emerge, it is difficult to formalise and describe the many alternative relationships that can begin to exist between the exchanges of products, services, ideas and money. Many of the new ideas for the ways in which environmental design can be used will not be instantly recognised, and need to be explored for what they are: the relationships between people that occur when a major cultural shift is taking place.

Here we come to the heart of humanity, and the truths that drive us all. Great designers undeniably manifest these elements. They bring form to aspirations, as well as actually making things fit for purpose. How would we know that a Hollywood star was divine if the gowns she wears did not reveal such a truth? Today, the designer is no longer an alchemist who is required to make sackcloth look like the golden fleece, it really must *be* the golden fleece, because we have to appreciate the true (non-financial) value of the constituent materials. The great designer is one who can dream and make that dream into reality. Anyone who says that this alchemy can be dispensed with really does not understand the unique nature that inspires the abilities of the textile and fashion designer: the one who really does make the dreams real.

1.2   Strategies for fashion and textile design

1.2.1   What do we need? How can we get it sustainably?

Our primary needs are air, water, food, shelter, warmth, safety, companionship, sex, procreation and, hopefully, happiness and contentment. To provide these necessities, we rely on identities that are upheld by those around us. Those identities are immediately apparent in what we wear. We have little time to perceive anything other than what we can apprehend in a momentary glance. Our costumes dictate our roles, and because we are all in competition for a share of resources, the way we present ourselves can help or hinder our passage through life. The expectations of our roles, social status and behaviour are hard to override. Imagine the hoodies helping old ladies across the road, ladies in suits pole dancing, clergymen picking fights, or facially tattooed police officers: this could only be an art piece to point out the power of daily costumes.

Textiles are used to make our chameleon skins. Textiles and clothes are psychologically important: uniforms of our particular tribe of hunter-gatherers, armour for our warrior nature, camouflage to hide behind, colourful, tactile, inventive, defensive, protective, decorative, socially and 'tribally' definitive, theatrical, smart, seductive, playful, as well as practical. Taken for granted, textiles are present everywhere in our daily lives. Even if they are eco-friendly, textiles still have to beautiful, fit for purpose and affordable. Or can we possibly allow these commonly held conventions to visibly change, if it means we can survive? How necessary are the uniforms? Can they be made differently?

Fast fashion has brought a democratising element to options and choices that we have never had before. Not only can we have the fashionable clothes we want, now, but we can have enough of them to be able, financially, to afford to throw them away rather than even to bother washing them! For example, at a government-backed conference on sustainable clothing Tom Fisher from Nottingham Trent University and Tim Cooper from Sheffield Hallam University described how cheap clothing often leads to short life spans: 'During recent research we met one woman who each year bought fourteen tee-shirts for £2 each for her holiday, wore one each day and then discarded them. She liked the idea of wearing something clean and fresh each day, and discarding them meant she had space for duty-free purchases on the way home.' We mix fast fashion with a few expensive designer items, a few cleverly bought fake designer items, and some vintage or second-hand collectable finds. The one we buy to dispose of, the other we have 'invested' in. If we continue to need 'fast' or at least inexpensive fashion, we have to ask how it can be made in an entirely environmentally acceptable way, throughout the whole life cycle.

1.2.2   Designer-initiated life cycle assessment

In order to understand the specific environmental implications of any given design project, and then begin to change any textile or fashion process for the better, companies conduct an LCA. This means looking at every possible material, process, use and finally the potential for reuse or recycling when the item in question is no longer wanted or needed. In itself it is also an experimental process that throws up as many questions as answers. Everyone is on the first steps of a long and changing path.

Designers and students certainly need to know what to look at in order to conduct LCAs to a greater or lesser extent, because this is a way of developing an understanding of, and confidence to change or advise on, alternative and beneficial changes that can be made. (An alternative way for a designer to engage with environmentally friendly practice would be by choosing to use and develop environmentally friendly materials and processes that have already been identified, and start with these.)

Tracy Bhamra and Chris Sherwin comment (jdr.tudelft.nl),

design ... is not an exact science ... it is a spontaneous and intuitive process ... designers work largely with visual imagery and stimulus, using creative and non-prescriptive design processes ... having little use for 'hard data' ... eco-design is ever strengthening its scientific and technical roots in striving for more precise and accurate environmental data, which may directly contradict the requirements of designers themselves.

This does not mean that designers cannot make their own independent enquiry, in fact it was these very comments that spurred me, as a designer myself, to try to bridge the gap. There are indeed many variables, and at the time of writing very few design websites and tools available that make information clearly accessible and comprehendible from a designer's perspective. But there are inspirational new ideas and sources that if presented in the right way will fit into the process of design and become an additional source of inspiration rather than a problem. Existing sources of help will be detailed at the end of this chapter, but designers should feel emboldened to invent their own methods to understand the environmental implications of the materials and processes that they are working with. This is what is meant here by individual designer-initiated LCA systems. Use the best information and processes that you have access to in order to do what you can, and keep questioning, improving and strengthening your own practice.

Governments may have legal requirements for accurate LCAs in certain circumstances. These are strict, time-consuming and expensive. Companies like Marks and Spencer have developed their own methods of assessment too, they are able to 'encourage' their suppliers to comply because they have economic leverage. In the fast-moving world of fashion, assessments can currently be too time-consuming in practice to be conducted for every garment in a collection. This is one reason why there has not been a sudden increase in alternative 'ecological' choices, and why the real pioneering retailers such as Howies are few and far between. Ideally, working to build a framework of acceptable materials and practices, and reaping rewards by building brand loyalty based on integrity and transparency of supply chains, is sensible. There are many opportunities to be the very first designer to present product lines or new ideas enabling consumers to engage with eco-textiles and fashion design.

The results of every assessment can also be variable. A designer has to be prepared to weigh the pros and cons themselves, but this is par for the course because it is a way of testing the validity of ideas, and also of making decisions that can incorporate mitigation. An example of mitigation might be that although the use of oil and lots of energy to make polyester seems bad, in the phase of use, polyester uses less energy than natural materials. If that polyester is then recycled efficiently, it could end up using less carbon in its whole life cycle than organic cotton. Sportwear manufacturer Patagonia, systematically recycles its polyester garments by 'taking them back' from their customers when they are worn out, sending them to the Japanese polyester recycling textile company Tiejin.

Of course there are other issues too, there is rarely a single right answer, but there are many better alternatives. It is the very consideration of alternatives that is life blood for a designer. Alternatives can inspire unexpected new ways of doing things, some of which might involve exciting and unanticipated design elements, or an easier way of making something. Designers who do conduct these assessments are pioneers, and it would be useful to create more resources such as those in specialised university departments like TED (Textile Environmental Design) at Chelsea College of Art and Design, London, where results and decisions made as a result of designer-initiated LCAs can be made available to others, since this is also a subject that is itself still evolving.

1.2.3 Designer life cycle assessment interpretation: underlying philosophies and strategies of eco-design

Many of the following eco-textile and fashion design strategies have been developed by looking at environmentally friendly architecture and product design which are more advanced, often because of the need to conform to legislation. Analysis tools are available, but these need to be interpreted to find appropriate parallels that may be of use for fashion and textile designers. There can seem to be an overwhelming amount of complexity, but there are founding principles that all products conform to, and because of this, the strategies and philosophies that are developing do offer alternatives. The core concepts that seems to underpin and support many solutions are 'cradle to cradle' and 'biomimicry'.

Cradle to cradle and biomimicry (or biominetics)

There are only two basic materials in nature:

(a) materials that grow, these are biodegradable and they re-grow;
(b) materials that are finite, these do not grow.

The materials that grow are 'renewable' and those that do not are 'non-renewable'. The core concept is that: in nature there is no waste. One process leads to another in a dynamic balance that is the result of cause and effect. William McDonough and Michael Braungart, the co-authors of the seminal book *Cradle to Cradle* say that 'waste is food' because every part of nature feeds on the remains of a previous form (McDonough and Braungart, 2002).

All materials will remain true to their basic nature. In manufacturing, we intervene and temporarily harness the natural order, but there are repercussions, mainly caused by our untrue belief that waste can be just thrown away. Out of sight may be out of mind, but when something is buried, for example, all the constituent parts will continue to work according to their own inherent nature, in combination with the new conditions that surround them. Natural materials biodegrade, can release methane and create nutrients for new growth and soil matter. Manmade non-renewable materials do not biodegrade or break down, and some are poisonous. Their inherent value is difficult and sometimes dangerous to retrieve if it is buried in mixed landfill.

For years the idea of getting rid of poisonous by-products from manufacturing just required their dilution into the wide world. Paracelsus, the father of toxicology, noted that benign substances can be poisonous in excessive quantities, but that toxic substances can be rendered safe in dilution. Textiles utilise many chemicals in the form of pesticides, fertilisers, mordents, chemicals, dyes, etc. that have accumulative effects. There is now a tipping point, where bioaccumulation threatens us and means we have to be far less messy, profligate and wanton. Now we have to understand and take control of the inherent nature of every chemical we use, understand its impact, then value and respect its unique qualities and apply them accordingly. This is a revolutionary call to grow up, and to reinvent the industrial revolution, by mimicking nature in a way that maintains a benign balance, not mixing it up in a way that disturbs global ecosystems and poisons us.

Biomimicry and the appropriate recycling of renewable and non-renewable resources

Cradle to cradle describes a system that calls for the appropriate redesign of all systems to enable all materials to be used in continual cycles of use and reuse. This system is inspired by mimicking the processes of nature. To complete the cycles, the end of the product life must rejoin the beginning of a new product life. This is the point at which the circle closes. In practice this means that renewable materials must be composted appropriately. If methane is produced as a by-product of decomposition, then it can be captured and used as a form of sustainable energy. Fabric treatments, dyes and pigments on natural materials should also be able to biodegrade harmlessly so that the resulting compost can be used to nourish renewable crops. This revolutionary environmental design strategy is known as a 'closed-loop system'. If the materials are renewable, the loop is closed when they are composted, because the old nourishes the growth of the new.

An industrial example of this is Climatex® fabric, from the Swiss company Rohner Textil AG. This is the world's first fully compostable industrially produced textile range. Developed in 1995 by Michael Braungart, co-author of *Cradle to Cradle* (McDonough and Braungart, 2002), the blend of ramie, wool and polyester is a substitute for cotton, and the new yarn dyeing plant reduces water use by 30%. Climatex® is biodegradable, and all processing waste is used for felt or garden mulch. Using closed production processes, the water entering the mill is less clean than when it exits the mill. (Michael Braungart has been asked for a clarification on the use of polyester in this otherwise compostable fabric. It obviously has important wear properties, but would not decompose and would remain as fibrous filler in compost. This seems to contradict the requirement of pure cradle-to-cradle systems to use materials singly in order to reclaim the non-renewable resource. A personal belief is that in small quantities this mixing is currently seen as allowable, particularly if the mitigation is that the fabric is fit for purpose only because of the durable qualities of the polyester.)

The environmental data for the production of Climatex® are extensive, and could be used as a model to inspire other manufacturers.

1   It meets the criteria for use required by the upholstery industry, for wear, aesthetics and full colour availability.
2   It meets the criteria of the environmental legislation in Germany with regard to dye, chemicals, heavy metals, poison, recovery of chemicals and production risks.
3   The raw materials are renewable, grown ecologically by closely monitored suppliers.
4   Constant research and updating of production systems occurs, with monitoring of chemicals where they have to be used, inside and outside the company itself.
5   Ecological accounting and constant upgrading, including regular recertification and independent inspections towards the objective of elimination of all waste and wastewater, noise reduction and the addition of alternative energy systems (www.climatex.com, environmental data, 2007).

Many fashionable products are currently designed for limited use. This is known as 'design for obsolescence'. It encourages people to throw things away so that they have to buy more things from the manufacturer. Taken to its extreme form this has become 'fast fashion', where people buy a cheap garment to wear only once before throwing it away. The global economic success and growth that this encourages is an ecological disaster.

In order to 'close the loop', the constituent parts of the product have to move through a new system into a new beginning or a new 'cradle'. It is this 'new system' that offers the new world of possibilities for a creative designer. After all, 'new design' can be as much about systems as it is about objects. In a service-driven economy, a key area for designers is to creatively develop ways for us all to keep money moving without necessarily relating it to repetitive sales of more and more new clothes. There may be a way in which this fast disposable fashion could enter easily back into the production loop for reuse or recycling. This could rely on the availability of very efficient take back systems, close to where the textiles and garments are processed and reprocessed. The fibres made from non-renewable materials such as oil would be valued for their irreplaceable and myriad performance values. To create an ideal environmental scenario, the energy source for this should be carbon neutral. Currently, energy is environmentally expensive because electricity is overridingly created from burning gas and oil.

Honouring the embedded energy

At present, even if the fast fashion garment is put into the recycling system, the 'embedded energy' within that garment has been entirely lost. Embedded energy means all the resources and energy used in the production of the garment. This includes that used in growing and harvesting and other synthetic fibre production, then spinning, weaving, knitting, sewing and finishing, including the production chains of zips, buttons, threads, beads and decorative embellishments. There are also air-miles used to transport the garment between countries and continents to find the cheapest factories for every part of the process, transport to the retailer, and then costs of disposal. (Importantly, all these aspects of production identify crucial aspects that need to be addressed within an LCA.)

The only way to change this system is to make it possible to retrieve all the materials, and in some way also justify, retrieve and honour the embedded energy. This means changing the way that particularly energy- and resource-rich products, particularly non-renewable products like oil, are made, so that they can be disassembled to reclaim and reuse those resources.

Design for disassembly

This is a new and important core strategy where the designer ensures that the constituent materials can be reclaimed for reuse at the end of the product's life. Fundamentally, this is easiest when all constituent materials to be used are in the simplest forms possible, because as soon as non-renewable materials are mixed, they become impossible to separate economically or ecologically, because they are contaminated and degraded. A textile example of this would be poly-cotton, where the polyester can only be retrieved by burning away the cotton, using excessive energy and wasting the cotton. An example of one design interpretation of design for

Eco-Friendly Apparel Design: 'The Dowry Dress'

Design Goals: to redesign the life path of a garment in order to better fit its purpose and increase usefulness

Design Results: a wedding dress that reflects the bride and reforms into keepsakes marking the life of the marriage

Phase 1: Design Embodies the Bride

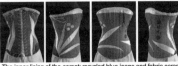

The inner lining of the corset: recycled blue jeans and fabric scraps

- **The Bride's Sense of Style:** bride can codesign her dress
- **The Bride's Memories:** recycling her favorite old pair of jeans for corset lining *(see above - also counts as her something blue)*
- **The Bride's Personality and Name:** scraps from the corset fashion fabric used to create appliquéed design on the lining that reflects her personality; Chinese characters symbolize her name
- **The Bride's Life:** reversible corset that she can wear again and silk veil becomes a scarf to wear on anniversaries
- **The Bride's Gift:** adjustable fit and easily removable decorations allow the bride to pass on her wedding corset to a loved one

Phase 2: Bride Sanctifies the Cloth

bobbi+mike, photographers   *bobbi+mike, photographers*

Skirt trained, veil lying down along the train   Skirt bustled, veil tied to hands

Phase 3: Cloth Reformed as Life of the Marriage Grows

Skirt formed from a cleverly draped, unfashioned length cloth that becomes keepsakes marking the milestones in the marriage

Dress on Exhibit at Cornell University

Step 1: Deconstruct the Skirt
- Remove center back seam and hems
- Detach from the waistband section

Step 2: Cut-out Patterns
- Pattern shapes designed for little waste
- Pattern layout to maximize cloth use for this cloth is sacred and must be conserved *(see diagram at the bottom)*

Step 3: Construct Keepsakes

Embellishments utilize other eco-friendly materials *(see captions below)*

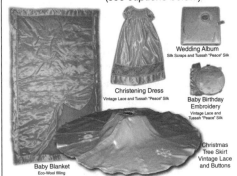

Wedding Album
Silk Scraps and Tussah "Peace" Silk

Christening Dress
Vintage Lace and Tussah "Peace" Silk

Baby Birthday Embroidery
Vintage Lace and Tussah "Peace" Silk

Christmas Tree Skirt
Vintage Lace and Buttons

Baby Blanket
Eco-Wool filling

Skirt Diagram - Layout for Keepsakes

| 28" | 38" | 49" | 9" | 14" | 14" | 14" | 14" | 28" | = 208" |

Baby Blanket
Christmas Tree Skirt
Embroidery Canvases
Christening Gown
Christening Gown Sleeves
Photo Album Cover 1
Photo Album Cover 2
Photo Album Cover 3
Photo Album Cover 4
Christmas Tree Skirt

1.2 Kathleen Dombek, wedding dress (published with kind permission from Kathleen Dombek).

disassembly is seen incorporated into Kathleen Dombek's wedding dress (Fig. 1.2), which can be changed again and again into new items that can change to meet the changing lives of the bride and groom. She also engages ideas of durability and emotional value in this work.

Factor 10 Institute

This principle of not mixing non-renewable materials and of being able to reclaim them should be the basis of a redesign of the way we manufacture things. If it is accomplished without disturbing the global economy, or even by enhancing profitability, it will have the desired effect of slowing environmental degradation. For those designers and students who need to know what will happen if we do not change our manufacturing systems: the Factor 10 Institute has calculated that we currently use 30 tonnes of non-renewable resources for every tonne of products, and that the trend is increasing; so much so that we would need the resources of over two more earth-like planets to maintain the current trend. This means that we have to get ten times more use out of every product. The system 'requires the simultaneous and even handed consideration of economic, social and ecological consequences of every impending decision' (www.factor10-institute.org).

Dematerialisation and design for durability

One of the ideas that is evolving to respond to the findings of the Factor 10 Institute is called 'dematerialisation', which essentially means getting more out of less. There are a number of design strategies that deal with how to put this into practice. Durable design looks not only at what the product looks like, feels like and how it functions, but then develops whole new systems of interaction with the customer that will add value and enjoyment, and create satisfaction and longer cycles of use. A number of companies exist that have begun to develop new systems as well as products. These systems include purchase, collaboration, use, hire, sharing, swopping, repair, reuse and eventually appropriate forms of recycling. Many of these activities can become economically viable businesses, where customers are often far more actively involved in developments of shopping experiences. Not least of these is a recognition of the value of shopping as entertainment. My own shop is already interactive, we offer to teach people specific textile skills, using our eco-friendly yarns, our varied expertise, eco-haberdashery and vintage hand tools. How and where the designer functions is also flexible. The methods for design and the materials and processes – as well as challenges to current relationships between time, motion and costing – offer many opportunities for creative lateral thinking.

1.3   Strategies for textile and fashion designers: recycling and reuse – beginning to close the loop

Recycling and reuse are particularly important and need to be the first problems that are solved within the whole environmental design puzzle. This is where the loop can close and where the wanton loss of our valuable materials and resources can be prevented. Currently, there is an abundance of materials available for creative recycling, the problem is that the systems to support the addition of value to discarded textiles and clothes are not abundantly advanced. Adding value to waste textiles by restyling them, or even simply sorting them is time and labour intensive and is the most expensive element. Accurate sorting should be a priority, and consideration needs to be given to where and why sorting facilities are sited. Sorting needs to happen where waste is created and where people need jobs, namely in large conurbations. Governments could decide to support the increased costs of employing and training staff, rather than relying on market forces which are preventing growth in this sector because it is impossible to pay even the minimum wage for such a time-consuming process. Market forces can not be relied on to make the best environmental choice, but only to obtain the largest immediate cash reward.

Governments also needs to subsidise the buildings in which the sorting and subsequent storage and addition of value can take place. This needs to be accomplished with coherent, non-party political forward planning, so that sorting systems can easily harmonise to take in new, more highly evolved waste streams that are working towards closed-loop harmonisation on a wide scale. The beginning will therefore be a period of transition where only some of the embedded energy and resources within the 'recyclate' can be utilised. (In other words, to kick start the future economic viability, let creative designers get their hands on the materials easily and affordably.) Success and the first profitable returns crucially depend on design companies. It therefore makes sense to also encourage the growth of new small design companies who can 'feed' from this source, while developing the sector. This could occur in much the same way as the encouragement of new enterprises in the 1980s. Again government could play an essential role in setting the system in motion.

There are two main categories for contemporary and future recycling.

1   Eventually, the largest, most-advanced companies will utilise and develop systems that will give them strategic control of material non-renewable resources. Some strategies would give future companies even more of this power, including custodianship. Large-scale management

of natural, renewable materials that biodegrade could also be controlled by the larger companies, mainly because of the economics of scale.

2   Until that time, the renewal of the sector will begin with companies who utilise the discarded value inherent in the global system as it currently exists.

The first category aims to maintain and control value; the second benefits from the lack of control of material resources as evidenced by profligate waste (in this case value means economic value, but also the ecological value, even if this is inadvertent).

Three successful design strategies have been identified that have been evolving naturally using a small fraction of the abundance of materials within the waste stream. There is plenty of room for all the potential to develop, and for new relationships of supply, demand, support and collaboration. The emerging aesthetic of environmentally friendly textiles and fashion is fascinating. There are web-links throughout the next section with examples from existing design companies.

1.3.1   Design-centricity

Design-centric recycling is where the designer is selective in choosing items or materials from a waste stream for reuse, remodelling and adaptation. It requires designers with a good eye, individuality, creativity, perhaps slight eccentricity, quirkiness, and the power for reinvention and reinterpretation. Its success is aspirational, intelligent, to do with recognising quality and enduring value. It can also be slightly exclusive, stylish, 'with it', but is not necessarily fickle because it is also at the cutting edge of creativity, where trends start.

Vintage in all its guises and uses plays a major part in the success of design-centric reuse and remaking strategies. A very important element is inherent in the received aesthetic, but also in the uniquely chosen physical properties of more unusual waste; for example bicycle inner tubes, which are used to very different effect by designers Barley Massey and Nani Marquina. The key is that the materials are specifically selected. This selection occurs by cherry picking from sources such as waste textile merchants, online auctions, jumble sales, antique and second-hand dealers, or by eureka moments where an available material inspires a creative mind to make a particularly innovative and relevant product. These constantly re-evolving interpretations of 'vintage' also go on to influence the mainstream contemporary design trends.

Textile and design students and professionals often create their own original looks based on past styles, while design training and professional practice has always included knowing how to conduct research into historic movements. Designers are familiar with the development of new ranges through sketchbooks, story boards, colour and pattern development, essays and fabric searches. Contemporary designers already revisit retro styles in their collections, and also analyse the actual cut of vintage pieces. These methods are little changed from the way that dressmakers have worked with their clients throughout the ages. Originally fashions changed, and were embodied in the same valuable fabric year after year. Oblongs were carefully cut and pleated into the new styles, then unpicked and put away every year to be remade again later. Fabric was in and of itself, far more valuable than we believe it to be today.

Beverley Lemire points out in *The Business of Everyday Life: Gender, Practice and Social Politics in England c1600–1900*, that until the middle of the nineteenth century, everyone in society, rich or poor, was able to evaluate the worth of textiles, which were used as currency, and used, reused and bartered even when they became rags (Lemire, 2006). Our current systematic inability to husband our resources is a relatively recent phenomenon which has its causes in the power of the world financial markets to override our respect and proper husbandry of goods and services. Lemire tells us that before 1850, there was very little money in circulation in the form of bank notes or actual coins. A good pair of shoes or some dress fabric really was an 'investment', and was commonly used as currency as well as clothing. There would be little desire to put a Primark T-shirt into the bin if you could use it to pay your fuel bills! What are the modern equivalents and possibilities? This is where the challenge to designers begins, and this is where many new business models can stimulate new economic growth.

Designers learn how to manipulate and master their materials in order to express practical and creative ideas which are then made by manufacturers. Designer-makers apply hands-on experimentation, dexterity, creativity, vision, imagination and flexibility to their fields of 'craftsmanship'. Technically proficient designers and designer-makers collaborate with good technicians too. Some materials are inspiring in themselves, others are transformed by creative intervention. Artists such as Tracy Emin involve varying levels of textile-based hand skills in their artwork to express more conceptual observations.

Within my concept of 'design-centricity' this is pretty much where the reuse and recycling stops: with concentrated forms of creative brilliance. This brilliance does not dissipate into manufacturing processes, but recognises and retains, enhances, preserves and plays with materials that it simultaneously confers value upon. Counterintuitively, these are not all exclusive and expensive products. This work is available at local levels whenever makers and artists are able to afford access, through galleries, shows, market stalls, exhibitions and fairs, as well as online sites.

1.3.2 Product-centricity

Product-centric recycling allows manufacturing systems to engage with textile waste streams. It can occur because materials and processes are affordable, available, sustainable, repeatable, in quantity, allowing the addition of value and profit. At the moment there is enough waste to enable some companies to recycle existing waste streams of textiles so efficiently that even variability can be integrated without compromising the requirements of larger scale production methods. Worn Again (www.wornagain. co.uk) utilises huge quantities of repeating fabrics to make its shoes, and Ecoganic uses faux sheepskin made from 100% recycled polyethylene terephthalate (PET) (www.hipandzen.com), which means that there is no remaining vestige of an original aesthetic to affect or influence the new design. Patagonia used PET for its fleeces and has now developed an extremely advanced system of take back where it fully recycles its polyester garments into new garments, the Common Threads Recycling Program utilises the ECOCIRCLE™ recycling system from Teijin in Japan (www. patagonia.com and www.teijinmonofil.com).

Currently the way that the waste stream can be utilised as a source of materials is different from the way that it should evolve. In future, the production systems need to evolve so that manufacturers will have more control over the life cycles of their textiles, clothes and other products. Eventually, it will be easier for textiles to go back more efficiently into the production stream. The systems that will make this change possible include 'design for disassembly', 'take back' and systems of use that would include 'products of service' where items are hired, returned and reused, and which can be upgraded.

Typically, materials in the product-centric group are very abundant and aestheticly random. Currently, the variation in materials requires that all companies using large quantities of waste have production methods that have been developed specifically to include sorting, cleaning and cutting. Small- and large-scale businesses use different methods, and the products themselves differ according to the economies of scale. For example, Worn Again can make thousands of shoes from 99% recycled materials, while smaller companies can use repetitive components and/ or make multiples of small saleable items such as fashion accessories. Examples include handbags made from ring pulls (www.escama.com) and handbags made from silk selvedge embedded in polyurethane (www. riedizioni.com). Traditional quilting and appliqué are even smaller scale, labour-intensive methods of making treasure from the smallest scraps (www.auburn.edu/academic/other/geesbend), while rug making (progging, hooking, pegging and plaiting) is a thrifty way to use more shabby and worn textiles.

The industrial side of the textile waste stream makes fibre that is used for wadding for furniture, underlay and fibreboard. It could be invigorated as evidenced by my own project to spin recycled yarn on a commercial scale once more, so that new basic materials, fabrics and yarns become available. This part of contemporary recycling is closest to the models that need to evolve if we are to be able to husband all our resources efficiently in closed-loop systems.

The keys to success in product-centric systems are that the materials are inexpensive, easily available (often at a local level) and that businesses are sustainable because they can exhibit healthy trading models that allow for wholesale, retail and also, most importantly, enough profit to be vertical units that can compete against globalisation and cheap imports. The very existence of such healthy product-centric systems could be the salvation of a lot of manufacturing capacity within the UK. This is absolutely essential in ecological terms because ultimately, every country needs to be able to accomplish appropriate self-sufficiency at local levels to deal with the soaring costs of energy and ensuing limitations on transport.

1.3.3 People-centricity

The first aspect of design within people-centricity is respect for what people need and how those needs are met, both in products that last, and in systems that support proper husbandry of materials and resources. By replacing the systems of debt bondage in order to make AND enjoy the choices we are able to make as responsible adults, it is not solely about profitability, but quality of life, products and systems that work harmoniously. All these things are defined as much by the way things are done as by what they are when they are complete. Relationships that are not mediated solely through objects, but that are enjoyed as part of daily life.

People-centric design is durable, loveable, valued and used. Things can be repaired, hired, swopped, shared, remanufactured, adapted, customised, upgraded, restyled, embellished and accessorised. In her thesis of 1999, where she identified that the largest environmental impact of textiles occurs in laundering, Dr Kate Fletcher (2008) suggests that garments could even be modular, so that bits can be taken off to be cleaned or replaced. Historically cuffs and collars on men's shirts were detachable for cleaning.

People-centric design involves creative thinking about how to service existing textiles. People-centric design processes creatively find ways to preserve textiles. It makes sense to work with and encourage good quality, workmanship and classic, lasting designs. Jonathan Chapman (2005) writes in his book *Emotionally Durable Design*, that we will cherish what we love. Often the items we care about most are very personal, individual, about ourselves and not necessarily designed to impress others. People-centric design can be achieved by noticing the quality of every activity of living, so that everything we do and create is not compromised or sacrificed on the altar of market forces. There are movements that already work in this way: slow design; permaculture farming methods; respect for local knowledge, skills and resources; and a respect for all talents within the community's own human resources. These skills get marginalised and lost whenever systems refuse to recognise and embrace the possibility for compatible diversity in all things. The greatest example for this is homogeneity of the high street. Now that those shops are empty, we can fill them with the products and services of our new sustainable revolution.

Interactivity can be manifested by co-designing, by wardrobe surgeries to remake your clothes, by commissioning new designers, artists and craftspeople, and by hiring, swopping, mending, and sharing. Sharing includes freeshare, and skill swopping, and is a modern form of barter, which works best in relatively small communities and flourishes when people are able to get to know each other personally.

1.4   The designer empowered

A referencing eco-textile glossary has been written (in wiki form so that it can continue to grow: http://ecotextileglossary.wetpaint.com/) because although the issues may seem complex, everything is interrelated so even a small action will have an effect. Use it to begin to answer specific questions.

1 Start by asking simple questions in order to identify which eco-options are possible.
2 Let a multi-dimensional awareness develop naturally with time and experience.
3 Allow your subconscious mind to process and include new 'sustainable' information within the mix of your design inspiration, market trends and your technical expertise.
4 It is better to get one thing right at a time than to fail or become dispirited by trying to change too much in one go. Remember, this is a whole system that has been evolving for centuries.

Bear in mind the following questions.

1 When you specify the production path for your design, can you factor out any big impacts such as excess use of resources, energy or transport?
2 When you design, will your product be a beloved classic that people want to treasure forever?
3 Will it preserve traditional skills? Will it use existing facilities? Will it provide or destroy employment?

4 How can you work more closely with your customers to give them exactly what they want?
5 Can you make sure that when your textiles are worn out they will have honoured the embedded energy and resources that made them, and will be able to be reused or recycled easily and appropriately?
6 How can all these ideas, and more, be accomplished in a way that maintains a healthy, growing economy as well as a healthy planet?

1.4.1 Use the following strategies to examine your particular production route

Can they help you to understand and alter anything along the way?

- *Efficiency and resource conservation*: ensures that the textile production is not taking out more than can be replaced, is not wasteful, and may even be contributing a replenishment of resources.
- *Design for disassembly*: asks you to be concerned with how a textile is going to be reused or recycled at the end of its useful life. It is essential not to mix or contaminate materials, particularly non-renewable ones, which you need to reclaim in order for them to be recycled.
- *Closed loop*: ensures that the end of the product's life leads to appropriate reuse or recycling. Non-renewables can be reclaimed and reused, renewables can be reused and eventually recycled by composting.
- *Design for durability*: asks you to value your materials, and make a textile or garment that will last at least long enough to honour the embedded energy and resources that have been used to make it. You could develop business plans that keep you in touch with your textiles and the customers who acquired them. Charge a little more for ongoing maintenance, spare parts, mending, swopping, collecting, sharing, altering, embellishing.
- *Emotionally durable design*: when you love something you have produced and so does your customer, it lasts longer, is cherished and treasured. It doesn't need to be replaced, but it might need to be maintained. It might be collected, traded, or mended, but probably not thrown away.
- *Product of service*: a modern-day twist on hiring rather than buying. A way of extending the way that needs and desires can be satisfied without the requirement for ownership; ensuring that everyone has access according to their need and desire when there are very limited resources. Sharing, hiring, borrowing, swopping, upgrading; everyone can have their day with a Prada bag!
- *Take back*: goods are returned to the manufacturers at the end of their cycle of use, either to recycle or to dispose of appropriately. If the materials are valuable or non-renewable, this is a way of maintaining access to raw materials with which to make new products.

- *Reuse and recycling*: this is the key to all fully functioning cradle-to-cradle systems of production; this is where the loop closes. Working hard at this point will reap rewards, because it is a major source of raw materials. Reuse is where a product is used again as is, or slightly changed and adapted. Recycling is where the materials transform into a new form either by composting and feeding the next renewable cycle of a natural fibre, or, in the case of non-renewables such as polyester, by being returned to feedstock for repolymerisation.

- *Local sourcing*: maintains and honours natural geographic and cultural diversity, enables specialism and is low on carbon emissions. An ideal is local self-sufficiency where possible and appropriate on a global scale. It does not exclude the possibility of global, specialist, individual trade if shop windows on the internet can use carbon neutral or more energy efficient transport systems.

- *Slow design*: an antidote to fast fashion, this is less stressful, and allows us to value materials and processes for their own sake. There are fewer 'deadlines' and more life. There is enough time for solutions and designs to evolve, for skills to develop and mature.

- *Tribal, ethnographic, regional, traditional design*: the industrial revolution sped up and relentlessly reproduced original technologies that had evolved slowly from generation to generation to meet needs and desires. The making of Harris Tweed, or tartan involved problem-solving skills in harmony with the use of local resources, efficiency, aesthetic expression and sophistication. These are the very images and allusions that fashion often emulates, but that market forces make uneconomic in their original form, while the workforce becomes redundant and the expertise is lost. But the originals are cultural treasures, whose creation gave dignity, identity and livelihoods. Not everyone wants to stack shelves or tweak computers. Can a new ecological economy revive and define traditional, tribal, regional and ethnic in a contemporary context?

- *Custodianship of skills and specialism*: needs to be maintained within local populations in order to be resourceful and resilient to change, but also to preserve the potential for self-sufficiency and the potential for future diversity at local levels. A high degree of accomplishment is fulfilling and satisfying. Schemes to recognise and encourage individuals and businesses arising from specialist skill bases are psychologically beneficial. This fosters a capable population.

- *Vintage, collecting, heirlooms*: husbanding resources, recognising and preserving value, enjoyment and entertainment. Economic vitality is enabled by passing things around, without expenditure of resources, allowing the culture to be reflected upon, maintained and vitalised. This is a way of sharing while maintaining economic movement.

- *Education, skills, training*: some of the most fulfilling activities of life, which should continue throughout life, refresh, contextualise, enliven, preserve, pass on, share and generally activate any community no matter how diverse, in challenging, yet non-combative ways. In the creative arts, this is the springboard of imagination and, for me, the zest of life. How we make things, how we meet, all the possible encounters, mediated through all the objects in our lives, these are activities of living a sustainable and fulfilling life; it is the quality of the encounter that holds the real value, not the speed of accumulation on a balance sheet.

1.4.2 Exercise: combine the previous strategies with issues in the life cycle stages below

Can they help you to understand and alter anything along the way? (The resulting ecologically friendly textiles and clothes need to be useful and beautiful. They need to be fit for purpose and affordable.)

- *What category of raw material?*
  Synthetic. Polymerised natural (cellulosic). Natural.
- *Raw material processing/harvesting methods?*
  Melt spun. Polymerised. Manufactured/farmed. Water use. Energy. Toxicity. Emissions. Land degradation. Waste. Appropriate locality. Appropriate regulated source (oil and non-renewables). Colouration. Biodiversity protected. Incorporation of recycled materials. Use of pesticides and fertilisers. Health and Safety for workers. Greenhouse gas emissions.
- *Yarn and cloth production?*
  Water use. Energy. Toxicity. Emissions. Waste. Appropriate locality. Infrastructure investment (machinery, factory, offices, transport fleet). Indigenous skills and expertise honoured. Transportation between processes/division of processes between localities and/or continents. Chemical usage: preparations, colour, finishing, dyes, printing inks. Airborne or waterborne pollution. Industrial waste created by rejected dye lots and production mistakes. Contained/closed production systems. Filtering and reclamation of chemicals, dyes and finishes. Greenhouse gas emissions.
- *Garment construction, finishing, embellishment?*
  Energy, toxicity, contamination between renewable and non-renewable raw materials, including finishing and embellishment. Informative labelling. Offcut waste and processing waste. Appropriate land use. Airborne or waterborne pollution. Transportation between processes/division of processes between localities and/or continents. Greenhouse gas emissions.

- *Packaging, presentation?*
  Origin of packaging, renewable or non-renewable, local or transported. Type of printing inks. Biodegradability. Use of machinery/energy in packing production methods of packaging. Toxicity, pollution. Excessive. Protective. Recycled. Recyclable. Reused. Reusable. Labelling. Greenhouse gas emissions.
- *Dispatching?*
  Transport. Energy. Packaging/protecting. Local or global distances. Toxicity/pollution. Greenhouse gas emissions.
- *Service/sales, industry infrastructure?*
  Where customer and textile meet. Local or global. Transport. Energy. Infrastructure investment (machinery/computers, shops, warehouses, offices). Trade and retail shows. Advertising. Public relations. Magazines. TV and film. Photography. Forecasting. Transport fleet. Greenhouse gas emissions.
- *Use?*
  Laundering and drying. Energy. Toxicity. Greenhouse gas emissions. Practicality/identity/aesthetic. Added value. Extending use.
- *Reuse/recycling?*
  Embedded energy. Consumption of resources. Reclamation of resources. Return of products. Sorting. Storage. Adding value to the recyclate. Reprocessing. Transport. Toxicity. Repolymerisation. Remanufacture. Water use. Energy. Emissions. Land degradation. Waste. Appropriate locality. Appropriate conditions for aerobic and non-aerobic biodegradability. Energy production. Production of compost. Conservation of finite resources.
- *Disposal?*
  Collection. Transport. Energy. Greenhouse gas emissions (methane). Appropriate or inappropriate conditions for aerobic and non-aerobic biodegradability. Energy production. Pollution. Water contamination. Toxicity. Chemical leakage. Land use. Burning. Waste of resources. Loss of access to non-renewable resources. Airborne particulates. Sorting. Added value.

1.5   Sources of further information and advice

Books

Black, S. (2008), *Eco-chic: The Fashion Paradox*. Black Dog Publishing, London.
Dombek-Keith, K. (2009), *Re-Fashioning the Future: Eco-Friendly Apparel Design*. VDM Verlag Dr. Müller, Saarbrücken.
Faud-Luke, A. (2005), *The Eco Design Handbook: A Complete Sourcebook for the Home and Office*. Thames and Hudson, London.

Websites

www.anniesherburne.co.uk
www.ethicalfashionforum.com
www.katefletcher.com
www.kingston.ac.uk/design/SDRC (dept: sustainable design)
www.productlife.designinquiry.wikispaces.net;   also   www.extra.shu.ac.
   uk/productlife
www.redesigndesign.org
www.tedresearch.net (Textile Environmental Design)
www.tfrg.org.uk (Textiles Future Research Group)
www.ecotextileglossarywetpaint.com/?t=anon

Articles and papers

Well Dressed? The present and future sustainability of textiles in the UK
   (www.ifm.eng.cam.ac.uk)

Certification, labelling

www.oeko-tex.com/OekoTex100
www.pan-uk.org (pesticides action network)
www.skal.com
www.soilassociation.org

1.6   References

CHAPMAN, J. (2005), *Emotionally Durable Design: Objects, Experience and Empathy*.
   Earthscan, London.
Ethical Consumer (2002), June/July, p. 12. http://www.ethicalconsumer.org/
   FreeBuyersGuides/clothing/alternativeclothescompanies.aspx.
FLETCHER, K. (2008), *Sustainable Fashion and Textiles: Design Journeys*. Earthscan,
   London.
LEMIRE, B. (2006), *The Business of Everyday Life: Gender, Practice and Social Poli-
   tics in England, C. 1600–1900*. Manchester University Press, Manchester, UK.
MCALPINE, T. (undated), (Cranfield) Eco design, vol. 5, no. 1.
MC DONOUGH W. and BRAUNGART, M. (2002), *Cradle to Cradle: Remaking the Way We
   Make Things*. North Point Press, New York.
PORRITT, J. (2007), *Capitalism as if the World Matters*. Earthscan, London.
ROSZAK, T. (1992), *The Voice of the Earth: An Exploration of Ecopsychology*. Phanes
   Press, Michigan, USA.
TRANBERG HANSEN, K. (2000) *Saluala: The World of Second Hand Clothing and
   Zambia*. University of Chicago Press, Chicago, USA.
WOOLRIDGE, M. (2006), '*The clothes line*' BBC Radio 4, 21 March 2006.

2

Sustainable cotton production

L. GROSE, California College of the Arts San Francisco, USA

Abstract: This chapter reviews prevalent perceptions about cotton's ecological impacts and brings additional considerations to the discussion. The benefits, risks and limitations of various farming systems in addressing the sustainability of cotton are then discussed. Complementary strategies to bring sustainable cotton to scale are presented. The chapter ends by linking cotton production to emergent trends in sustainable fashion.

Key words: organic cotton, integrated pest management, biological integrated pest management, best management practices, transitional cotton, genetically modified cotton.

2.1   Introduction

Cotton is arguably the most publicly scrutinized fiber in the world. It is through this scrutiny that we have come to develop a lens through which to examine all other fibers and define a playing field for sustainable textiles as a whole. The scale and complexity of cotton has drawn the attention of the industry and consumers beyond the materiality of products into the supply chain that produces, processes, trades and delivers them. This has encompassed a range of related industries in addition to textiles, including agriculture and the synthetic chemical industry.

From the platform of organic cotton, which is now well understood and accepted by mainstream markets, our scrutiny of this ubiquitous fiber is deepening and broadening and with it our understanding of what comprises a truly sustainable textile product and industry. This chapter briefly explores the range of cotton's impacts and suggests corresponding strategies for the future of cotton and for sustainable fashion as a whole.

2.2   Cotton basics

Cotton is one of our favorite fibers and represents almost 38% (ICAC, 2008) of the world's textile consumption, second only to polyester, which recently took the lead. Cotton is produced in approximately 90 countries worldwide (IIED, 2004), the largest producers being China and the United

States; with Pakistan, India, Uzbekistan and West Africa also being significant producers. These six major countries account for 75% of our global cotton supply (IIED, 2004). Cotton cultivation supports about 30 million farmers worldwide, 80% of which live in developing countries, working as smallholders (BCI, 2008).

The total area dedicated to cotton production accounts for 2.4% of arable land globally and this has not changed significantly for about 80 years. In that time yields have tripled, primarily owing to the use of chemical pesticides and fertilizers, expansion of irrigation, mechanical harvesting, precision farming and genetic modification of seed (Baffes *et al.*, 2004; Clay, 2004). These capital-intensive inputs have been focused primarily in developed nations, which account for 90% of global cotton fiber supply.

2.3   Global ecological effects

Chemical use on cotton has been well documented and publicized. We now know that cotton accounts for an estimated 25% of global insecticide consumption and 11% of the world's pesticide consumption (Clay, 2004). These chemicals have caused a wide range of environmental problems including severe human health problems, water and air pollution, insect and weed resistance, depleted soils and loss of diversity to name a few. Although chemicals are most often cited in relation to its production, cotton has also become known as a 'thirsty' crop. It evidently requires 7000–29 000 liters of water to produce 1 kg of cotton fiber (Clay, 2004). In retailing terms, this is enough fiber to produce one pair of jeans.

The standard industry approach for marketing 'sustainable fashion' assumes that the ecological impacts of products must be communicated succinctly to a consumer who has very little technical or industry knowledge. However, reducing information on cotton (or any fiber for that matter) to over-simplified statements results in narrow definitions and erodes our capacity to 'hold the tension' between effective marketing and the complex realities in the field. If, as an industry, we have to harness market forces to provide a mechanism for converting cotton fields to more sustainable practices, then we must equally cultivate the capacity for consumers to do more than simply consume. We must enable them to become well-educated, informed and active participants in real solutions to real challenges.

Expanding one's perspective to include the larger systems at work that influence farm practice, as well as focusing on the specific regional challenges in cotton cultivation, is the first step. This chapter begins to look at both scales and invites further debate.

2.4   Economic systems

Anyone who 'scratches below the surface' of our favorite fiber, is quickly drawn into the systems and institutions that perpetuate cotton's production, processing, promotion and consumption. These systems affect the 'sustainability' of cotton as much as, if not more than, its actual cultivation, since they direct cultural behavior and relationships throughout the supply chain. Cotton is a publicly traded commodity, with 30–35% of the global crop being traded annually (IIED, 2004). Cotton futures are traded on the stock market where the prevailing price for the fiber is based on the dynamics of supply and demand. Consequently, the grower has no control over the price he receives for his cotton. Cotton farmers worldwide are under pressure from declining prices. There are many forces that come into play to cause this decline, but in general they can be summarized as: technological improvements leading to increased production, efficiencies and growth in fiber supply, competition from chemical fibers, stagnant per capita demand and domestic policies (Baffes *et al.*, 2004).

Cotton is also subject to various marketing and trade interventions (Baffe *et al.*, 2004) in the form of subsidies, tariffs, quotas and taxes. At least eight countries have consistently supported cotton production, including Brazil, China, Egypt, Greece, Mexico, Spain, Turkey and the United States. Perhaps the most notorious intervention is the complex US government subsidy program, which at 3.5 billion dollars in 2005 (Chapman *et al.*, 2006) is almost double that provided by the second largest interventions from China (Baffes *et al.*, 2004). US subsidies were first instituted in the 1930s depression years. Originally intended as a temporary measure to minimize the financial risks inherent in farming, subsidies enabled farmers to survive bad years while keeping cotton and food crops in plentiful supply for the rest of the nation. However, the US farm program has since strayed from its original purpose. Today, much of the subsidy, based on acreage regardless of income, flows to large landowners, primarily in the southeastern United States, many of whom do not make their livelihoods from agriculture at all (Chapman *et al.*, 2006).

Farm subsidies comprise a complex set of overlapping financial relationships, which are difficult to explain, even for an economist. Simply put, subsidies guarantee farmers a fair price above and beyond the commodity market rate. When the commodity price falls below the cost of production, the subsidy makes up the difference capped at a certain price maximum. US cotton subsidies, it is argued, undermine free trade and encourage overproduction, increasing cotton fiber supply, which, under the 'laws' of supply and demand, depress world prices below what is economically tolerable for farmers in poorer countries.

Direct subsidies, however, are only one part of the complex global economic picture for cotton. There are many policies in place that influence farm costs and world prices to indirectly encourage fiber production, including local government incentives and quotas, government support for farm equipment and chemical inputs, World Bank loans and public–private loan schemes. In Turkey, for example, the GAP water project (otherwise known as the Southeastern Anatolia project) is projected to irrigate 1.7 million hectares of land and though not yet finished has already influenced the tripling of cotton production in the last decade (Barker, 2007). The effects of increased fiber supply on the commodity price can be seen clearly from the following example. In 1998 China faced increased population and greater need for food crops. To encourage farmers to grow soybean instead of cotton, the Chinese government purposefully put more than 200 000 metric tons of cotton on to the world market, and the increased supply dropped the price of cotton globally by 13% (Clay, 2004).

Whether international governments and organizations provide direct or indirect subsidies or influence the supply and demand economics of the global trading system, it is questionable that these actions benefit the farmers. Many smaller farmers, both in rich and poor countries, lease the land they cultivate and take out loans at the beginning of the season to cover costs of materials, labor and rent. They pay the loans back once the crop is harvested. Low prevailing commodity prices for the crop greatly hinder the ability of all farmers to settle their debts. Moreover, farmers around the world are pressured to maximize yields per acre in order to gain the highest return on investment and pay back loans. This creates an economic treadmill that fuels adoption of technologies and practices that push natural resources beyond ecologically sustainable thresholds. Squeezed on price, growers worldwide are forced to cut corners. Whether it is farmers in developing nations using cheaper, more toxic chemicals (Clay, 2004), or growers in the United States, refraining from making investments in more efficient and more expensive irrigation equipment (Fickett and Williams, 2008), many of the ecological problems associated with cotton production can be traced back, or at least linked, to national and global economic policies.

Changing these policies is a complex challenge since each region has a different set of mechanisms at play and vested interests at risk. However, perhaps the most challenging hurdle of all is that our global textile industry is the greatest beneficiary of low cotton fiber prices. The corresponding lower processing costs enable our companies to deliver cheaper goods at retail to the customer at record low prices, feeding consumption at ever-increasing speeds. This is evidenced by the value of the budget clothing market in the United Kingdom, which has grown by 45% in the last five years; twice the rate of the normal clothing market (Shah, 2008).

2.5   Farm systems

Our global cotton farms are either in the hands of a small number of large mechanized farms (Australia, United States and Brazil) or millions of smallholders (China, South Asia, West Africa) (IIED, 2004). Farm size, chemical inputs/outputs, energy efficiency and production efficiency differ widely from farm to farm and region to region, as do their related environmental impacts. In developing countries, farms can be as small as 0.5 hectares, and up to 20 hectares. These are often family run and are usually part of a mixed crop rotation system. Food and grains are grown generally for family consumption and animal feed, while additional income for the family is provided through cash crops, one of the most popular of which is cotton. In general, larger farms are situated in three main countries: United States, where they range from 200 to 800 hectares; Brazil, where cotton farms can be as large as 2000 hectares; and Australia where they can be as large as 15 000 hectares (IIED, 2004). Continuous cropping is the dominant system for these larger farms and is designed to generate as high an income as possible to the owner (often an absentee landlord) through large economies of scale. There is a large degree of mechanization in these developed nations since most farms are too large for the cotton to be picked by hand and wages are high, so costs of mechanization and chemical inputs are, by comparison, relatively low (ICAC, 2005b). Smaller farmers in these highly capitalized farming environments find it increasingly difficult to compete with the efficiencies of scale of the corporate growers. In the United States, for example, the number of cotton farms decreased from 43 000 in 1987 to 24 805 in 2002 and during this same period the average cotton farm doubled in size (Freese, 2007).

2.6   Water

The specific local or regional impacts of cotton cultivation differ widely according to a number of factors including climate, natural resources available, pest complexes, chemical and water inputs/outputs, access to capital and farm production efficiency. Looking at the two most cited impacts of cotton production – water and chemical use – from a regional perspective, some interesting options and 'sustainable' strategies emerge.

Cotton's volume of water consumption frequently tops the headlines as a major issue. Yet cotton's effect on water takes several forms: drawdown of natural water bodies for irrigation (inputs), contamination of fresh water from fertilizer and pesticide runoffs (outputs), and water management. Water management affects soil quality particularly when salinization occurs. Salinization is the process by which water-soluble salts accumulate in the soil. This is a resource concern because excess salts hinder the growth of crops by limiting their ability to take up water. The causes of salinization include the natural presence of soluble salts in the soil, a high water table, a high rate of evaporation or low annual rainfall. It is estimated that 4% of the world's total arable land is abandoned owing to former intensive cotton cultivation with soil salinization being the main reason (Kooistra and Termorshuizen, 2006).

The most well-known case of cotton's impacts on water is in Uzbekistan where water drawdown, contamination and poor irrigation practices all contributed to a well-documented social and ecological disaster. Surface water diverted for cultivation of cotton in the Araal Sea Basin reduced the sea to a fraction of its former size and the once-thriving fishing villages in the region are now surrounded by desert. More to the point, pesticides and fertilizer residues on the Araal Sea bed blow into surrounding communities. The population in this region suffers from chronic poor health as a result of exposure to agricultural chemicals and unsafe drinking water (EJF, 2007). Furthermore, the soil in this area has been so intensively farmed that it has now degenerated beyond its capacity to support future cultivation (IIED, 2004).

The case of the Araal Sea region represents the impact of cotton cultivation on water systems at its very worst and has contributed to cotton's global reputation as a thirsty crop. It is perhaps due to the extremity of this case, that countermeasures proposed by some marketers have been equally extreme: eliminating cotton production altogether or replacing all cotton fields with hemp, for example. While provocative statements might highlight differences and benefits between fibers to potentially influence market share, they do little to promote deeper understanding. Finding pragmatic and practical ways to achieve optimal water use is the imperative. Then, communicating the details to the end customer to provide education on real issues will help shape the market.

Real solutions to water challenges demand knowledge of the particular region where the fiber is sourced. Local climate, regional natural resources, access to technology and general farm practice are just a few of the contributing factors. Consider West Africa, for example, where seasonal tropical storms bring 32–50 inches (80–125 cm) of water to the cotton crop (IIED, 2004; Toulmin, 2006); and Texas, where almost all cotton is dry farmed. In Brazil, 50% of cotton production comes from rain-fed farms. In these areas, the drawdown of water from local sources is not such a critical issue, despite our perception to the contrary.

Responsible water management is not only defined by rain-fed areas. For example, in Israel, the cotton crop is irrigated, but water scarcity and cost has spurred technical innovation resulting in the most efficient watering systems in the world. Similarly, in California, where farmers face drastically increased water costs, a variety of solutions are emerging, including timely 'water deficit' and subsurface irrigation. In fact, in many areas, cotton is a moderate water user, accounting for less water consumption than perennial crops such as grapes, almonds, pistachios and stone fruits, and considerably less than field crops such as alfalfa (D. Munk, 2008).

2.6.1 Water options

A list of the options available to the grower for reducing water inputs is given below. While each of them may be appropriate in one area with a particular set of circumstances and conditions, they may be equally inappropriate for different areas and conditions. All strategies offer opportunities for marketing the end product. Knowing the source of the cotton fiber in our products and the regional impacts specific to that area is a prerequisite for accurate and authentic marketing.

Increased costs for water

Where water cost is low, over-irrigation may result (as in Uzbekistan). High water costs, however, will usually prompt the conservation to keep the costs of production low (as in Israel). However, increasing water costs as a blanket strategy may over-burden farmers and particularly smallholders beyond what they can financially bear (IIED, 2004). This is already apparent in California where the cotton crop has declined from more than 1 million acres (405 hectares) to less than 300 000 acres (121 500 hectares) over the last decade. This is in part due to the low commodity price, which no longer supports the cost of production, and in particular due to increased water costs (M. Fickett and F. William, 2008).

Rain-fed cotton

Rain-fed cotton offers an alternative to irrigated and diverted water supplies. However, rain-fed cotton tends to produce irregular fiber quality owing to the inconsistency of watering, and yields tend to be 50% of that of irrigated fiber (D. Munk, 2008). Rain-fed cotton is also in relatively short supply, representing 27% of global cotton production.

Changing cultural practices in the field

Shallow soil cultivation, mulching, minimal or zero tillage and organic production all improve soil structure and higher water retention is attained as a result. However, yields may also be adversely affected in some cases.

Deficit watering

Cotton benefits from stress more than other crops, since if it becomes too leafy, it produces less fruit (bolls) and fiber. Timely deficit watering, withholding water from the plant at non-critical times and supplying water to it at critical times, can suppress leaf growth and encourage the fruiting cycle. This both reduces water loss through evapotranspiration in the leaves and increases fiber yield (D. Munk, 2008).

Irrigation systems

Highly efficient irrigation systems such as those employed in Israel tend to be expensive to set up and maintain. These systems are therefore only possible to implement if funding is available, or if the price of the fiber allows for some discretionary capital investments. Longer-staple cottons command a higher premium and so producers may be more able to offset the cost of investment.

Subsurface drip irrigation

These systems are also expensive to install and maintain, and growers usually make the investment over a variety of crops in addition to cotton. Subsurface delivery of water reduces evaporation off the field and can be adjusted according to soil moisture content and other environmental considerations, including weather. Continuous maintenance of irrigation equipment is necessary after set-up so technical support is important, especially for smallholder farmers.

Alternate furrow watering

Alternate furrow irrigation can reduce some water losses by limiting evaporation from the field when the cotton plant is small in the early part of the season. Reduced losses through deep percolation are sometimes also effected. However, this is soil and location dependent; in short, results are time and location specific.

Traditional breeding of germ plasm for drought tolerance

Classic breeding of seed for drought tolerance in addition to fiber quality may take 10–20 years. Marker-assisted research allows geneticists to conduct genetic crosses in as little as five years (Allen, 2008).

Although all of the above options address drawdown of regional water bodies (inputs), contamination caused by farm outputs, still warrant attention. Hazardous pesticides associated with global cotton production are known to contaminate rivers in the United States, India, Pakistan, Uzbekistan, Brazil, Australia, Greece and West Africa (EJF, 2007); while in Australia and India, cotton irrigation also contributes to salinity, water logging and groundwater pollution (IIED, 2004). Chemical use reduction best addresses these issues.

2.7   Chemicals

Cotton is highly susceptible to pests, especially in humid areas. In order to protect their crop and yields, farmers use a variety of techniques, including chemical sprays, to deter and kill insects. These sprays set in motion a variety of dynamics that alter the natural food chain in the field. When sprays are used, it is not only target pests that are affected, but also their predators (beneficial insects). Furthermore, some insects flourish once the target pest is absent, causing a phenomenon called 'secondary infestation'. Farmers spray more chemicals to bring infestations of the secondary pests under control and overuse of the pesticide leads to genetic resistance in the insects. When the resistant pests return to feed on the crop, farmers often resort to stronger chemicals and spray more often, creating a continuous cycle where the more chemicals are sprayed, the more resistance builds. In Pakistan's Punjab region, for example, pesticide use on cotton has increased from 665 tons in 1980 to 47 592 tons in 2002 (IIED, 2004). Although use of pesticides on cotton now represents 50% of Pakistan's total pesticide consumption there has been no increase in yields (IIED, 2004).

The development of resistance in each insect is complex and determined by a variety of environmental and cultural conditions as well as the physiological characteristics of the pest. Insect lifespan, tolerance, volume and frequency of spraying, and length of exposure all contribute to potential resistance, so once the continuous spraying cycle is set in motion it is difficult to slow or stop. To date, almost 500 insects and mites have developed resistance to specific insecticides.

Besides chemical controls for insects, additional agricultural chemicals are applied to control weeds (herbicides) and plant pathogens (fungicides). Continued use of these chemicals also fosters genetic resistance. It is estimated that 48 weed species have developed resistance to one or more herbicides and more than 100 plant pathogens are resistant to fungicides (Clay, 2004).

Over the years, agricultural chemicals have been developed to be 'narrow spectrum', to more precisely target the problem pest or plant. They have also been modified to be less persistent in the environment and to have less toxicity to humans. However, cotton pesticides are still among some of the most toxic available for use in agriculture. Temik (chemical name:

aldicarb), for example, is classified as '1a extremely hazardous' by the World Health Organization and is widely used on cotton. In the United States, 85–90% of Temik sales are attributed to cotton production (Clay, 2004).

Farming systems also influence the impact of these toxic substances on human health. In developed nations, chemicals are applied mechanically from tractor spray rigs or airplanes called crop dusters. In these countries, pesticide use and storage is regulated and standards are enforced. In poorer countries, smallholder farmers have less access to technology and the farm system relies more on hand labor. Chemicals are applied by hand often from backpack sprayers, with the applicator held in front of the person walking into the just-sprayed areas. Pesticides are often stored at home in common living areas. Protective clothing is rare and regulations are lax.

2.7.1 Alternatives to chemicals

Just as with water strategies, pragmatic solutions to chemical use in cotton cultivation demand knowledge of the regional conditions and pest pressures where the fiber is being sourced. Only when cultivation conditions are understood can the appropriate chemical reduction strategy be effectively implemented. Conversion of cotton acres from chemical-intensive systems to biological systems is the imperative for our industry. Marketing worldwide data on cotton chemical use while using organic cotton from regions using no chemicals helps expand the market for organic, but we must redirect marketing from being the driving force to being the by-product of real solutions to real problems.

The following farming systems are examples of different management practices that can be implemented to improve the sustainability of cotton production and reflect differences in philosophical positions regarding conceptions of nature and environment. This sometimes presents the systems as opposites or being in conflict, whereas they are in fact all tools or technologies working toward the same end. While presenting them in opposition helps convey the differences between them (and there are clear differences), it does not necessarily aid understanding of the issues they aim to address. The following section of this chapter attempts to describe the different philosophies and the similar goals of organic, integrated pest management (IPM), biologically intensive IPM (BioIPM) and genetically modified (GM) farming systems and invites further discussion.

2.7.2 Organic production

Organic cotton production began in the United States and Turkey in the late 1980s and early 1990s, followed by Egypt, Uganda, India and Peru. The production of organic cotton has grown from 6 countries in the 1992/1993 crop season to 24 countries in 2006/2007 (Ton, 2002). India, Syria, Turkey, China, Tanzania, United States, Uganda, Peru, Egypt and Burkina Faso are the top 10 organic cotton-producing countries in order by rank. India recently overtook Turkey's long-time standing as the number one producer, and now accounts for 55% of organic cotton production worldwide (Organic Exchange, 2008).

The organic cotton market has grown from 2075 metric tons in 1992/1993 to 145872 metric tons in 2007 (Organic Exchange, 2008) and now represents 0.2% of the global cotton production. It is hoped that this increased market demand will create a 'pull-through' effect to convert chemical-intensive farming systems to organic. In fact, it is already enabling a more robust organic cotton infrastructure to be developed throughout the supply chain.

Limitations of organic

Organic cotton farming is very well suited to smallholdings where inter-cropping with other crops is standard practice and where the hand labor required is readily available. It does not easily lend itself to larger farms in developed nations where capital-intensive farming systems (use of chemi-cals and machinery versus labor) prevail and labor costs are high. In these systems, where efficiencies are already optimized, the additional labor costs are, with a few exceptions, more than the market can bear.

The primary focus of organic systems is reducing toxicity and thus reduc-ing chemical use. Organic production does not track optimal water use and practices, nor does it track responsible water extraction from the environ-ment, nor salinization or contamination of local water bodies from agricul-tural run-off. However, there are documented benefits associated with organic production. Farmers in Israel have reported that cotton grown under organic conditions requires 30% less water than cotton in conven-tional systems and that this may be related to better soil structure and higher water retention (IIED, 2004). Organic cotton farmers in Texas report similar results. The organic system does not track labor issues, nor does it ensure fair prices to farmers or fair wages for farm workers.

Organic cotton is primarily produced in areas where pest problems are manageable. In regions where the pesticide treadmill effect is most severe and the checks and balances of natural systems, such as predator popula-tions, have been completely degraded; switching immediately to an organic system represents too great a risk of crop failure and threatens the already tenuous farm finances.

Although the market demand for organic cotton is increasing, it repre-sents a very small percentage of the fiber inventory in the supply chain.

Preparation for an organic crop begins before planting in the spring, more than 9 months before harvesting. In order to allow time for spinning, processing and shipping, projections from the retailer must reach the grower at least 18 months before delivery of garments. Companies usually judge sales projections a few weeks or months in advance of key retail periods and moving this forward creates financial risk, especially in erratic economic times. The absence of advanced contracts from retailers and the industry means that a disproportionate financial risk of converting to organic lies on the shoulders of the farmer. In their 2007/2008 Farm and Fiber Report, the Organic Exchange notes that the lack of contracts and delaying or breaking of contracts results in caution by growers and slows the growth of organic production (Organic Exchange, 2008).

Organic cotton cultivation represents 0.2% of the world's cotton supply. Organic fields are, therefore, surrounded by conventional acreage. Cotton has a 9-month growing season and is often the last crop standing in an area where a variety of other crops are grown. As the surrounding fields are harvested, pests migrate to the cotton for food and shelter. It is therefore of utmost importance that biological systems are expanded region-wide, to help buffer the negative effects of conventional farms and optimize the success of organic farms. This is particularly challenging in developing nations where smallholdings are extremely diverse. But it is just as challenging in developed nations where farms tend to be large and neighboring farms may be owned by the farmer, or by absentee landlords. Leased land can flow from one person one season to a different farm entity the next. Developing relationships in farming communities to influence cultural practices is therefore as crucial as converting actual land to biological systems.

Persuading a conventional farmer to switch from a chemical-intensive system to an organic system is the equivalent of asking a Western medical doctor to switch to Chinese medicine and acupuncture; it is a fundamentally different system, requiring years of practical training, experience and technical support through the required 3-year transition. The perceived risks of shifting to a fully organic system are quite real. Reduced yields or crop loss from pest infestations not only increase the pressure on farmers when repaying loans. Reduced yields may also mean the difference between having and not having enough income on which to live. This economic pressure, in addition to unpredictable fluctuations in organic cotton demand, is the main hurdle to persuading a conventional farmer to change his growing practices.

These barriers can be overcome by building new business relationships and supply chain structures that mitigate risks for all sectors. It is in these new relationships, connecting farmer to market and market to farmer, that organic cotton has made the greatest contribution to the textile industry.

For today's consumer is becoming more interested in where their cotton comes from and how it is grown, and the current supply chain does not easily accommodate this information nor does it readily supply it. There are many examples of co-operatives, collaborations and partnerships that have been developed to bring organic cotton to market. A few examples are given below:

1   BIORE, established in 1991, is one of the pioneers. The mission guiding the company is: 'To meet the customer needs and produce the best product ... while helping to end poverty by respecting farmers as a partner in the supply chain.' Working with growers in Tanzania and India, a five-year contract and purchase guarantee secures both the supply and the demand for the organic fiber. 'Fair price' is established as a 15% premium on the average market price of the past five years. This formula 'flattens out' the price for the retailer and steadies the return for the farmer. With reduced volatility, both sectors are able to make further long-term commitments, and the flow of fiber through the supply chain in strengthened.

2   In addition to translating the demands of the customer into market access for the farmer, Bio Re Tanzania further strengthens the supply chain by buying the seed cotton, ginning and selling the lint, and providing training and advisory service to the farmers. The Bio Re project now employs a total of 236 employees in its Swiss production management center, and regional offices. Between India and Tanzania, more than 10 000 farmers have been mobilized to grow organic cotton. Retail customers include Co-op, Monoprix, Rewe, Leclerc and Mammut, to name a few, and Bio Re yarns are shipped to the European Union, Japan, United States, China and North Africa.

3   Gossipium is another pioneering project, headed by Textile technologists Abigail Garner and Thomas Petit. The company grew from working with the farmers service center Agrocel, based in Kutch, Western India. Having developed a formula for trading organic cotton while also providing a fair price to farmers, Garner and Petit then connected producers to local textile mills and established a retail store in the United Kingdom. A website, mail order catalogue and wholesale distribution channel further diversify the routes from farm to market.

4   The American bed linens wholesaler, Coyuchi, partners with Rajlakshmi Cotton Mills in India, a mill in which farmers are 10% owners. The farm project, Chetna Organic, receives support from Solidaridad and ICCO (Interchurch Organisation for Development Cooperation) in Holland, and provides a premium of 45% over conventional cotton prices. The partnership has increased from a few hundred

growers to several thousand (C. Nielson, 2009). Both Rajlakshmi and Chetna have fair trade in addition to organic certification.

There are countless additional examples. Spearheaded by non-governmental organizations (NGOs), the private sector, farmers themselves or a combination of all three, each project has its own structure and means of meeting the needs of specific growers and markets. Vertical manufacturer Pratibha Syntex also partners with Agrocel; The NGO Helvetas assists projects in Tanzania and Kyrgystan; Woolworth South Africa collaborates with ComMark and local farmers; the US-based farm group, Texas Organic Marketing Cotton Co-operative, negotiates its own terms directly with retailers. And so the list continues. These efforts are as diverse as biological systems themselves, yet all combine fair trade or fair price with organic, and all are committed to long-term partnerships as a means to mitigating risks from farm to market. Collectively they indicate a significant movement: a shift in culture throughout the value chain.

Transitional organic production

Transitional organic cotton is registered in an organic certification program and grown to strict organic standards, but is going through a mandatory three-year period before being allowed to carry an 'organic' label. This is a critical risk period for the farmer, since there is no 'safety net' of familiar chemicals that can be used in case of infestation, and yet the fiber cannot carry a premium in the marketplace. Recognizing the transition period as a hurdle to converting farmers and acres to organic, several companies have chosen to purchase this fiber as part of their 'sustainability' efforts. Esprit and Patagonia did this on a small scale in the early 1990s, and, most recently, Wal Mart committed to transitional cotton on a much larger scale. Developing transitional cotton products represents a shift in focus for the industry; a shift away from marketing as the driving force for a sustainability program to real ecological goals and partnerships. The collaboration and communication of these transitional cotton programs may also create a model for additional biological systems.

2.7.3 Biological systems

The strict standards and enforcement for organic certification are clearly necessary to reassure the industry and consumers that the fiber is authentically organic. However, these strict criteria also limit accessibility. If we are to reduce the global impact of cotton production we must ask critically: 'Is organic cotton the only available tool to achieve this?' Developing partnerships as described above takes concerted effort, leadership, money and time. If the perceived risks outweigh the ability of partnerships to support greater conversion of fields to organic cotton cultivation, then we must act as an industry to expand our 'sustainable' cotton options. Developing these options independently of market demand is imperative, for degradation of natural systems continues irrespective of market trends.

IPM, BioIPM and best management practices (BMPs) all offer the farmer an opportunity to test biological controls, with the safety net of being able to spray softer chemicals if an infestation threatens failure of the crop beyond economic thresholds. Because of this safety net, IPM, BioIPM and BMP have more potential to convert acres and farmers to biological practices than the organic system, which has a narrow set of criteria for certification. Simply put, these options are scalable. The following is a brief synopsis of each system.

Integrated pest management

IPM emphasizes the growth of a healthy crop with the least possible disruption to agricultural ecosystems and encourages natural pest control mechanisms (EJF, 2007). IPM is an ecosystem-based strategy that focuses on long-term prevention of pests and their damage through a combination of techniques such as biological control, habitat manipulation, modification of cultural practices and use of resistant varieties. Pesticides are used only after monitoring indicates they are needed according to established guidelines, and treatments are made with the goal of eliminating only the target organism. Pest control materials are selected and applied in a manner that minimizes risks to human health, beneficial and non-target organisms, and the environment (University of California Davis, http://www.ipm.ucdavis. edu/IPMPROJECT/about.html).

The most significant program aiming to engage farmers in the developing world in IPM cotton production is the Food and Agriculture Organization (FAO)-EU IPM Program for Cotton in Asia (EJF, 2007). This program operates in six countries: Bangladesh, China, India, Pakistan, Philippines and Vietnam. To date, more than 100000 farmers have graduated from IPM schools established under the scheme (EJF, 2007). IPM has also been implemented by a number of larger producers in Brazil, United States and Australia. It has been show in IPM systems worldwide that chemical use can easily be reduced by 50% (San Philippo, 2009).

Much of the IPM in developed countries includes transgenic seeds. In fact, the International Cotton Advisory Committee's (ICAC's) Second Expert Panel on Biotechnology of Cotton viewed genetic modification as a foundation or valuable component of IPM systems (ICAC, 2004). However, IPM schemes were operating long before GM seed was first commercialized in 1996, so it is not a given that genetic modification is a necessary element for a successful IPM program.

The Sustainable Cotton Project and biologically intensive IPM

Founded in 1996 to help reduce the toxic impacts of cotton in rural California, the Sustainable Cotton Project (SCP) helps growers in the San Joaquin Valley to convert practices from chemical-intensive to biologically based farming systems. SCP's BASIC program (Biological Agricultural Systems in Cotton) operates without the use of genetically modified seed and reports consistent chemical savings of 50–73% in comparison to conventional fields in the same region (Gibbs and Grose, 2008). The BASIC program is referred to as 'biologically intensive IPM' (BioIPM), because it has a full emphasis on naturally occurring biological systems as a first resort for controlling pests. BioIPM fully engages conventional farmers in recognizing and managing biological systems in the field, rather than simply substituting one class of chemicals for another more benign class. Because GM seed is not used, the results of using the biological systems become very clear and the farmer begins to trust them fully. Some farmers even go on to grow certified organic cotton.

The SCP provides technical information through their own certified organic pest control advisor, and links farmers to technical information provided through University of California (UC) co-operative extension advisors and UC IPM advisors. Growers enroll on the program before planting and team with mentor farmers. Biological controls match organic systems and include:

- inter-planting with alfalfa or other beneficial habitats such as corn, sunflowers, sorghum, blackeye beans);
- intensive scouting to monitor pests and beneficials;
- early releases of beneficials;
- limiting or eliminating pesticide applications in the spring and/or using soil fertility and nutrient monitoring;
- partnering with UC technical experts to customize approaches for each farm.

In addition, when pest infestations do threaten the survival of the crop, growers are required to use 'softer chemicals'. The program disallows the use of the top ten most toxic chemicals (based on PAN's *Bad Actor* category list, PANNA, 2008).

The results of the BASIC program are monitored through a statewide law. California has the largest pesticide regulatory system in the United States with more than 400 inspectors overseeing local enforcement, operating from agricultural commissioner offices, county by county. Farmers must obtain site-specific permits from their local commissioner to buy or use agricultural chemicals. The request for a permit is evaluated according to environmental and public safety parameters, the volume of chemicals used is documented and aggregated under California Environmental Protection Agency (CALEPA) State Department of Pesticide Regulation (DPR) supervision, and then the data are made publicly available. This same system tracks chemical use on the BASIC fields and the data are then compared with county averages on conventional fields in the same region.

Since it has been shown to maintain yields very close to conventional fields, BASIC has been more successful in enlisting farmers than organic, which has many more economic risks associated with its production; organic yields can be as much as 40% less than conventional yields in California. In 2006, there were just two organic cotton farmers in California, cultivating 250 acres (100 hectares). In contrast, there were 22 BASIC growers farming 2000 acres more widely.

The Better Cotton Initiative (BCI) and best management practices (BMPs)

The BCI's BMP program is designed to improve the production practices of as many farmers as possible, rather than excluding farmers who do not employ practices consistent with the BCI's production principles. This approach enables more conventional farmers to transition into biological systems, thereby expanding the total area of land with BMP (BCI, 2008). The defining feature of the BCI's BMP is its emphasis on water management and labor in addition to reduced chemical use. BCI production principles include:

- minimizing the use and impact of pesticides;
- optimized use of water and care for water availability;
- conservation of natural habitats;
- water must be extracted legally with neither groundwater nor water bodies used for irrigation being adversely affected;
- soil management practices maintain and enhance the structure and fertility of the soil and include minimum tillage, cover crops and rotation crops;
- nutrients are applied to minimize risk to the health of workers;
- labor rights include voluntary overtime with pay, no child labor, etc;
- preservation of fiber quality is a priority.

In addition to developing these goals, the BCI also aims to strengthen the culture of BMP through knowledge sharing and skills development; negotiating more favorable terms of business and lending, encouraging group investment in equipment, marketing cotton in the best interests of the grower and extending rural lending to enable farmers to invest in long-term sustainability. Monitoring and evaluation is also planned to demonstrate improvement over a specified period of time. Although this approach is complex given the different farm systems, cultural practices and pest/

weed complexes worldwide, the BCI has developed a global network of companies and organizations to advise and support the program. Ecological and social hot spots in cotton production have been identified, based on acute problems with pesticides, labor and water management. Uzbekistan, Pakistan, China, Turkey and Australia are current priorities (IIED, 2004). The BCI firmly places ecological and social touchstones as the measure of progress, and marketing as a secondary consideration.

Limitations of biological systems

Some of the challenges to IPM, BioIPM and BMP are quite similar, the most profound being the lack of market incentives. While organic cotton currently commands a premium price, BioIPM, IPM and BMP are little known in the market. Australia's Commonwealth Scientific and Research Organization (CSIRO) reported that a study had not revealed any evidence of a market for IPM or BMP cotton (BCI, 2008). This is primarily because the term and label 'organic' is perceived as being the pinnacle of sustainable agricultural practice by the market, and any other systems or approaches are perceived as 'less than' or compromised (see Fig. 2.1). The Sustainable Cotton Project has developed a trademark, Cleaner Cotton™, to market its BioIPM cotton, which together with a 'locally grown by family farmers' message may provide a model for marketing BioIPM, IPM and BMP in other regions.

Currently, there is no legislation to 'shape' the market towards better practice. Although BioIPM, IPM and BMP can all reduce costs in terms of chemical inputs, 'softer' chemicals are generally more expensive. With the low prevailing commodity price for cotton, there is no excess capital to buffer the financial risk taken by the grower. Any associated costs for implementing these programs therefore falls disproportionately on the shoulders of the producers (IIED, 2004). Forward contracts, partnerships and new relationships could encourage farmers to take the risk and implement these systems. In addition to lack of market awareness, there are also in-field hurdles to implementing biological systems. Though producers in developed countries may have access to agricultural support from universities, qualified technical support for smallholders in poorer nations is weak or lacking.

Since the market is focused on organic systems, foundation funding is generally directed to those programs working on organic cultivation. As a result, biologically based programs work with very limited budgets, which threaten their long-term viability. Although university agricultural extensions provide technical support in developed nations, dissemination of information to the farmers is a challenge without the support programs, which work with farmers in the field on a daily basis. Their presence

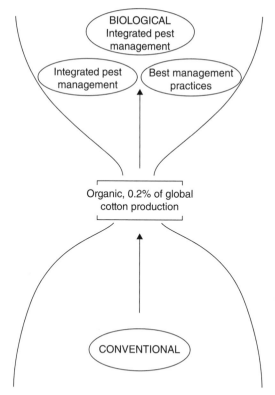

2.1 Market bottleneck caused by organic cotton. Organic cotton is seen as the pinnacle of sustainability in the market and creates a bottleneck, restricting market incentives for farmers transitioning to biological systems. (Credit illustration: Denise Ho, Academy of Art University, San Francisco, California.)

helps to maintain the farmer's focus on and engagement with biological controls. If technical support is not present, farmers may slide back into 'easier', more familiar (chemical) production methods (M. Gibbs, 2008).

Most organic production is concentrated in poorer nations where costs for hand labor are cheaper. This further compounds market challenges in developed nations, since imported organic is often less expensive than the locally grown biologically based cotton.

The benefits of IPM, BioIPM and BMP, and the changes in the biological balance of the field (such as the slower emergence of resistant pests), are more apparent over large areas. In smaller areas, benefits are not so apparent, since conventional growing practices of a nearby field can cause pests to migrate to neighboring biologically based crops. Verification, documentation and monitoring of chemical use globally are complex, given the large number of smallholders, the variety of farming systems in play and the different pest and weed complexes in each. SCP's BASIC program is able to measure and make data available through a California state-mandated reporting system and this may provide a model for other regions (see earlier paragraphs on SCP in Section 2.7.3).

Many IPM programs and the BCI's BMP are 'technology neutral', and embrace the prevailing regional use of GM seed to maximize farmer participation. The BCI's BMPs include water use and management criteria as well as labor standards. Weighing the wider benefits of IPM and BMP against the inherent risks of GM seed needs to be done on a case-by-case basis (see Section 2.7.4 below).

2.7.4 Genetically modified cotton

The adoption of GM cotton has been rapid from its introduction in 1996 and cotton growers have adopted biotechnology at a faster pace than growers of any other crop (Cantrell, undated). Ten countries now allow GM cotton to be grown: Argentina, Australia, Brazil, China, Colombia, India, Indonesia, Mexico, South Africa and the United States (ICAC, 2006). There are reportedly seven million farmers using biotech cotton, more than 85% of which are resource-poor farmers planting pest-resistant cotton (ICAC, 2004).

In the 2005/2006 crop season, 28% of the global area of cotton production was planted with seed having GM traits of herbicide tolerance and insect tolerance, representing about 38% of the bales exported globally (ICAC, 2006) and in the United States where the technology was first developed, it is estimated that 83% of the total 2007/2008 cotton crop was GM (Freese, 2007). This technology is engineered to add several desirable qualities into the cotton plant, including higher agronomic performance, fiber quality, stress and drought tolerance, etc. (Cantrell, undated), but thus far only two traits have been commercialized: pest-resistant cotton (Bt) which constitutes 50% of global GM cotton; and herbicide-tolerant cotton (Ht) which represents 16% of global GM cotton. There are also 'stacked' varieties of GM cotton that are both herbicide tolerant and insect resistant, these represent 37% of global GM cotton.

Insect-resistant cotton is genetically engineered to produce a single toxin derived from a soil bacterium, *Bacillus therungensis* (Bt), which kills certain insect pests when they eat the plant. Bt cotton specifically targets the budworm and the boll weevil family of insects including budworm, pink bollworm, beet armyworm and cotton bollworm (Gould and Tabashnik, 1998). Studies show that in several areas around the world, Bt technology has been very effective in reducing losses due to bollworm infestations in particular. The most dramatic reductions in the environmental impact cotton production through the use of biotech cotton have occurred in the United States, Australia and China, who were all early adopters of the technology (Cantrell, undated).

Ht cotton is genetically engineered to survive the application of a broad-spectrum herbicide that would otherwise kill a non-engineered plant. This allows farmers to use the herbicide to control a variety of weeds, without having to shield the cotton crop from exposure to the chemical. It also simplifies weed management and reduces labor costs. Monsanto's cotton, engineered for use with the company's Roundup® herbicide, is the most dominant herbicide-tolerant GM cotton, representing 96% of the Ht market, and is called Roundup Ready® cotton. The active ingredient in the Roundup® herbicide is glyphosate. The reported benefits of using Ht transgenic cotton are reduced use of more toxic herbicides and reduced tilling of the soil to disturb weed populations (otherwise known as conservation tillage). Conservation tillage is desirable because it reduces soil erosion from wind and water, particularly on sloped land, increases the soil's water holding capacity, reduces soil degradation, and reduces water and chemical run-off. It is estimated that Ht is responsible for the expansion of conservation tillage practices on approximately 60% of the US total cotton acreage (USDA-NASS, 2000). Reduced till also has associated economic savings for the farmer, since it requires less labor and fewer tractor passes resulting in lower fuel consumption.

The genetic modification debate is a heated one and the available facts and figures for and against can be confounding to companies and designers trying to make educated decisions about sustainable product development. The following sections give a few examples.

Insect-resistant (Bt) cotton

In China's Yellow River Valley region, severe bollworm infestation developed into a crisis associated with yield loss and hazards from repeated spraying of pesticides. Use of Bt technology in these circumstances resulted in a 60–80% reduction in pesticides use and helped revive cotton production in this region (Pray *et al.*, 2002). Chemical use reductions of this volume are clearly remarkable, yet other sources contradict these claims. Kooistra and Termorshuizen (2006) report that although Bt cotton initially helped revive cotton production in the Yellow River Valley region, infestations of lygus bug, red spider mites and whitefly are still prevalent and Connor (2006) reports that within a few years, growers in this region found themselves using more insecticide than non-transgenic cotton growers, owing to secondary pest problems. Furthermore, farmers pay a higher price plus a technology fee for GM seed, anticipating benefits from reduced expenditures on insecticides. When pesticide use actually increases, so do costs. The

ICAC (2005a) reports that with prevailing planting seed prices, the net benefit from the adoption of Bt technology does not unambiguously favor adoption of Bt cotton.

Similarly, in India, Bt cotton has reportedly reduced pesticide use by 60% and increased net yields by 29% (Manjunath, 2004). Yet scientists counter that yields depend on such a complex variety of inputs – including water, timely fertilizer applications and general management of problem pest populations by the farmer – that it is near impossible to attribute increased yields solely to the use of transgenic seed. Reports show that although Bt cotton has helped farmers reduce yield losses due to damage from boll-worms, it has not been effective against other pests. It is also noted that in areas that do not have heavy bollworm infestations, the impact on yield of Bt cotton is negligible (Freese, 2007).

Despite the mountain of scientific evidence that argues points for and against GM technology, there are some areas where both sides of the debate concur. It is these areas that can perhaps guide us to draw some rudimentary conclusions.

Evolving genetic resistance

The Bt toxin is currently used in moderation as a topical spray by both organic and IPM farmers and is considered safe for humans and wildlife. Unlike broadspectrum insecticides, which kill a whole host of insects, including the natural predators of the target problem pest, the spectrum of insects that Bt affects is quite narrow and the negative effects on beneficial insects have been rare and exceptional (Gould and Tabashnik, 1998). Bt can therefore be used in combination with biological controls, which has helped to establish it not only as a staple of organic farming systems, but also as a highly valuable complement to the more widespread IPM systems. In contrast to the spray, the Bt toxin in transgenic cotton is present continuously in every cell of the plant. This in turn continuously exposes the insect to the toxin and, it is argued, is highly likely to foster resistance.

ICAC (2005a) notes: 'All sectors of the cotton industry, including pesticide companies and biotech technology owners, agree that it is only a matter of time before cotton pests evolve resistance to the Bt toxin. The threat is real and acknowledged by everybody'. There is recent evidence from Bruce E. Tabashnik, insect researcher at University of Arizona, that pests have already developed resistance to the Bt toxin formed in plants. According to Tabashnik's research, published in *Nature Biotech* (Tabashnik, 2008), cotton bollworms found between 2003 and 2006 in the United States display significantly lower levels of sensitivity to the toxin compared with in previous years. To combat insect resistance, experts generally acknowledged that 'some level of pre-emptive resistant management is required' (ICAC, 2004), and others state that 'regulation, approval and monitoring (of GM technology) should be rigorous, transparent and continuous for the life of the technology' (Cantrell, undated).

Pre-emptive resistance management

However, resistance management plans are a challenge to develop for several reasons. Firstly, the causes of resistance itself can vary from pest to pest since there are many ecological and genetic factors that affect the rate at which an insect will evolve resistance to the Bt crop. The number of generations per year exposed to Bt in both transgenic crops and sprays, the percentage of the population that was exposed and the mortality of the individual pests are just a few of the relevant factors that must be considered (Gould and Tabashnik, 1998). Furthermore, since each pest differs in terms of population biology and physiological response to Bt cotton, no single resistance management plan could be optimally suited for all pests.

One resistance management strategy for biotech crops recommended by the US Department of Agriculture (USDA) is 'refuge areas'. This requires a 25 acre (10 hectare) section for every 100 acres (40 hectares) planted to Bt cotton, to be planted with conventional cotton and sprayed using an insecticide other than Bt. This provides a habitat for pests not exposed to Bt which are then available to mate with pests that have been exposed, thereby reducing the probability of genetic resistance. However, due to differences among pests in different areas no single plan can suit all of the major cotton growing regions; the scale of refuge that is most effective varies from pest to pest and is dependent upon a host of conditions, including adult movement and mating patterns as well as larval movement. Moreover, in some areas multiple pests occur (Gould and Tabashnik, 1998). Expert opinions on the 'safe' scale of refuge area vary widely. Gould and Tabashnik (1998) recommend that at least 50% of the cotton acreage on any southern US cotton belt farm should be planted with non-Bt cotton, while in Australia the Bt cotton acreage is limited to below 20% of all cotton grown.

Pest resistance strategies present particularly daunting challenges in third-world countries. As the Second Expert Panel of Biotechnology of Cotton (ICAC, 2004) states: 'GM demands local knowledge of the eco system, the weed complex, and pests and pathogens specific to a given bioregion, especially in smallholdings, where the plantings are diverse and pest populations are tremendously varied'. In South Africa, for example, the government and Monsanto recommended a refuge area of 20%. However, adherence to the strategy was reportedly poor, especially among smallholders, who had difficulty grasping the concept (Bennett *et al.*, 2003). Jayaraman *et al.* (2005) warn that poor refuge practices call for serious attention with the expanding area of Bt cotton. Some 85% of the adoption of Bt transgenic cotton is in third-world countries and further growth is planned, yet a proliferation of biosafety plans is clearly impossible to administer.

Herbicide-tolerant cotton

Ht cotton initially benefited farmers by reducing labor costs for weeding. However, in the absence of multiple weed management strategies, and with reliance on this single herbicide, genetic resistance to glyphosate has rapidly developed. This is particularly apparent in the United States where the technology was first adopted. Resistant weeds include: pigweed in Georgia, which survives almost 12 times the normal application rate of Roundup®; hairy fleabane, which has shown resistance in 12 states (Laws, 2006); and horseweed, which shows resistance in California (M. Ficket and F. Williams, 2008). All told, there are now 12 herbicide-resistant weeds across several states in the United States (Allen, 2008). The immediate response from farmers faced with resistant weeds is to spray more herbicide, and this has resulted in increased volume of glyphosate on cotton. In California, for example, where Ht cotton represents 61% of the cotton crop, glyphosate volume increased 200% between 1996 and 2005 (PANNA, 2008).

The long-term strategy for weed resistance involves a broader selection of weed eradication techniques – such as hand cultivation, double gene technologies in the cotton itself and use of a wider variety of chemicals. However, since Roundup Ready® cotton has been so widespread, research into new classifications of herbicides has been dampened and there are very few alternatives available (Mueller *et al.*, 2005; Yancy, 2005). Choices therefore tend to be older more toxic chemicals. And again, these additional strategies increase costs to the growers, who were originally sold GM technology with anticipated reduced input costs. In light of the evident resistance, the cotton industry now recommends that a range of methods and techniques for weeding be employed, rather than a single technology (ICAC, 2004). In short, GM requires the variety of techniques that IPM offers to ensure its long-term effectiveness.

Indirect benefit of genetically modified cotton

An indirect but potentially significant benefit of transgenic technology is the apparent 'improved populations of beneficial insects and wildlife in the cotton fields' where GM cotton is used (ICAC, 2004). In areas that are so degraded by intensive chemical use that the checks and balances of natural and biological systems are absent, immediate conversion to organic production is practically impossible. In these cases, GM cotton may provide a short-term stepping stone to biological and IPM systems. There are apparently a number of field studies that have shown increases in beneficials (Fitt and Wilson, 2002). However, these regional short-term benefits must be weighed against the large-scale, systemic and long-term risks. Simply put, the more that weeds and pests are exposed to GM crops, the faster the evolution of resistance. As contributors to the Second Expert Panel Report on Biotechnology of Cotton (ICAC, 2004) acknowledge: 'we can never know everything about a new technology, nor definitely predict long term consequences'.

2.8   Conclusions

Currently the 'sustainable' fashion industry and market view organic cotton as the pinnacle of sustainability. At 0.2% of global cotton production, this perception creates a bottleneck in the market and restricts incentives for farmers to move to biological and IPM systems (Fig. 2.1). Furthermore, water issues and labor are not captured by organic standards and yet also represent a significant part of cotton's impacts. Simply put, organic production, is not so much the 'pinnacle' of sustainability, it is rather a 'pillar'. If certified organic is too narrowly defined, the market niche too small or conversion too slow to convert fields and farmers to organic practices, then we must as an industry support bringing additional tools through the supply chain. IPM, BioIPM and BMP are scalable and therefore amplify the ecological benefits.

Marketing IPM, BioIPM and BMP involves re-educating the consumer on why a variety of approaches to sustainable cotton production is a more effective strategy than a singular approach. It will take a concerted effort to present the approaches as complementary rather than in competition with each other (Fig. 2.2). However, as an industry, we have already succeeded in capturing the attention of the consumer. We have already educated them on issues reaching far beyond the product quality, style and materials and they now have the capacity to understand the complexities of sustainability beyond an organic label. Furthermore, our industry is one of the most effective communicators in the world, touching the everyday lives of people across all demographics and cultures. We simply need to apply our skills and resources in a collaborative effort to communicate a deeper level of understanding regarding cotton's production and processing.

Scientific data comparing GM and organic cotton is covered in Chapter 11. Here, the evidence seems to indicate that where biological systems are so out of balance and degraded that crops are under siege from pests, GM Bt used in moderation may provide a useful short-term stepping-stone to

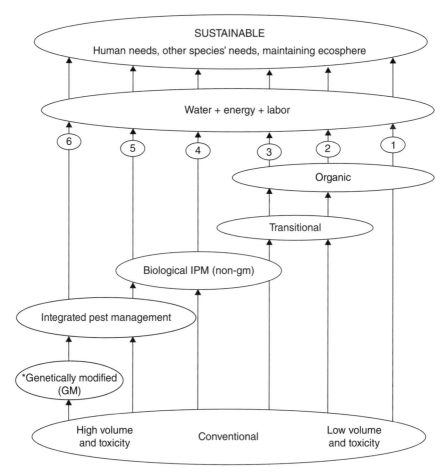

2.2 Expanded options for 'sustainable' cotton. Different farm systems (e.g. numbers 1 to 6 here) are often perceived as being in competition with each other, when they are in fact working towards the same goals. Organic production is one tool that is a stepping-stone to more sustainable practices in cotton, especially in areas of low chemical use. Additional biological systems broaden ecological goals through scalability and may be 'sustainable' in and of themselves. *GM and Bt cotton in particular may initially provide a stepping-stone to biological systems in highly degraded areas where chemical use and toxicity is high, but the risks of evolving genetic resistance in insects are real and acknowledged by everyone. (Credit illustration: Denise Ho.)

IPM where organic cotton production cannot be used. However, the effects of this technology are not yet known and may never be, and so the precautionary principle would be the best guide for GM systems or IPM systems that include genetic modification.

Given the limits of a micro-niche organic market on the one hand, and the wide-scale systemic risks of GM technologies on the other, best management practices proposed by the BCI, and BioIPM in particular, hold the most immediate potential for the future of sustainable cotton production. Biologically intensive IPM brings the farmers back into the field, thoroughly engages them in managing biological systems and may provide a stepping-stone to organic systems as market support builds or when legislation helps shape the market through incentives. BioIPM represents less risk of both reduced yields and increased costs to the farmer and is therefore scalable, converting more farmers and more acres faster than organic systems. BioIPM also provides a useful function as biologically intensive buffer zones between organic and conventional fields, mitigating pest migrations resulting from different cultural practices and harvesting times on neighboring conventional farms. BMP perhaps represent a more holistic approach since they address water and labor issues in addition to pesticide use. BMP and IPM that include GM crops may provide a stepping-stone into BioIPM systems, although they carry the inherent long-term systemic risks of GM technology.

Although organic cotton is criticized as a micro-niche in market and field terms, this detracts from the real contribution of organic cotton to the textile industry. For its promise can not be measured in volume alone. Perhaps the greatest contribution organic cotton has made is in forging new partnerships through the supply chain, fostering collaborative approaches to bring awareness of the ecological issues in production to the consumer and engaging all sectors of the industry to work together to build new markets and infrastructure. It is this shift in cultural practices from the field and throughout the supply chain where the future of fashion lies.

2.9   Future trends

Our current supply chain is based on cheap labor and cheap oil. It is a model that is increasingly outdated. With diminishing resources and corresponding increased costs, our industry must radically change the way it operates if it is to survive at all. Ironically, inspiration for new approaches may lie in the cotton fields themselves. In developed nations, where the newest technologies have been applied for decades, yields have increased tremendously and this has fuelled an ever-expanding textile industry. There is every indication that we have reached maximum yield potential in these areas. Farmers now carefully weigh the expense of cultivation inputs against the potential return in crop yields. For example, they often decide to hold back on irrigation, and sacrifice some of the yield to arrive at a place where the economics work (D. Munk, 2008). In other words, at the intersection where inputs, yield and costs meet, increased fiber volume is not necessarily the desired outcome.

Some of the more progressive companies in our industry are coming to similar conclusions about their own businesses. Given higher oil, material and labor costs, they are realizing that the constant drive to increase the volume of product throughput is no longer a relevant approach; it's an old model. Companies are now seeking to decouple growth from consumption (Jones, 2008). Regional manufacturing pods, reduced supply chain length, local to market production, extending product life, zero waste flat patterns, integral knitting (Rissanen, 2008), closed-loop recycling (patagonia.com, 2008) and product cycles and services (Fletcher, 2008) are just some of the newer concepts being deliberated. The emergent business models resulting from these inquiries challenge our industry's perception that we constantly need to increase fiber production. Stability, exchanging resources business to business, and income generation through a variety of means and services, are the new measures for success.

In the short term, cotton will only be part of the agricultural landscape if it is economically viable for the farmer, since a range of crops competes with cotton economically for a place in the field (D. Munk, 2008). New relationships and partnerships through the supply chain need to be developed stabilizing prices for the farmer and the industry, thereby securing supply. If these partnerships do not occur, farmers will simply grow other crops for more secure markets, and cotton will become a smaller percentage of overall world textile production.

In the long term, partnerships and collaborations across all sectors of the cotton industry, from farm gate to end user, are necessary if we are to support our farmers worldwide, secure fair returns for everyone through new business models, restore the natural environment on which our industry depends and stabilize our industry as a whole.

2.10 References

ALLEN W (2008), *War on Bugs*, Chelsea Green Publishing, Vermont, USA.
BAFFES J, BADIANE O, NASH J (2004), Cotton: market structure, policies and development issues, Paper presented at the WTO African Regional Workshop on Cotton, Benin, 23–24 March.
BARKER D (2007), *The rise and predictable fall of globalized industrial agriculture*, report from The International Forum on Globalization (IFG), San Francisco, USA.
BENNETT RM et al. (2003), in International Cotton Advisory Committee (ICAC) (2004), *Report of the Second Expert Panel on Biotechnology of Cotton*, USA, p. 32, ISBN-9704918-6-7.
BCI (BETTER COTTON INITIATIVE) (2008), *Global Principles and Criteria, Version 0.5*. http://www.bettercotton.org (accessed August 4, 2008).

BIOSAFENET (2008), *Research on the Development of Resistance in Six Pests. Cotton Pest Develops Resistance to Bt Plants: 'Evolution in Action'*, supported by Federal Ministry of Education and Research, accessed on February 8, 2008 at http://www.gmo-safety.eu/en/news/618.docu.html.

CANTRELL R (undated), *The Role Of Biotechnology in Improving the Sustainability of Cotton*, Cotton Inc., USA, www.cottoninc.com.

CHAPMAN D, FOSKETT K, CLARKE M (2006), How your tax dollars prop up big growers and squeeze the little guy, *The Atlanta Journal*, October 1, http://www.ajc.com/metro/content/metro/stories/cotton1.html.

CLAY J (2004), Cotton, Chapter 12, in *World Agriculture and the Environment*, World Wildlife Fund, Island Press, Washington DC, USA, pp. 283–304.

CONNOR (2006), in Freese B (2007), *Cotton Concentration Report*, Center for Food Safety (CFS), International Center for Technology Assessment (CTA), Washington DC, USA, p. 15.

EJF (ENVIRONMENTAL JUSTICE FOUNDATION) (2007), *The Deadly Chemicals in Cotton*, Environmental Justice Foundation in collaboration with Pesticide Action Network UK, London, ISBN 1-904523-10-2.

FITT G, WILSON LJ (2002), as cited in International Cotton Advisory Committee (ICAC) (2004), *Report of the Second Expert Panel on Biotechnology of Cotton*, USA, p. 31, ISBN-9704918-6-7.

FLETCHER K (2008), *Sustainable Fashion and Textiles: Design Journeys*, Earthscan, London.

FREESE B (2007), *Cotton Concentration Report*, Center for Food Safety (CFS), International Center for Technology Assessment (CTA), Washington DC, USA.

GIBBS M, GROSE L (2008), New cotton category comes to Market, *Pesticide News*, PAN*UK*, London.

GOULD F, TABASHNIK B (1998), Bt cotton resistance management, in Mellon M and Rissler J (Eds), *Now or Never*, Union of Concerned Scientists, Washington DC, USA.

ICAC (INTERNATIONAL COTTON ADVISORY COMMITTEE) (2004), *Report of the Second Expert Panel on Biotechnology of Cotton*, USA, ISBN-9704918-6-7.

ICAC (INTERNATIONAL COTTON ADVISORY COMMITTEE) (2005a), Concerns, apprehensions and risks of biotech cotton, *ICAC Recorder*, March, **XXIII**(1).

ICAC (INTERNATIONAL COTTON ADVISORY COMMITTEE) (2005b), as cited in Kooistra K and Termorshuizen A (2006), *The Sustainability of Cotton*, Report 223, Science Shop Wageningen UR, the Netherlands, p. 19.

ICAC (INTERNATIONAL COTTON ADVISORY COMMITTEE) (2006), as cited in Cantrell R (undated), *The Role Of Biotechnology in Improving the Sustainability of Cotton*, Cotton Inc., USA, p. 2.

ICAC (INTERNATIONAL COTTON ADVISORY COMMITTEE) (2008), *Cotton's share to decline in 2009*, USA, http://www.icac.org/cotton_info/tis/organic_cotton/documents/english.html (accessed November 14, 2008).

IIED (INTERNATIONAL INSTITUTE FOR ENVIRONMENT AND DEVELOPMENT) (2004), Research for IFC Corporate Citizen Facility and WWF-US, *Better Management Practices and Agribusiness Commodities. Phase 2: Commodity Guides*, IIED, London.

JAYARAMAN KS, FOX JL, JI H, ORELLANA C (2005), Indian Bt gene monoculture: Potential time bomb, *Nature Biotechnology*, **23**(2), 158, as cited in ICAC (2005), *The Performance of Bt Cotton Hybrids in India*, March 2002, p. 12.

JONES H (2008), *Nikes Journey*, in Ecollection at MAGIC, Las Vegas, February 13.

KOOISTRA K, TERMORSHUIZEN A (2006), *The sustainability of cotton*, Report 223, Science Shop Wageningen UR, The Netherlands.

LAWS (2006), in Freese B (2007), *Cotton Concentration Report*, Center for Food Safety (CFS), International Center for Technology Assessment (CTA), Washington DC, USA, p. 23.

MANJUNATH (2004), in Kooistra K and Termorshuizen A (2006), *The Sustainability of Cotton*, Report 223, Science Shop Wageningen UR, The Netherlands, p. 22.

MUELLER *et al.* (2005) and Yancy (2005), in Freese B (2007), *Cotton Concentration Report*, Center for Food Safety (CFS), International Center for Technology Assessment (CTA), Washington DC, USA, p. 25.

ORGANIC EXCHANGE (2008), *Organic farm and fiber report*, Organic Exchange, O'Donnell, Texas, USA.

PANNA (PESTICIDE ACTION NETWORK NORTH AMERICA) (2008), *PAN pesticide database*, http//www.pesticide info.org (accessed August 21, 2008).

PRAY *et al.* (2002), as cited in Cantrell R (undated), *The Role Of Biotechnology in Improving the Sustainability of Cotton*, Cotton Inc., USA, p. 4.

RISSANEN T (2008), Creating fashion without the creation of fabric waste in Hawthorne J and Ulacewitz C (Eds), *Sustainable Fashion, Why Now?* Fairchild Publishing, Washington DC, pp. 184–206.

SAN PHILIPPO D (2009), The GM cotton debate: science or ideology? *Eco Textile News*, December 2008/January 2009, **20**, 20–21.

SHAH D (2008), View, *Textile View Magazine*, **82**, Metropolitan Publishing, The Netherlands.

TABASHNIK B (2008), Field-developed resistance to Bt crops, *Nature Biotechnology*, **26**(2), 199.

TON (2002), as cited in Organic Exchange (undated), The worldwide development of organic cotton cultivation: progress and future challenges, IFOAM textiles paper.

TOULMIN C (2006), From harvest to high street, *Pesticide News*, December **74**. PAN*UK*, London.

USDA-NASS (2004), as cited in Cantrell R (undated), *The Role Of Biotechnology in Improving the Sustainability of Cotton*, Cotton Inc., USA, p. 5.

3

Sustainable wool production and processing

I. M. RUSSELL, CSIRO Division of Materials Science and
Engineering, Australia

Abstract: This chapter reviews the history and development of wool as a modern textile fibre and describes the systems that have been developed to overcome the inherent variability in the fibre production process. Wool is a natural and renewable protein fibre, with a complex physical micro- and nano-structure and a complex chemistry. Together these allow development of textile garments with unique comfort, performance and appearance characteristics. However, 'natural' does not automatically equate to 'sustainable' and the wool industry, like most other mainstream textile fibre industries, is examining its environmental performance. For Australian Merino wool, this is not being done as a marketing tool, but as a means to identify and confront the main environmental challenges and to direct future research. Three main production, processing and garment scenarios representative of Australian wool supply chains have been chosen for examination using life cycle assessment (LCA). Overall conclusions from the current study are that: biogenic methane is the major contributor to the carbon footprint, and the Australian industry is intensively developing tools to reduce enteric emissions; garment laundering accounts for most water consumption, and much energy (heating water and electrical drying); more research is needed to address the gaps in our background environmental knowledge, especially for electricity and energy production in Asia where the bulk of textile processing is currently performed.

Key words: wool, production, processing, EU eco-label, life cycle assessment (LCA), energy, CO_2-e.

3.1   Introduction

Wool is defined by the *Shorter Oxford English Dictionary* as the 'fine soft curly hair forming the fleecy coat of the domesticated sheep (and similar animals)'. While several animals produce similar protein-based fibres – such as mohair and cashmere (from goats), alpaca and angora – wool obtained from sheep is by far the main protein fibre in common use. Historically, wool is one of mankind's oldest fibres, valued for its natural warmth and water repellence. Sheep were one of the first animals to be domesticated by man around 10000 BC (International Wool Textile Organisation, 2009), and they were mainly valued for their meat and for

63

their milk (for immediate drinking or for storage as cheese). The sheep skins were converted to leathers, often with the fleece preserved to increase warmth. The wool from these early sheep was poor, often coarse and uneven, although there were some finer fibres that were suitable for spinning. The proportion of spinnable fibres was increased over the next several thousand years by selective breeding and this led to the development of manufactured clothing. With its longer fibre length and residual wax, wool was easier to spin than vegetable fibres, and could be more readily dyed. One report suggests that the tribes of northern Europe were spinning and weaving animal fibres before 10000 BC (International Wool Textile Organisation, 2009).

The process of selective breeding of sheep to increase the value of the wool continues to the present time, driven by the value of the wool fibre in international trade. Wool and cotton were important raw materials that drove the industrial revolution in the eighteenth century in England. Various breeds of sheep were exported from Spain and other European countries from around 1800 to what were to become the major wool-producing countries: Australia, New Zealand, South Africa, China, Uruguay and Argentina. Selective breeding continued to develop animals that were adapted to the new climates and conditions and to optimise the production and quality of the fibre and the meat. As an example, the first Merino sheep that were introduced to Australia in 1797 (11 years after the first settlement) for wool production produced a fleece weighing just 1.5–2 kg each year. The current average cut of greasy wool is almost 5 kg. Sheep are able to thrive in rough, barren and arid regions, or in high altitudes or high temperatures where other animals can not survive. Sheep can utilise weeds and vegetation that other animals will not eat and can tolerate quite high salt concentrations in their drinking water.

There are around 40 different breeds of sheep in the world producing around 200 wool types of varying standards and end uses. An important requirement within the breeding programs was to develop sheep with a uniform fibre diameter over the whole body of the sheep so that all of the wool could be harvested and processed uniformly. Cashmere remains an exception where a dual-layer coat persists; however, specialised equipment has been developed to separate the coarse hair from the ultra-fine and highly valued cashmere fibre.

3.2   Wool uses

It is largely the fibre diameter of the wool that determines the value and the end uses of the harvested fleece. British breeds produce mostly coarser quality wool (fibres are around 30 micrometres (or microns) in diameter or more). This wool is highly suited for products such as carpets, blankets and hand-knitting yarns. Finer wools (with fibres of 20 microns in diameter or less) produce soft and luxurious fabrics that are highly suited to next-to-skin wear, and that can be used for applications that range from active sports-wear to elegant evening wear. For fine wool garments it is important that wool is uniformly white and contains no dark fibres that would limit the range of garment colours that can be produced. This is less important for the coarser fibres where naturally coloured fibres can be tolerated or may even enhance the 'natural' character of the products.

Different countries have tended to specialise in different sheep types and different wool fibre diameters. Australia has specialised in finer wools pro-duced mainly by Merino sheep (in June 2002 the Australian flock was composed of 85.1% Merino, 10.4% crossbred and 4.5% other breeds) (Aus-tralian Wool Testing Authority, 2009). In fact the Australian Merino is not a single homogeneous breed. Four main strains of sheep have been devel-oped in response to the environment and the need to optimise both wool and meat production. Australia is increasing its relative production of finer wools. In 1993/1994, only 8.8% of the wool clip was finer than 19 micron, compared with 30% in 2003/2004. In 2002/2003, Australian wool accounted for 48.5% of the global total of wool used in apparel (Australian Wool Testing Authority, 2009). A recent estimate is that Australia dominates the luxury apparel end of global production – producing 85% of the world's apparel wool and 95% of the world's wool of <19.5 microns (Gray, 2009; Lyons, 2008). Even though consumers invest US$80 billion per annum in wool apparel in the OECD (Organisation for Economic Co-operation and Development) countries alone (Lyons, 2008), wool remains a minority fibre in a textile world dominated by synthetics and cotton.

As the economic returns from both meat and wool can vary annually, a significant number of Merino ewes are mated with rams from English breeds of sheep such as Border Leicester to produce offspring with good meat characteristics, i.e. good carcass, high fertility, robust constitution and good milk production (important for rapid lamb growth). These sheep produce crossbred wool, coarser than Merino wool, but still valuable. This strategy provides farmers with the option to rapidly shift their farm output from wool to lamb meat in response to economic and climatic factors, especially with further breeding of the 'Border/Merino' ewes with 'Downs' breed rams to improve meat production. Sheep can therefore be seen as dual-purpose animals and this complicates any environmental assessment of wool produc-tion owing to the fact that the environmental inputs (land, water, fuel, fertiliser) and environmental outputs (methane, urine, faeces) need to be allocated between the products from the animal (meat, wool, skin). In addi-tion, few farms produce sheep as a sole product. Most produce some beef and a variety of crops, and many use the crop stubbles as a food resource for sheep and other animals (Australian Wool Testing Authority, 2009).

3.3   Consumer trends and environmental impacts

There is little doubt that consumers in Northern Hemisphere countries are developing a strong preference for sustainable and ethical textiles. There is growing awareness that all manufactured goods have an environmental impact and that the textile industry has a disproportionate effect. Cooper (2007) demonstrates that the chemical and textile industries are large users of water and large emitters of contaminated process waters. Of these two industries, most of the water used in the chemical industry is for cooling, while, in textiles, most water usage is process water which is discharged with processing contaminants.

This will become more important in a future dominated by global warming, water shortages and peak oil, and by environmentally aware retailers and consumers. AWI (Australian Wool Innovation) Consumer Insights research shows a mass trend toward a lifestyle of health and sustainability – and this extends to apparel (Lyons, 2008). The LOHAS (lifestyles of health and sustainability) market represents 'one-in-three' Americans (Karp, 2008) and is expanding. The LOHAS consumers are seeking goods and services that support their desire for health, environment protection, social justice, personal development and sustainable living.

Europe is exerting increased pressure on global supply chains, initially because it is collectively the world's largest market, but more recently because of its environmental leadership. Kanwar (2008) noted that 'If you manufacture globally, it is clearly simpler to be bound by the toughest regulatory standard in the supply chain. As China and other sourcing countries lean towards the European approach, US companies are also beginning to work to EU rules.' REACh legislation (Registration, Evaluation, Authorisation and Restriction of Chemical substances; EU Regulation, 2006) and other EU standards will continue to have an impact on global supply chains that seek access to the European Union. The EU eco-label for textiles is a powerful standard for environmental good practice in the production of textiles although it is poorly understood by industry and remains a poor marketing tool.

It has already been noted that the wool and cotton industries led the way into the industrial revolution and the modern textile industry was among the first to adopt globalisation. Through the 1990s there was progressive movement of the textile factories from traditional processing countries to less developed countries, attracted by lower wage rates, progressive assistance packages and by less stringent (or less monitored) environmental standards. Unfortunately, the sheer size and pollution loads from areas where the textile wet processing mills are concentrated have led to new pollution events (Reuters, 2008). China is responding to recent high-profile pollution incidents with stricter controls on effluent volume and pollution loads, with increased pollution monitoring and plant closures. It is reported that 10000 factories across all sectors have been closed (Woo, 2008).

Retailers are also responding to consumer pressures. Retail giant Wal-Mart has signalled that by 2009, factories where their goods are manufactured should be identified, and by 2012, 95% of all products should be sourced from factories with highest rankings for environmental and social standards (Anon., 2008a). Buyers are increasingly seeking supply chain transparency and hard evidence to support environmental claims. There is increasing awareness of 'Greenwash' both by consumers and regulators (Greenwash is the practice whereby disinformation is disseminated by an organisation with the intent to present an environmentally responsible public image). Regulators are increasingly insisting that environmental claims should be accurate, verifiable and in context, and that the basic information to support the claims should be easily accessible. In general, vague and non-specific claims that are too general in nature to be of any use to consumers should be avoided and this includes terms such as 'environmentally friendly', 'green', 'non-polluting' and even 'sustainable' (ACCC, 2008; Defra, 2000; Mohr, 2005).

Global consumption of textiles is large. The textile industry has grown, not only in line with global human growth, but also as a result of per capita consumption. Around 1900, humans used around 2 kg of textiles per head, whereas by 2010 consumption will exceed 10 kg per head (Bide, 2008; Lenzing Group, 2008). As with most other consumable goods, per capita consumption is weighted to the wealthiest countries, with the USA and UK consuming as much as 10 times the quantity of textiles per head as in developing countries.

The environmental problems are associated not only with resources used in the initial manufacture of the textiles but also in the use of resources to clean and dispose of the garments. Most textile garments finish their life in landfill, a diminishing resource in most developed countries. It has been estimated that 1.2 million tonnes/year of textiles are sent to landfill in the UK (Draper et al., 2007), while two million tonnes/year are sent in Japan (Akihiro, 2008). This mass is split evenly between clothing and other textile products. There are increasing calls for greater recycling of textiles as this consumes much less energy than the manufacture of new textiles; however, problems with identifying useful end products for all of this recycled material remain difficult. Even in the UK which has reasonably well developed recycling systems, only around 25% of garments are recycled (Waste Online, 2004).

A second estimate for the combined waste from clothing and textiles in the UK is about 2.35 million tonnes (38 kg/person). Of the 330000 tonnes of recovered textiles, 200000 tonnes are exported while 100000 tonnes are recycled within the UK. These recovered clothes are given to the homeless, sold in charity shops or sold in developing countries in Africa, the Indian sub-continent and parts of Eastern Europe. Over 70% of the world's population use second-hand clothes. Incineration is used for 10% of clothing while 60% is sent to landfill (Alford *et al.*, 2006). The EU has indicated that there is likely to be legislation to encourage greater recycling of textiles, initially in the UK and France, but then across the EU (Anon., 2008b; Defra, 2008; Paillat, 2008).

Because wool is a relatively expensive fibre, there is higher demand for discarded wool garments and processing wastes which are sold to specialist firms for fibre reclamation to make yarn or fabric. Incoming material is sorted into type and colour to minimise re-dyeing. The material is shredded into 'shoddy' (fibres). Depending on the end uses of the yarn, other fibres are chosen to be blended with the shoddy by carding before spinning. Products may be garments, felt and blankets. In anaerobic landfills, wool, cotton and other natural fibres degrade rapidly while oil-based textiles degrade extremely slowly, but all potentially produce methane. Methane is a potent greenhouse gas and unless it is recovered, the global warming potential is greater than from incineration.

3.4   Wool fibre: structure and properties

Wool is a naturally produced protein, biodegradable and renewable, and is therefore well placed to take advantage of the emerging 'green' trend for textiles. Wool has a number of advantages as a textile fibre and these arise as a result of its complex chemical and physical structure. The wool fibre forms in follicles in the sheep's skin, initially as a structure of elongated living cells, but these cells dry and harden (keratinise) as the fibre emerges. An outer layer of overlapping, flattened cuticle cells provides a tough water-resistant coating to encase spindle-like cortical cells that contain a nano-structure of crystalline, water-insensitive filaments within a high-cystine, water-sensitive matrix (Fig. 3.1). This complex histology provides a fibre with unique properties. While the surface is water-, dirt- and stain-repellent, the fibre interior is highly moisture absorbent (higher than all other common fibres) and this provides good wear comfort. The fibre is breathable and fabric structures can be engineered so that garments can be either warm or cool. The filament/matrix structure ensures that, when moisture is absorbed, the fibre diameter increases but the length and strength are less affected. The crystalline filaments also ensure that the fibre is highly elastic so that it does not wrinkle under most wear conditions and garments have excellent drape and shape retention. Garments can be formed and permanently set under the influence of heat, moisture and if needed, reducing agents.

The internal chemistry of the fibre is complex as it is based on a protein structure with more than 20 amino acids. This ensures that the fibre is

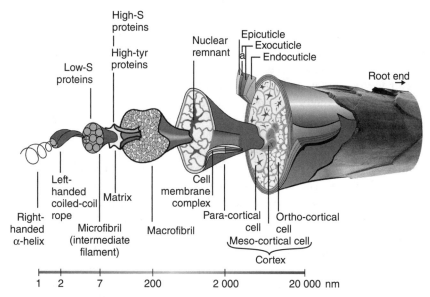

3.1 Expanded schematic of wool fibre structure (© CSIRO Materials Science & Engineering Textile and Fibre Technology Program Graphics by H.Z. Roe, 1992 & B. Lipson 2008, based on a drawing by R.D.B. Fraser, 1972).

readily dyed, that it absorbs odours, and that it is naturally fire-resistant and anti-static. The overlapping scales on the surface of the fibre provide a differential friction effect that allows fabrics to compact or felt in water. This can be desirable to achieve effects in finishing, but it can lead to garment shrinkage if uncontrolled. A variety of shrink-resist finishes have been developed that can be applied at loose fibre or at garment stages so that machine-washable, easy-care garments can be produced.

3.5   Wool and ecolabels

While wool is natural, it is important to note that 'natural' does not automatically mean environmentally friendly. Wool, like all other textiles, needs to develop a defendable environmental profile in response to the growing environmental awareness of consumers in developed countries. Of course, only the hardest 'green' consumers are prepared to purchase garments purely on their environmental profile, to achieve body covering with minimum environmental cost. For the vast majority of consumers, purchasing decisions are made on appearance, functionality, fashion, quality, performance and price. Of these, price seems to be the single item where consumers seem most prepared to compromise. Surveys in Europe and the USA show a progressive increase in the numbers of consumers prepared to pay more for ethical and sustainable textile goods (Anon., 2008c). Some commentators believe that good environmental credentials are now a 'given', however the problem remains that it is still very difficult to identify goods manufactured using good environmental practices in the market place. Ecolabels are important in this area, however not all textile ecolabels identify environmentally preferred goods. Some textile ecolabels might deal only with labour issues, while some, such as the 'human ecology' labels, deal only with the absence of toxic chemicals in the final manufactured product. These types of ecolabels are silent on environmental practices through the supply chain. This is often not understood well by either consumers or, unfortunately, many retailers and textile processors. The International Wool Textile Organisation (IWTO) has recently defined 'eco-wool' textiles as goods that meet the EU eco-label criteria at all stages of the supply chain, from raw wool to finished product (deBoos, 2008).

The EU eco-label is an important textile processing standard that covers all stages of textile manufacturing, from raw fibre to finished product. It contains criteria that exceed 'environmental best practice' for textile processing as defined in the EU BREF document (EIPPCB, 2002) and the Integrated Pollution Prevention and Control legislation (IPPC, 96/61/EC). The EU eco-label is a Type 1 ecolabel (as described by Internation Standard ISO 14024). These labels are based on criteria independently established across the life cycle of a product and certified or audited by an independent third party. The criteria in Type 1 ecolabels are set and assessed independently, they are readily available and they cover the total processing sequence. Specific criteria are applied to those processing stages with the greatest environmental impact. Type 1 ecolabels provide the greatest degree of transparency and the greatest assurance that the product has been manufactured with minimum environmental impact.

Other international Type 1 textile ecolabels have usually taken their criteria from the EU eco-label. Some of these derived Type 1 standards (but not the EU eco-label itself) may add some social elements (fair pay, safety, child labour, etc.). Type 1 ecolabels adopt a supply chain approach where information on all processing additives is available to downstream processors. This information flow is often missing in textile supply chains. A British Government report has identified the EU eco-label as the most robust measure of textile sustainability after considering 207 other standards, databases and product lists globally (Anon., 2008d), and the UN Environmental Programme (UNEP) is advocating its use in South Africa and India (Ferratini, 2008/2009).

There is increasing awareness that environmental discharges from a wet textile process such as dyeing arise from two sources:

(a) the compounds deliberately added to the process by the dyer;

(b) the materials already present on the fibre when it is received. The identity of these materials is often outside the control or knowledge of the processor receiving the fibre and they may contribute a greater pollution load than the process additives. They may be difficult to remove and they may be toxic or poorly biodegradable.

The general issue of supply chain communication has been raised as an important element of good environmental practice in the EU BREF document (EIPPCB, 2002) and by Cooper (2007). The problems are most severe at dyeing since dye liquors are difficult and costly to treat because of their volume and temperature. Wool has a long supply chain compared with most other fibres and the fibre may change ownership several times as it progresses from farm to garment. Because the manufacture of EU eco-label garments begins with identifying compliant batches of raw wool, garment manufacturers have been deterred by the task of identifying and then commissioning the processing of the specific wool lots. An important development has been the opening of a database in the EU eco-label website that allows late-stage manufacturers to find partially processed fibres, tops or yarns that have been manufactured in compliance with the EU eco-label (EU, 2007). At least one supply chain has moved in this direction. It is also expected that the adoption of the EU eco-label by the international wool industry as its definition of eco-wool will also assist this process within wool supply chains.

3.6   Life cycle assessment (LCA) studies

In 2006, the Commonwealth Scientific and Research Organization (CSIRO) and the Australian wool industry began a small project to perform a preliminary LCA on Australian wool. Other wool-producing countries are performing partial studies (Barber and Pellow, 2006; Kelly, 2008). Australian wool is used for the production of many types of garments and processing occurs in many different countries. In order to simplify the study and to allow identification of the major environmental pressure points, three typical Australian wool supply chains manufacturing common apparel products were selected:

• fine wool grown in a high-rainfall climate by a specialist producer, processed in Italy into lightweight next-to-skin knitted garments;

• medium-micron wool produced on mixed enterprise farms, processed in Asia into men's suits;

• slightly coarser wool, produced in Australia's arid pastoral zone, processed in Asia into an outerwear knit.

All garments were assumed to be sold and used in Europe.

The study covered the total production, transport, processing, garment care and garment disposal sequences. The project adopted the ISO 14040 as a framework, and the main environmental impacts focused on the usage of water and fossil fuels and on emissions of greenhouse gases. The project was conducted in order to identify and rank the main environmental challenges – were the major emissions of greenhouse gases and usages of water associated with on-farm fibre production (methane emissions from sheep, CO_2 emissions from trucks and tractors, N_2O emissions from fertiliser), with transport, with processing, or with garment care?

The aim of the study was not for tactical marketing or inter-fibre comparisons; it is acknowledged that all fibres have their own unique pressure points. Rather the study was conducted as a strategic planning tool, to ensure that future investments by the Australian wool industry in the environmental area addressed the major issues. Comprehensive modelling of all inputs and outputs associated with the provision of a product or service is potentially extremely complex. The task is simplified considerably by the availability of calculation and database software that includes much of the 'background' data on environmental impacts associated with the provision of general services such as transport, electricity (in various countries) and fertiliser manufacture.

This study was conducted with the SimaPro software package (Pré Consultants bv, The Netherlands) to perform overall calculations and to provide the background LCA data through its associated databases. The most comprehensive resources are available for services provided in Europe through the EcoInvent 2.0 database (May 2008 version). The Australian distributor of the SimaPro software (Life Cycle Strategies, Victoria, Australia) has modified some of the European data with assumptions appropriate to Australia and has provided a reasonably comprehensive Australian database.

3.6.1 On-farm production of wool fibre

The process of LCA is generally poorly developed for agriculture, especially as it applies in the arid Australian environment. The most detailed studies and methodologies have been developed for European farming but these are often poorly applicable to Australia. As an example, in European agriculture, water is generally plentiful and soils are often wet. This has an impact on such factors as emissions of nitrous oxide (N_2O) from animal urine and from use of nitrogenous fertilisers. In Australia, water is a scarce resource and access to water is highly contested. At present there is no means to value the scarcity of water in an LCA study, however tentative steps are underway.

In this study we only considered drinking water supplied to the sheep from contestable sources such as dams (holding stored run-off water captured from the property in wet seasons and prevented from draining to a waterway) and underground bore water, with associated overheads (evaporation and transport losses). Other brief LCA studies also adopted this approach (Laursen *et al.*, 1997). Consistent with many other agricultural LCA studies, rainwater landing on the property (an important resource for plant/fodder production) and dew condensation were not considered, even though the water in and on plant matter are an important water resource for sheep. Water stored in the soil ('green water') is a valuable resource for the production of plants and for retaining carbon and nitrogen in the soil; Australian farmers actively manage this resource (where possible). The estimates of the volumes of drinking water depend on climate (mainly temperature), food intake (quality and quantity), animal size, distance to water and factors such as lactation. Sheep are extremely hardy animals and in temperate climates they can survive for prolonged periods on dew and plant moisture alone. However, this study assumed a plentiful drinking water supply and, as a result, estimated values of water usage probably erred on the high side. The volume of urinary water returned to the soil was subtracted from the total water consumed on the basis that N_2O was generated when the urine reached the soil, and that the urinary water (with nutrients) was retained on the farm and contributed to plant growth. Sheep respire large volumes of moisture but this can not be captured on the farm and is regarded as lost to the system. Unfortunately there are no agreed protocols.

The estimates of water use obtained in this way were more than two orders of magnitude lower than worst-case estimates previously made. These previous calculations were based on the assumption that sheep were raised solely for wool production on irrigated pastures (Meyer, 1997). In fact it is now extremely rare for sheep to be run on irrigated pastures in Australia, and, when it is done, it is invariably for meat production (fattening of lambs).

The methodology for carbon accounting in sheep production is well developed because of Australia's international greenhouse gas reporting commitments. A small part of the carbon emissions on farm are as CO_2 itself, from use of farm equipment, electricity, transport and, indirectly, for production and transport of fertilisers. A very small adjustment in the carbon balance included the carbon fixed in the wool fibre product from the farm.

A larger proportion of on-farm carbon emissions arise from the production of the potent greenhouse gases, methane and N_2O. On a mass basis, both are many times more potent than CO_2 itself. Total emissions of greenhouse gases are usually expressed as CO_2 equivalents (or CO_2-e), where the mass emissions of other gases are multiplied by factors according to their global warming potential. The principal emissions of methane from sheep production are from enteric digestion processes. It is possible that methane could be produced from faeces, but this is regarded as negligible under Australia's dry farmland conditions. The volumes of methane from sheep depend on the size of the animal, the diet (poor diets produce more methane), lactation and numerous other factors. The principal sources of N_2O production are from urine when it is deposited on soil (along with other nutrients), as well as from the use of nitrogenous fertilisers in crop production. In general, superphosphate is the only fertiliser used on high-rainfall pastures in Australia for intensive sheep production, and no fertilisers are used in the arid pastoral zones. Nitrogenous fertilisers are not used directly on pastures for sheep (they are too expensive), however they are used on grain crops in mixed farming enterprises and farm animals are allowed to graze on the stubbles once the grains have been harvested. Again the emissions of N_2O from nitrogenous fertilisers depend on rainfall. In the dryland farms used for wheat production in Australia, emissions are lower than in Europe and other wet climates. The current study allocated some of the N_2O emissions from grain production to the sheep according to the protein value of the residual grain and stubble. Nitrogenous fertilisers are often used in the production of fodder used to supplement the feed of sheep in drought periods and again the associated N_2O was assigned to the sheep (as well as transport and energy used to produce the fertiliser) based on the farm expenditures on fodder.

Sequestration of carbon on the farm was not included although many Australian farmers plant trees to fix carbon and to lower the water table to reduce soil salinification. Agricultural soils are capable of sequestering large amounts of carbon and nitrogen. While this was considered in this study, no allowance was made as:

• under the greenhouse gas accounting rules, sequestered carbon must be 'additional, permanent and verifiable', and these conditions are not yet met;
• these activities are not relevant to the product's carbon footprint.

Generally the allocation of environmental inputs and emissions to the sheep was made on economic factors based on overall farm expenditure and receipts, as it is economic returns that decide whether the land is used for animal production or for cropping. An initial economic allocation was based on the overall farm receipts for all farm products to obtain a 'per sheep' share of inputs and impacts. A number of other environmental inputs and outputs (such as drinking water and methane emissions) were already available on a 'per sheep' basis. A second economic allocation was performed based on farm receipts for sheep meat and for wool to divide the 'per sheep' inputs between the meat and the wool products. Again it is the economic returns that decide whether the sheep is produced for either wool or meat. While ISO 14040 regards economic allocation as a poorly preferred option, with a product such as wool, the price per kilogramme of wool varies more than tenfold depending on the micron of the wool. Crossbred sheep are usually produced for their meat and often lead short lives; sheep with fine wool are often retained for several productive years, by which time the meat has relatively poor value. It is only economic allocation that reflects these differences.

3.6.2 Early-stage wool fibre processing

In comparison with cotton and synthetic fibres, wool undergoes several processing stages before it can be effectively used in a textile factory as combed, aligned fibres for blending or for production of yarns. The wool is manually shorn from the sheep and the main fleece wool from the back, sides and chest of the sheep is then separated (or skirted) to remove the stained and burr-contaminated wools from the rear and underside of the animal. This poorer wool still finds valuable uses in wool supply chains. The fleece wools are then sorted into consistent lines of wool (similar micron, fibre length, appearance and style) by wool classers before the wool is baled and transported to a central store. These lines of wool (weighing around 1 tonne on average) are then sampled and objectively tested (for yield, micron, vegetable matter, etc.) as these factors determine the amount of clean wool in the sales consignment, its value and its likely processing characteristics. These presale evaluations are necessary to counter the fleece-to-fleece and flock-to-flock variability in a natural product. In Australia most of the wool is sold at 'open cry' auctions mainly on the basis of the objective measurements. The wool is purchased by buyers whose job is to acquire (at minimum cost) the wools that will allow a processing lot (20–50 tonnes) to be compiled and that will, after initial processing, meet specifications for fineness, variation in fineness, fibre length, variation in fibre length, colour, etc. in the combed tops. Predictive computer programs are available to assist this process.

It is only when the wool is ready to be washed or 'scoured' (either in Australia or overseas) that the specific lines of wool that were purchased are retrieved from their original place of storage. Wool to be shipped overseas may be 'dumped' (three bales are highly compressed together) to increase the packing density in shipping containers. This reduces the energy overheads in shipping the container as well as the wool. Transport is by truck and ship to the overseas seaport (China or Italy) and then by truck to the mill.

The total consignment is mixed in a variety of ways to ensure that the finished consignment is consistent from start to finish. Scouring of the wool is performed at 60–70°C in a continuous counter-current operation in a scouring line usually with six scouring bowls. In a typical scouring operation, non-ionic detergent is used and around 10 L of water are used per kilogramme of greasy wool. The most energy-efficient scours use covered scouring bowls, heat the water using direct gas firing of the bowls and recover heat from the discharged liquors; however practices vary widely. Steam is a less efficient heating medium but unfortunately in some countries, such as China, where coal-fired boilers are common, it may be the only option. It is reported that 90% of industrial boilers in China are coal fired, many of which operate at less than design efficiency (Li and Sun, 2007).

The scouring operation is designed to remove contaminants from the wool. A typical bale of Australian wool may contain only 65% of wool, the remainder is wool wax (around 30% of this is recovered from the scour and sold to be refined into lanolin), dirt, sheep sweat salts and other skin contaminants. Effluents from the wool scour are therefore highly contaminated. While there are high-energy methods such as evaporation that can be used to clean these effluents, in fact there are lower energy options that can remove most of the contaminants in-line and recover valuable by-products (compostable organic material, potassium fertilisers) (Bateup et al., 1996). Within any supply chain there will be environmentally good operators and there will be environmentally poor operators and this complicates the comparison of generic textile supply chains. The uniform and cleaned mat of wool from the scour is squeezed to remove excess water and dried in efficient drum driers usually fitted with moisture meters to ensure that the wool is not over-dried and damaged (but also to save energy).

There are two main streams in wool processing. The woollen system uses short wool fibres and the wool passes through a woollen card where a series of wire-coated rotating drums align the fibres. A final set of rubbing tapes split the fibre web into 'slubbings' that are subsequently spun into yarns with little further drafting. These yarns are typically coarse and hairy and are used for tweed-type fabrics and hand-knitting yarns. The worsted system is used for fine woven and fine knitted fabrics. Scoured wools with long fibre lengths pass though a shorter worsted card where rotating wire-covered cylinders begin to align the fibres. The wool scouring process unfortunately entangles some of the wool and there is significant fibre breakage in the worsted card. This reduces the average fibre length. Most of the vegetable matter is also removed. The carded web emerging from the card is further blended and gilled (combed) with steel combs in four or five stages. The shortest of the broken fibres in the now well-aligned wool are removed in a rotary comb, a process that separates the long fibres from the short broken fibres ('noil') that may comprise 6–8% of the incoming fibre. This fibre is lost to the worsted process but finds ready use in the woollen system. The highly aligned long fibres are given a final gilling before typically being rolled into large balls ('tops') ready for blending or spinning.

These early processing stages are functionally equivalent to the extrusion (primary spinning) processes used with synthetic fibres. As with synthetic fibres, various lubricating oils are added on the fibre to minimise fibre breakage in mechanical processing. While the operations in the woollen and worsted processing sequences are complex and specialised, the fibre throughputs are high and the energies required per kilogramme of fibre are generally lower than the energies for the synthetic fibres, even when the on-farm energies and transport are considered.

3.6.3 Fabric production

All textile fibres, once in the mill in the form of clean, aligned staples, are typically spun, woven or knitted, dyed, finished and made up into garments. There are environmental similarities in all of these sequences for all fibres. The differences between mills are probably greater than the differences between fibres. There will be modern mills that use state of the art, efficient, low liquor ratio equipment and there will be mills that use greater resources. As an example, a Finnish study found that the water usage across six large textile mills dyeing and finishing mainly synthetics and cotton ranged from 140 to 370 L/kg of fabric. Energy uses ranged from 55 to 124 MJ/kg. The mill with the highest water usage also consumed the most energy. Discharges of COD (chemical oxygen demand, a measure of oxygen depletion in water) ranged from 61 to 393 g/kg of fibre (Kalliala and Talvenmaa, 2000). In general, the main factors that differentiate the environmental footprints of different fibres are the energy, water and resource usages to produce the combed, aligned fibres.

There are a few studies that provide processing data (energy and water use) for processing operations, but most show that between-mill resource usage varies over large ranges. It is difficult to combine much of this data into a coherent processing sequence because of the high variability within the individual processing stages. As an example, British Textile Technology Group (BTTG) (1999a) cites an energy use for winch dyeing of 5.7–16.9 MJ/kg and a water use of 28–293 L/kg. Additionally, it is often difficult to know specifically what is included and excluded in the data that are available. As an example, is the energy for air conditioning or effluent treatment included in the mill operations? A streamlined LCA for selected cotton and polyester garments (Collins and Aumônier, 2002) performed for Marks and Spencer solved the problem with the range of data by using the best environmental data for each stage, effectively assuming environmental best practice for all processing stages.

The current study used specific process stage data from specific early-stage wool processing mills, including data from a South African study evaluating life cycle impact assessment methods for the production of twofold wool yarns (Brent and Hietkamp, 2003). For wool dyeing and finishing, because of the variety of processes encountered and the associated uncertainties, overall mill data were used. These data were poorly available and the results from only a single woven fabric mill and from a single knitted fabric mill were used. However, for the woven fabric mill, all activities were captured including air conditioning, effluent treatment and operation of staff canteen and toilets. The fact that the mill had achieved EMAS (Eco-Management and Audit Scheme) certification probably meant that it was among the better environmental operators.

Another major difficulty was the lack of background data available within LCA software packages for operations in Asian processing countries. While most of the standard packages include a wide choice of operational sequences in Europe (for example, country-specific data for supply of electricity), there are few corresponding sources for Asia. The LCA practitioner may be forced to select electricity data from another country where coal-fired power generation is predominantly used, but that may poorly reflect the coal mining, transport and air quality impacts in China. The men's suit and outer knitwear scenarios assumed garment manufacture in China, necessitating transport of the garments to Europe by container ship and surface transport.

3.6.4 Environmental impacts in use and on disposal

The Marks and Spencer streamlined LCA study demonstrated that the water and energy resources required to clean and maintain a garment in wear can exceed the resources required to manufacture the garment initially. In fact, for the polyester trousers and the men's cotton briefs, the garment washing and care phase consumed 76% and 80%, respectively, of the total energy in the garment life cycles. The main contributors to energy use were high-temperature washing, garment drying and ironing. The Marks and Spencer study demonstrated the environmental benefits of lowering the wash temperature. While some modern wool garments are capable of withstanding hot washing and tumble drying, most consumers are not prepared to take the risk, even if there are assurances on the care labels. This current study assumed that consumers would use a gentle warm wash cycle and air-dry the wool garments. This had the effect of substantially reducing the in-use care resource usage for the knitted garments.

This study found little information on the lifetime of wool garments or on the number of cleaning cycles. Wool has good odour absorbing properties and this reduces the frequency of laundering. A garment lifetime of 20 washes was assumed. The men's suits were assumed to be dry cleaned four times per year for three years or 12 dry cleanings with the energy and resource usages taken from the Marks and Spencer streamlined LCA report

(Collins and Aumônier, 2002) and from BTTG reports (British Textile Technology Group, 1999b).

UK default figures for disposal of garments were essentially used, however wool is a relatively expensive fibre and in Japan and in the UK schemes have been developed to recycle the fibre. For the purposes of this study, 40% of wool garments were assumed to be recovered/recycled (slightly higher than for other fibres), with 7% incinerated and 53% disposed to landfill (Alford *et al.*, 2006).

SimaPro models for incineration of combustible wastes and for disposable of degradable materials such as paper and cardboard to European landfills were used to model the end-of-life behaviours of these wool garments. In fact the end-of-life disposal caused very little change in the overall picture for these wool garments. The recovered/recycled fibre and garments were assumed to be environmentally neutral as substitution of recovered fibre to replace new garments manufactured from virgin fibre cannot be demonstrated.

3.6.5 Overall results

Figure 3.2 shows the relative proportion of CO_2-e for the three wool production and use scenarios as calculated by SimaPro. These emissions are calculated per kilogramme of garment and therefore take account of mass losses through the processing chain. For each supply chain, around 2 kg of greasy wool is required to produce 1 kg of garment and this increases the apparent contribution of on-farm emissions. The main mass loss is at scouring where wool wax, dirt and sheep sweat salts are washed from the wool.

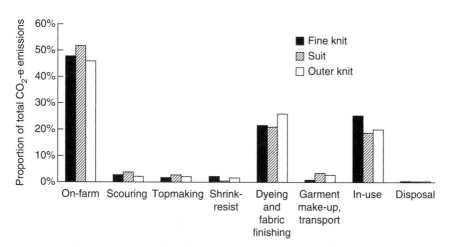

3.2 Emissions of CO_2-e through the wool garment life cycle.

In this simple model, losses through processing have been considered as lost to the specific supply chain, but in fact fabric losses are usually valuable clean fibre which is sought after as raw material for other wool products. Examples are losses of noils in topmaking (these are used in the woollen system for manufacture of knitwear) and fabric offcuts in suit manufacture and in cut-and-sew garments.

The on-farm emissions of CO_2-e form around half of the total emissions and include biogenic methane (the largest contribution) as well as N_2O from urine and, for the sheep/wheat-zone farm, N_2O emissions from the use of nitrogenous fertiliser on crops. Energy used on the property as well as in production and transport of fertilisers (for the high-rainfall zone and sheep/wheat-zone farms) and fodder are also included.

The other major stages for emission of CO_2-e are in dyeing and finishing and in garment care (washing and dry cleaning). The emissions from dyeing and finishing will be roughly similar for all fibres, however the laundry operations for the knitted wool garments in the model are low-energy scenarios (warm wash, air dry). For other fibres, the CO_2-e emissions from garment care would probably be higher. Emissions in sea and land transport are relatively low and are included in the scouring and garment make-up components.

In order to more clearly show the relative contributions of animal-related emissions from fuel, electricity and fertiliser production, SimaPro is able to estimate usages of fossil fuels through the supply chain. Figure 3.3 shows that the energy used on-farm and in the early stages of fibre preparation is low in comparison with the fossil fuels used in dyeing and finishing and in garment care. The higher energy use on-farm for the men's suit scenario in

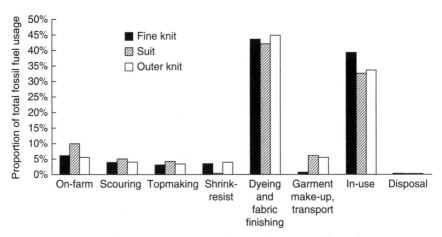

3.3 Fossil fuel usages through the wool garment life cycle.

the mixed farming scenario is largely associated with manufacture (in Australia) of nitrogenous fertiliser used on the crops. Again surface transport is a minor component of fossil fuel use.

This study has clearly shown the relative importance of enteric methane in contributing to the carbon footprint of Australian wool. The problem is not unique to sheep but is common to all ruminant animals including cattle. There is substantial scope to reduce livestock emissions and research is underway both in Australia and internationally (Trivedi, 2008). It has been found, for instance, that substantial differences exist within breeds of cattle in the amounts of emissions, independent of diet.

While researchers have developed methods to achieve small reductions, it is hoped that genetic and management tools being developed by sheep, beef and dairy industries and by researchers sponsored by the Australian and New Zealand governments will yield larger reductions. Researchers have recently found that kangaroos in Australia produce very little methane, yet are a ruminant. The differences relate to the gut microorganisms, but translation of this observation into practice is likely to be difficult.

Figure 3.4 shows the relative water consumption through the three wool production and processing supply chains. The major water usage is at the garment laundry stage. While the men's suit is assumed to be dry cleaned, water is used in electricity generation. In the on-farm scenarios, water is used in the production of nitrogenous fertilisers (men's suit), while sheep in the hot and dry pastoral zones have a high intake of drinking water. The mill used to provide the overall data for the men's suit scenario was a relatively high user of water, largely because the multiple processes used to finish high-quality woven wool fabrics can be water-intensive.

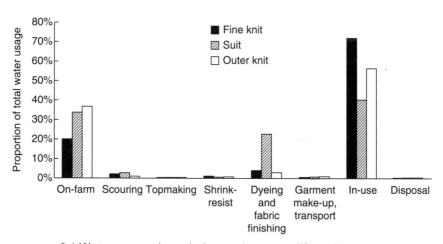

3.4 Water usages through the wool garment life cycle.

3.6.6 Energy use and the carbon footprint

The current study did not perform fundamental and first-principle energy studies for other fibres; however in asserting that wool uses low energy over its life cycle but has high greenhouse gas emissions, it is important to consider the relationship between energy and carbon emissions. For textile processing, data on energy usage are relatively readily available. Greater complications occur in estimating the associated CO_2 emissions that arise in the production of the energy, whether from coal, electricity, natural gas, etc., in the different countries where processing is performed. In the current study, the SimaPro software was allowed to perform the calculations for CO_2 emissions based on the source of the energy using regional data where available. It is well known that the output of CO_2 depends heavily on the source of the energy. There are in-built inefficiencies associated with use of electricity compared with direct use of thermal energy. In addition, even with thermal energy, steam is a less efficient energy source than direct-fired gas heating in a mill. The source of electricity in a process will have a significant impact on the carbon footprint. Electricity generated from wind, solar or nuclear sources has low carbon emission factors, whereas brown coal, with its high moisture content, has a high emission factor. This means that a specific process will have different carbon emissions depending on where it is located.

Most of the textile reports that have provided information on total energy usage have not calculated CO_2 emissions, nor do they generally provide sufficient data on energy sources to calculate the associated carbon footprint. The energy required to produce 1 kg of polyester fibre has been variously estimated as 109 MJ/kg (Marks and Spencer Streamlined LCA study – Collins and Aumônier, 2002) and 112 MJ/kg (Franklin Associates, 1993) – made up of resin manufacture (97 MJ/kg) and from fibre production (15 MJ/kg). There are several other estimates but most are similar.

Few studies provide a sufficiently accurate breakdown of the fuel mix for textile production to generate accurate data on carbon emissions. The Franklin Associates study of a polyester blouse did provide an inventory of fuel types used, and, on the basis of the fuel mixture, the New Zealand Merino study (Barber and Pellow, 2006) calculated an emission of 6.9 kg CO_2/kg polyester fibre produced or 62 g CO_2/MJ of energy.

As a basis for comparison, a number of sources show that black coal generation of electricity produces 278 g CO_2/MJ of electricity (Anon., 2009). Quantities of ozone, sulphur dioxide, oxides of nitrogen and other gases as well as particulate matter are also produced.

Morris recently estimated an emission of 4.2 kg CO_2/kg polyester staple or about 40 g CO_2/MJ of energy (Morris, 2008). This probably reflects production in Europe where nuclear, gas and hydroelectric power are extensively used. Morris noted that France has the lowest CO_2 emissions because its electricity is mainly generated from hydro and nuclear resources. His report also observed that coal-based processing in China increased the carbon emissions per kilogramme of fibre, especially if transport energy was included.

Wiseman (1981) reported an energy for wool production figure of 38 MJ/kg while the New Zealand Merino study (Barber and Pellow, 2006) calculated an energy usage of 46 MJ/kg to produce wool top, half of which was on-farm. The New Zealand study calculated a CO_2 emission for production of wool staples of 2.2 kg CO_2/kg (corresponding to 50 g CO_2/MJ of energy). This value of 2.2 kg/kg wool appears to have been used by Morris in his comparison of fibres, but the New Zealand study ignored enteric methane and N_2O emissions.

The carbon footprint of commodities is the subject of a British Standard, PAS 2050, and this initiative is being supported by major retail chains. Many commodities are viewing this development very seriously and it is anticipated that carbon footprints of commodities will begin to have an impact on market access (Mello, 2007). Energy ecolabelling is also being proposed for textiles in Europe (Anon., 2008e). Many farmers in Australia are planting trees to offset their on-farm carbon emissions and The Merino Company in Australia has developed a scheme to source ZeroCO2 Wool™ for markets that are seeking a carbon-neutral product.

3.7   Outcomes

LCA studies are data intense and outcomes depend on the assumptions made, both in the foreground and in the background. At a practical level, the task of the LCA practitioner is to evaluate the importance of a particular assumption to the outcome. If the result is important to the outcome, the practitioner must spend time obtaining the best data. If the outcome has a minor effect, less time needs to be spent on improving data quality and accuracy. However, it is important that the assumptions are clear and transparent so that the user of the study can decide if the overall outcomes are appropriate.

Some of the commercial LCA programs have different strengths, e.g. GaBi has a very detailed textile finishing database, which would be very useful for benchmarking European processing sequences (GaBi, 2009). However, because of the variability of the environmental practices of different mills, there is little substitute, if detailed comparisons of specific products are to be performed, for studies to be conducted for all stages of manufacture at the mill level. Unfortunately there is a paucity of background LCA environmental data for Asian mills, where the bulk of textile processing is currently performed.

There are some overall conclusions from the current study of three wool production scenarios for the manufacture of typical Australian wool products.

- Biogenic methane is the major contributor to the carbon footprint, and the Australian industry is intensively developing technologies to reduce enteric emissions.
- Garment care during normal wear accounts for most water consumption, and much energy (heating water and electrical drying).
- More research is needed to address gaps in our background environmental knowledge, especially for electricity and energy production in Asia.
- The environmental performances of the different potential participants in any supply chain vary widely – there are opportunities to optimise the environmental performance for specific products by selecting the participants in the supply chain. Unfortunately the data to make such a selection are not generally available.

3.8   Sources of further information and advice

- Further information on the properties of wool and its applications in garments and in technical textiles can be obtained from the CSIRO website in a series of information sheets (CSIRO, 2009). Links to other wool information sites (Australian Wool Innovation, Australian Wool Testing Authority) are also available.
- The suppliers of the major LCA software packages provide a great deal of background information and even training packages on the use of their software.
- The Society of Environmental Toxicology and Chemistry (SETAC) has published several books on the theory and practice of LCA.
- The monthly trade magazine, *Ecotextile News*, is an important source of environmental information on trends, issues and news releases.
- The EU BREF document (EIPPCB, 2002) is a major resource that defines good environmental practices for textile processing. It was written using European data that were obtained between 1999 and 2001 and reflects that viewpoint; however it remains a valuable benchmarking document.
- The EU eco-label textile-specific website is found at http://ec.europa.eu/environment/ecolabel/product/pg_clothing_textiles_en.htm. The site links to many other useful articles, including to the current suppliers of ecolabel textiles. The background material describing the reasons for setting the specific criteria are valuable.

3.9   Acknowledgements

This study was supported by funding provided by the Australian Government to CSIRO and by Australian wool growers through funds administered by Australian Wool Innovation (AWI). Information provided by colleagues in CSIRO, especially as part of the Agricultural Sustainability Initiative, and from the AWI Project Advisory Team and the AWI Project Manager, Dr Paul Swan, is gratefully acknowledged.

3.10   References

ACCC (AUSTRALIAN COMPETITION and CONSUMER COMMISSION) (2008). 'Green marketing and the Trade Practices Act', http://www.accc.gov.au/content/index.phtml/tag/greenmarketing/, accessed 24 April 2008.

AKIHIRO O (2008), 'Tech-upgradation leads to mass scale textile recycling', http://www.fibre2fashion.com/news/association-news/jcfa/newsdetails.aspx?news_id=61061, accessed 11 February 2009.

ALFORD JM, LAURSEN SE, DERODRIGUEZ CM and BOCKEN NMP (2006), 'Well dressed, the present and future sustainability of clothing and textiles in the United Kingdom', http://www.ifm.eng.cam.ac.uk/sustainability/projects/mass/UK_textiles.pdf, accessed 10 February 2009.

ANON. (2008a), 'Wal-Mart's green new deal', *Ecotextile News*, **19**, November, 15.

ANON. (2008b), 'Textile waste plan', *Fashion Business International*, June–July, 4.

ANON. (2008c), 'US consumers still buying "green"', *Ecotextile News*, **15**, June, 7.

ANON. (2008d), 'EU eco-label tops sustainability study', *Ecotextile News*, **19**, November, 12.

ANON. (2008e), 'Energy eco-labelling', *Fashion Business International*, June–July, 4.

ANON. (2009), 'Environmental concerns with electricity generation', http://en.wikipedia.org/wiki/Environmental_concerns_with_electricity_generation, accessed 30 January 2009.

AUSTRALIAN WOOL TESTING AUTHORITY (2009), http://www.awta.com.au/en/Home/Education/Historical-Material/Brief-History-of-Wool-in-Autralia/, accessed 29 January 2009.

BARBER A and PELLOW G (2006), '*New Zealand Merino Industry: Merino Wool Total Energy Use and Carbon Dioxide Emissions*', The AgriBusiness Group, Pukekohe, Auckland, New Zealand.

BATEUP BO, CHRISTOE JR, JONES FW, POOLE AJ, SKOURTIS C and WESTMORELAND DJ (1996), 'Effluent management', in Proceedings of Top-Tech '96 Conference, CSIRO Division of Wool Technology, Geelong, Australia, 11–14 November, p. 388.

BIDE M (2008), http://www.colourclick.co.uk/colourclick/articleimages/China%20Conf%20Report_full.pdf, accessed 30 January 2009.

BRENT AC and HIETKAMP S (2003), 'Comparative evaluation of life cycle impact assessment methods with a south african case study', *International Journal of Life Cycle Assessment*, **8**, 27–38.

BRITISH TEXTILE TECHNOLOGY GROUP (1999a), '*Report 5: Waste Minimisation and Best Practice*', http://www.e4s.org.uk/textilesonline/content/6library/fr_library.htm, accessed 11 February 2009.

BRITISH TEXTILE TECHNOLOGY GROUP (1999b), '*Report 4: Textile Mass Balance and Product Life Cycles*', http://www.e4s.org.uk/textilesonline/content/6library/fr_library.htm, accessed 11 February 2009.

COLLINS M and AUMÔNIER S (2002), *Streamlined Life Cycle Assessment of Two Marks & Spencer plc Apparel Products*, Environmental Resources Management, February, http://www.satradingco.org/Reports/LCA_Final.pdf, accessed 30 January 2009.

COOPER P (2007), 'Clearer communication', *Ecotextile News*, **4**, May, 30–33.

CSIRO (2009), http://www.csiro.au/science/Wool-Textiles/Publications.html, accessed 10 February 2009.

DEBOOS A (2008), 'Wool industry tweaks standards', *Ecotextile News*, **15**, June, 28.

DEFRA (2000), 'Consumer products: green claims code', http://www.defra.gov.uk/environment/consumerprod/gcc/index.htm, accessed 11 February 2009.

DEFRA (2008), 'M&S leads – legislation may follow', *Ecotextile News*, **14**, May, 29.

DRAPER S, MURRAY V and WEISSBROD I (2007), 'Fashioning sustainability', http://www.forumforthefuture.org/files/Fashionsustain.pdf, accessed 10 February 2009.

EIPPCB (EUROPEAN INTEGRATED POLLUTION PREVENTION and CONTROL BUREAU (2002), *Integrated Pollution Prevention and Control (IPPC) Reference Document on Best Available Techniques for the Textiles Industry*, VD/EIPPCB/TXT BREF FINAL, November.

EU (2007), 'A database of green suppliers', *The Flower News*, **2**, 4, http://ec.europa.eu/environment/ecolabel/pdf/flower_news/2007_oct/flower_0207_en.pdf, accessed 30 April 2008.

EU REGULATION (2006), *Registration, Evaluation and Authorisation of Chemicals*, Regulation No. 1907/2006, http://eur-lex.europa.eu/JOHtml.do?uri=OJ:L:2007:136:SOM:EN:HTML.

FERRATINI S (2008/2009), 'UN develops EU eco-label plan', *Ecotextile News*, **20**, December/January, 36.

FRANKLIN ASSOCIATES (1993), *Resource and Environmental Profile Analysis of a Manufactured Apparel Product – Life Cycle Analysis (LCA): Woman's Knit Polyester Blouse*, Franklin Associates, Kansas, USA.

GABI (2009), 'New database on fabrics and processes in the textile finishing industry available', http://www.gabi-software.com/fileadmin/user_upload/database_textile_finishing_01.pdf, accessed 29 January 2009.

GRAY D (2009), 'The free ride on the sheep's back is over', *The Age*, 31 January, Business Day, 1.

INTERNATIONAL WOOL TEXTILE ORGANISATION (2009), http://www.iwto.org/WhyWool/WoolHistory.htm, accessed 29 January 2009.

KALLIALA E and TALVENMAA P (2000), Environmental profile of textile wet processing in Finland, *Journal of Cleaner Production*, **8**, 143–154.

KANWAR P (2008), 'The EU sets the standard', *SGS Consumer Compact*, **III**(4), 2, http://newsletter.sgs.com/eNewsletterPro/uploadedimages/000006/ConsumerCompactNewsletterJanuary2008.Pdf, accessed 30 April 2008.

KARP E (2008), 'I can eat it, I clean with it – now do I have to wear it?', *Ecotextile News*, **18**, October, 39–41.

KELLY B (2008), 'Life cycle analysis the key to meeting eco demands', *Wool Record*, January, 28.

LAURSEN SE, HANSEN J, BAGH J, JENSEN OK and WERTHER I (1997), '*Environmental Assessment of Textiles*', Environmental Project No. 369, Ministry of Environment and Energy, Denmark.

LENZING GROUP (2008), http://www.lenzing.com/images/lenzing_kk/NH_Brosch_EN_2008_SCREEN.pdf, accessed 30 January 2008.

LI L and SUN G (2007), 'Policies and measures on energy efficiency improvement and renewable energy development', www.oecd.org/dataoecd/15/52/38332844.pdf, accessed 11 February 2009.

LYONS B (2008), 'Positioning Australian Merino as an environmental and ethical fibre of choice: the challenges of life cycle analysis for a natural fibre', In RITE Conference, London, 7 October 2008.

MELLO F (2007), 'Top labels spy a stylish edge in carbon labelling', *The Age*, 17 September, http://www.theage.com.au/news/business/top-labels-spy-stylish-edge-in-carbon-labelling/2007/09/16/1189881340189.html, accessed 30 April 2008.

MEYER WS (1997), 'Water for food – the continuing debate', http://www.clw.csiro.au/publications/water_for_food.pdf, accessed 11 February 2009.

MOHR T (2005), 'Reputation or reality? A discussion paper on greenwash & corporate sustainability', www.greencapital.org.au/index.php?option=com_docman&task=doc_download&gid=12, accessed 23 April 2008.

MORRIS D (2008), 'The carbon challenge for textile fibres', *Ecotextile News*, **16**, July, 30.

PAILLAT B (2008), 'Eco-levy on clothes to boost recycling', *Ecotextile News*, **17**, August, 36.

REUTERS (2008), 'Lake pollution could hit deliveries', *Ecotextile News*, **6**, July, 10.

TRIVEDI B (2008), 'Kangaroos to the rescue', *New Scientist*, 20/27 December, 48.

WASTE ONLINE (2004), http://www.wasteonline.org.uk/resources/InformationSheets/Textiles.pdf, accessed 10 February 2009.

WISEMAN LA (1981), 'Oil and the textile industry', *Textiles*, **10**, 24.

WOO PN (2008), 'RITE Conference energises industry', *Ecotextile News*, **19**, November, 20.

4

Sustainable synthetic fibres: the case of poly(hydroxyalkanoates) (PHA) and other fibres

I. CHODÁK, Polymer Institute of the Slovak Academy of Sciences, Slovakia, and R. S. BLACKBURN, University of Leeds, UK

Abstract: This chapter describes structures based on the bacterial polyesters poly(hydroxyalkanoates) and fibres based on the synthetic polymer poly(caprolactone). The various materials and techniques available for producing these fibres are discussed along with their properties relevant to sustainable synthetic materials. Examples of other biodegradable polyesters are also given and the chapter concludes by describing possible applications of the technology and future trends in research directions.

Key words: poly(hydroxyalkanoates), poly(caprolactone), poly(hydroxybutyrate), copolymers, biodegradable polymers, synthetic polymers.

4.1 Introduction

Bacterial polyesters poly(hydroxyalkanoates) (PHAs), with poly(hydroxybutyrate) (PHB) (Fig. 4.1) as the first homologue, belong to the most interesting, but also the most controversial, group of biodegradable polymers. Advantages include production from fully renewable resources, rather fast and complete biodegradability, biocompatibility, and excellent strength and stiffness, which favours this material as a polymer of the future.[1] However, several serious drawbacks hinder its wider application, including rather high susceptibility towards thermal degradation, difficult processing related partially to thermal instability as well as to low melt elasticity, brittleness of the material resulting in low toughness (which increases further during storing due to an interesting phenomenon of physical ageing), and rather high price. These are the main reasons for low production volumes and an unsatisfactory number of applications.[2]

The low number of applications up to now seems to be also partially the reason for the high price of the polymer, creating a vicious circle where the applications are not developing because of too high a price, while the price is not decreasing due to a low volume of produced polymer. Since fibres can be considered as a new prospective product, successful development of PHA fibres would also contribute significantly to the general

PHB R=CH₃

4.1 Poly(hydroxybutyrate) (PHB).

4.2 Poly(caprolactone).

spread of products from PHAs. Moreover, it is generally believed that after drawing, the properties of PHAs will improve, including an increase in toughness.

Poly(caprolactone) (Fig. 4.2) is a synthetic polymer prepared mainly by ring opening polymerisation of caprolactone. The polymer is, similar to PHAs, fully biodegradable, although the rate of biodegradation is lower compared to PHAs. This, together with a low melting temperature (of about 60 °C), is the reason why the polymer is used mainly either as a component of polymer blend or as a matrix for biodegradable composites. Among the latter, its mixture with starch is possibly best known under the trademark *MaterBi* produced by Novamont, Italy.[3] Nevertheless, poly(caprolactone) has a number of interesting properties, such as good processability, high toughness and deformability, and good thermal stability. Therefore its wider application is expected in the future.

4.2   Poly(hydroxyalkanoate)-based oriented structures

4.2.1   Materials and techniques available

PHAs represent a number of materials with a broad range of properties. Generally, all types described in scientific literature are produced by bacteria, although synthetic routes are also known and used.[4] About 40 roots of bacteria exist that are able to produce polyester-type polymers.[5] Although the various bacteria differ in the conditions and the efficiency of the PHA production, they produce more or less the same products. The variation in the polymer produced is reached more by changes in the production conditions than by changing the root of bacteria; the most important from this point of view is the substrate for feeding the bacteria.[6] By sophisticated selection of the substrate, PHB homopolymer, its copolymers with higher PHAs,[7] even polyesters with branching[8] functional groups (epoxy,[9] aromatic structures[10] chlorine,[11] double bonds[12]) in the chain may be produced.

Table 4.1 A comparison of physical properties of PHB, copolymers of PHB with higher PHAs, polypropylene (PP), and low density polyethylene (LDPE)

Property	PHB	20V*	6HA[†]	PP	LDPE
Melting temperature (°C)	175	145	133	176	110
Glass transition temperature (°C)	4	−1	−8	−10	−30
Crystallinity (%)	60	ng[‡]	ng	50	50
Density (g cm^{-3})	1.25	ng	ng	0.91	0.92
E modulus (MPa)	3.5	0.8	0.2	1.5	0.2
Tensile strength (MPa)	40	20	17	38	10
Elongation at break (%)	5	50	680	400	600

* Poly(3-hydroxybutyrate-co-20 mol% hydroxyvalerate).
[†] Poly(3-hydroxybutyrate-co-6 mol% hydroxyalkanoates) = 3% 3-hydroxydecanoate, 3% 3-hydroxydodecanoate, <1% 3-hydroxyoctanoate, <1% 5-hydroxydodecanoate.
[‡] ng, negligible.

The properties of the first homologue, PHB are similar to polypropylene, as seen in Table 4.1.[2] While strength parameters (tensile strength, Young's modulus) and most physical properties (crystallinity, melting temperature, and glass transition temperature) are basically the same, the important difference concerns elongation at break and, consequently, toughness. While the ductile polypropylene breaks at elongation around 700%, PHB hardly exceeds 10%, with typical values between 1 and 3%; PHB copolymers with higher PHAs have higher elongation at break mainly due to much lower crystallinity. However, the preparation of these, while well mastered in laboratory conditions, results in much more expensive materials if industrial, large-scale production is considered. From this point of view, only PHB copolymer with poly(hydroxyvalerate) (Fig. 4.1, R = CH$_2$CH$_3$) could be considered to be suitable for some applications.

Considering fibres based on PHB or higher PHAs, no successful process was reported for preparation of PHB fibres by conventional fibre processing technology, i.e. melt or gel spinning with subsequent hot drawing. Therefore, more sophisticated procedures have to be developed to achieve reasonable draw ratios (DRs), resulting in production of anisotropic material with important improvement of properties.

To prepare fibres, it is usually advantageous to start with a polymeric precursor with molecular weight within certain limits. The molecular weight values should be as high as possible to achieve good drawability and high DR. On the other hand, it should increase only to the values acceptable from the point of view of processing (spinning); in the case of very high molecular weights, extremely high DRs can be reached resulting in ultra-high modulus, e.g. for polyethylene,[13] however, rather sophisticated preparation techniques, e.g. dry gel technology,[14] have to be applied.

In the case of PHB and generally PHAs, polymers with high molecular weight are also important concerning the rather high susceptibility of the polymer towards thermal degradation. Thus, starting with a polymer with high molecular weight may result in a polymer having a molecular weight still well above certain limits, even after rather demanding thermal treatment during processing and resulting decrease in the molecular weight due to thermal degradation. Fortunately, by a sophisticated selection of bacteria and preparation conditions, PHB or its copolymers with high M_w can be produced. Kusaka et al.[15] reported a preparation of P(3HB) with a final average molecular weight of between 1.1 and 11 million produced by Escherichia coli XL-1 Blue (pSYL105)[16] containing a stable plasmid harbouring the Alcaligenes eutrophus H16 (ATCC 17699) PHB biosynthesis geneoperon phaCAB. Two-step cultivation of the recombinant E. coli was applied for the production of high molecular weight P(3HB).[17] The molecular weight of PHB produced within the cells was strongly dependent on the pH of the culture medium.[17]

Apparently, as mentioned above, the most important reason to synthesise high molecular weight PHB seems to be a high susceptibility of PHAs towards thermal degradation. When starting with high molecular weight material, even after a substantial decrease in molecular weight, the polymer still has M_w high enough to be processed to secure a product with acceptable properties. Another option to avoid these shortcomings might be to develop modified processing procedures, leading to a thermal treatment as short as possible or proceeding at a processing temperature below the degradation limits. Thus, solid state processing was suggested as a viable alternative for PHB processing with low thermal degradation.[18] An extrusion of PHB powder at temperatures well below melting temperature was successfully preformed to products with improved mechanical properties. Compared with melt-processed PHB, the ductility (and consequently toughness) improved significantly.

4.2.2 Processing/preparation

Generally, fibre-forming polymers can be processed either by drawing the preformed amorphous material at temperatures above, but near, the glass transition temperature (T_g) or by drawing a crystalline material below but near melting temperature. Thus, the easiest way to produce fibres should be melt spinning and, subsequently, cold or hot drawing. More sophisticated procedures involve so-called gel spinning, which was successfully applied for the production of ultra-high modulus polyethylene fibres[13] and later for other polymers.[19] These basic techniques have been tested and modified also for the preparation of PHA fibres; other procedures have been suggested and investigated. Although the number of scientific papers dealing with PHB or other PHA fibres is much smaller compared with important synthetic polymers, the information on various spinning/drawing processes deserves reviewing, especially regarding differences between different processes and between properties of fibres prepared by different research teams.

Melt spun fibres

PHB melt spinning is, from several points of view, not as straightforward a process as for many other fibre-forming polymers. The problems include rather rapid thermal degradation of PHB at temperatures just above melting temperature, low melt elasticity, and slow crystallisation after spinning (which results in the formation of large crystallites leading to extremely brittle material). The brittleness increases during storing owing to an interesting phenomenon of physical ageing; some improvements can be achieved by an addition of plasticisers and nucleating agents; boron nitride was reported to be one of the most efficient nucleating agents for PHB.[20]

An important problem is the fact that crystalline PHAs and especially PHB are rather brittle materials, though the brittleness can be partially removed by compression moulding. The cold rolling of PHB at room temperature results in an increase in elongation at break from 8% up to 200%, if measured in tensile mode in the rolling direction, while no change (elongation 8%) was observed if measured in the perpendicular direction.[21] Such pre-deformed material could be drawn at 125°C leading to an increase in tensile strength by a factor of 5. A patent from 1984[21] claims the procedure of cold rolling and subsequent drawing at temperatures in the ranges of 50–150°C below the melting temperature of the respective polymer.

Melt spinning may lead to an improvement of some of the features mentioned above. Gordeyev *et al.* reported in 1977 that melt spinning followed by a pre-orientation may prevent the ageing process.[22] However, the melt spinning is not so easy; several methods have been developed and described, differing in many details of the process. Yokouchi *et al.*[23] and Nicholson *et al.*[24] reported the procedure of melt quenching below T_g and subsequent drawing. A more successful process seems to be the drawing of melt spun PHB immediately after spinning while the material is still hot, to obtain pre-oriented material. These pre-oriented fibres can be drawn to high DRs even after a few weeks of storing at room temperature, under conditions when bulk PHB would become extremely brittle as a result of physical ageing.[22]

Table 4.2 Mechanical properties of melt spun PHB fibres

Sample	Draw ratio	Tensile strength (MPa)	Young's modulus (GPa)	Elongation at break (%)
As spun	2	109	2.2	160
Hot drawn, 110°C	8	127	3.5	95
Annealed, 155°C, 1 h	8	190	5.6	54

A successful procedure leading to melt spun fibres with good properties was described by Gordeyev and Nekrasov.[25] The authors suggested dissolving PHB with a molecular weight of about 300 000 and $T_m = 180°C$ (determined by differential scanning calorimetry (DSC)) in chloroform and filtering the solution before spinning to remove impurities as well as high molecular weight portions, although it is not explained why higher molecular weight portions are recommended for removal nor what portion, if any, of what molecular weight was actually removed. The spinning and pre-drawing step (DR = 2) was performed in an extruder heated in four regions between 170 and 182°C. Hot drawing proceeded at 110°C and the DR achieved was about 8. Although only a modest rise of Young's modulus was achieved, an increase in tensile strength by a factor 4 to 5 was found when compared with undrawn bulk PHB. Moreover, the fibres were rather elastic showing an elongation at break of about 50%, as seen in Table 4.2.

More detailed studies on the melt spinning/hot drawing of PHB were published in Yamane *et al.*[26] Filaments, about 0.3 mm in diameter, were obtained by extruding. These fibres were drawn at 110°C immediately after melt spinning; the maximum DR achieved was about 6. The drawing immediately after melt spinning is important since after even short storing periods, PHB turns to a brittle material which is impossible to draw. A further requirement seems to be the presence of nucleating agents to increase the originally low nucleation rate of PHB. Boron nitride can be used as an additive; contaminants such as remnants of proteins and lipids from the culture media can also act as efficient nucleating agents.[26] The authors refer to further improvement of mechanical properties of the fibres by annealing after drawing.

It seems that to achieve the drawability, a certain structure of the polymer must be formed. The material has to be in crystalline form, but crystals must not be too large to be able to be deformed during drawing. Thus, the basic requirement for melt spinning/drawing seems to be the presence of nucleating agents on the one hand and the drawing or at least pre-drawing while the material has not developed a fully crystalline structure on the other.

Table 4.3 Mechanical properties (tensile strength (σ), elongation at break (ε) and sonic modulus (E)) of PHB fibres in relation to spinning speed (v) and draw ratio (DR)[27]

v (m min^{-1})	σ (MPa)	ε (%)	E (GPa)
2000	228	72	5.8
3000	281	48	7.1
3500	250	26	7.6

DR (m min^{-1})	σ (MPa)	ε (%)	E (GPa)	T_1/ T_2* (°C)
4.0	52	10	n.a	40/50
4.5	108	60	n.a	40/50
5.0	220	53	n.a.	40/50
5.4	178	71	5.2	45/60
5.5	263	60	5.6	40/50
6.4	310	45	6.8	45/60
6.9	330	37	7.7	45/60

*Temperature of the first and second godet in the production line.

High-speed melt spinning and spin drawing

Certainly, a demonstration of an ability to draw fibres is of primary importance for further development. The possible industrial production depends not only on procedures enabling spinning and drawing the polymer, to produce fibres with good properties, but also on efficiency of the production, which is limited mainly by the development of a procedure for high-speed melt spinning. Such a process was described by Schmack et al.[27] The spinning line consisted of an extruder, spinning pump, heated godets, and two winders, enabling speeds in the range of 2000–6000 m min^{-1}, comparable with the speed of production of synthetic industrial polymeric fibres. Exceptional attention was paid to the thorough drying of the PHB powder prior to spinning to minimise hydrolytic degradation. In spite of the extreme care regarding the moisture removal, the viscometric molecular weight of the PHB being 540 000 for virgin powder dropped down to 175 000 after spinning; this has to be attributed to thermal chain scission since the water content of the dried pellets was only 0.01%. The process described seems to be fast enough to be considered for efficient production of PHB fibres, especially since the mechanical properties of the fibres are satisfying, as seen in Table 4.3. The paper deals with the procedure in rather a detailed way, so that the effect of changing preparation conditions on the ultimate properties of the fibres can be estimated.

Gel spun fibres

Gel spinning is an example of a special technique for preparation of fibres with unique properties. The well-known commercially available fibres prepared by this method are those of ultra-high molecular weight (UHMW) polyethylene.[19] The gel spun material can be drawn to very high DRs, well above 100; fibres exhibit extremely high stiffness and strength, reaching the values of 3–6 GPa and 150 GPa for tensile strength and modulus, respectively.[13]

The typical procedure leading to gel spun PHB fibres is described by Gordeyev *et al.*[28] It involves dissolving the PHB in a suitable liquid (1,2-dichloromethane is recommended as the best solvent); a solution with a PHB concentration as high as possible should be prepared, which depends on the original molecular weight and is about 20 wt % for PHB with M_w about 300 000 g mol^{-1}. Then a solid gel is prepared by evaporation of a part of the solvent; at this stage the concentration of the polymer is about 30 wt %. The gel was extrudable at about 170 °C. The extruded gel is consequently processed in three stages.[28] For the first, so-called pre-conditioning stage, the fibre was wound on a speed-controlled drum – the optimal pre-conditioning DR was estimated by comparing Tex values of the extruded and drawn fibres, respectively. For further efficient drawing the optimal DR of the pre-conditioning step was found to be around 2. Continuous hot drawing between two rollers was performed at 120 °C as the second step – the total DR was around 10. Finally the fibres were stretched at room temperature to 180% of the length after the second step – they were fixed and annealed at 150 °C for 1 hour.

'As-spun' fibres from the stage 1 could be drawn via the necking process both at, as well as above, room temperature; the necking begins at a strain about 6–7%. The mechanical properties of the drawn fibres are shown in Table 4.4.

The drawability depends on drawing temperature to a certain extent, manifesting the highest DR (about 5) at 120 °C. Although the first step

Table 4.4 Tensile properties of gel spun fibres at room temperature[28]

Sample	Draw ratio	Tensile strength (MPa)	Static modulus (GPa)	Dynamic modulus (GPa)	Strain at break (%)
As spun	2	103	2.0	4.6	250
Hot drawn	10	332	3.8	5.8	104
Annealed	10*	360	5.6	7.5	37

*After hot drawing.

(pre-conditioned drawing after spinning) has to be performed soon after spinning, the pre-conditioned fibres, once drawn, can be stored for several months without losing the ability to be drawn at stage 2. Surprisingly, the drawn material exhibited rubber-like elastic behaviour. Presumably, drawing may introduce changes to the brittle bulk PHB similar to those of cold rolled polyester,[29] the effect on drawing is much more pronounced, obviously as a result of a much higher degree of chain orientation.

It was reported that the tensile strength of gel spun fibres is about double that of melt spun material of similar parameters.[28] This behaviour was attributed to a lower degree of thermal degradation due to lower thermal treatment during three-stage processing. However, this assumption was not directly proved, e.g. by comparison of changes in molecular weights. Comparison of mechanical properties of gel spun fibres from reference 28 with melt spun fibres prepared by the high-spin procedure shows certain differences, but generally the properties are similar, being dependent mainly on the DR. In the latter case the original molecular weight of powder, 550 000, dropped down to 175 000 after processing; similar molecular weights of the fibres can be expected after the less detrimental gel spinning process from the original molecular weight around 330 000. These considerations suggest that the procedure itself does not affect the ultimate properties of the fibres if the DR and molecular weight of the fibre-forming polymer is the same; the effect of the process consists of secondary phenomena, mainly the extent of thermal treatment which is less detrimental in the case of gel spinning compared with melt spun material. It must be stressed also that for gel spun fibres annealing leads to a certain increase in stiffness while tensile strength does not change.[28] It is important to note that mechanical properties do not change significantly during storing as demonstrated by strength and modulus values measured during a period of 120 days. The modulus increased by a factor of less than 1.2 while a decrease in tensile strength was observed to about 82% of the original value.[28]

Besides fibre formation, oriented PHB can also be prepared in the form of films. A patented procedure[30] refers to PHB with M_w higher than 500 000, which can be oriented at temperatures of 144–180 °C. As an example, PHB with $M_n = 6 \times 10^6$ was treated at 160 °C at DR 6.2 to produce oriented film with $T_m = 186$ °C, T_g value 2.2 °C, crystallinity higher than 90%, Young's modulus 1.7 GPa, tensile strength 80 MPa, and 70% elongation. The strength parameters are not extraordinarily impressive, but reasonable elongation indicates that the material may form a flexible foil which could be considered for packaging. In that case, strength and modulus are similar to polypropylene foils and are substantially higher compared with common foil-forming materials such as LDPE.

Other procedures

Kusaka *et al.*[15] described a preparation of PHB fibres with a DR higher than 6 via stretching solution-cast high molecular weight films in a silicon oil bath at 160°C. PHB with a molecular weight well above 10^6 was used. The mechanical properties of the fibres are shown in Table 4.5. Compared with solution-cast isotropic undrawn films, drawing results in higher tensile strength while the modulus is the same or slightly lower. Elongation at break is substantially higher and it is almost unaffected by ageing, especially if annealing of the fibre is performed after drawing. The elongation values, the absence of the ageing phenomenon, as well as a certain decrease in modulus seem to be of interest especially considering a clear increase in crystallinity due to drawing and a further rise resulting from annealing. Obviously, high crystallinity is not necessarily a reason for the high stiffness/low toughness of PHB; the changes in crystal morphology and also changes in the supramolecular structure of the amorphous phase seem to be much more important. Some authors reported a conformational transformation as a result of drawing (e.g. Orts *et al.* for polyhydroxybutyrate-valerate (PHBV) stretched films);[31] however, reflections indicative for helix to planar conformation were not observed for UHMW PHB by Kusaka *et al.*[15] A centrifugation spinning process for PHB fibre preparation was also demonstrated[32] to be an alternative to gel spinning. An entangled fibrous material was produced which resembles 'cotton wool'. The fibres were found to possess various surface irregularities such as pores with a diameter in the range 1–15 μm.

Table 4.5 Physical properties of solution-cast drawn ultra-high molecular weight poly(3-hydroxybutyrate) films with original molecular weight M_n = 6 000 000; M_w = 16 000 000

DR	t_{anneal}	t_{ageing} (days)	C (%)	σ (MPa)	ε (%)	E (GPa)
0	0	7	65 ± 5	41 ± 4	7 ± 3	2.3 ± 0.5
0	1 sec	7	65 ± 5	41 ± 3	6 ± 3	2.4 ± 0.3
6.5	1 sec	7	80 ± 5	62 ± 5	58 ± 1	1.1 ± 0.1
6.5	1 sec	190	75 ± 5	88 ± 8	30 ± 1	2.5 ± 0.2
6.5	2 hrs	7	>85	77 ± 10	67 ± 1	1.8 ± 0.3
6.5	2 hrs	190	>85	100 ± 10	67 ± 2	2.5 ± 0.2

DR = draw ratio; t_{anneal} = annealing time at 160°C; t_{ageing} = ageing time at room temperature; C = crystallinity (X-ray); σ = tensile strength; ε = elongaton at break; E = Young's modulus.

Copolymers of poly(hydroxybutyrate) with higher
poly(hydroxyalkanoates)

Most of the papers on PHA fibres deal with PHB. Although sometimes a copolymer of PHB with a low amount of poly(hydroxyvalerate) is also referred to as PHB; in the literature quoted in this review such inconsistency is not expected. The reason for not using PHBV may be due to the fact that the copolymer has a much lower crystallinity and consequently, also, its modulus is lower compared to PHB. Thus, the processing to fibres may be more difficult because of low crystalline content and also the properties of the fibres may be expected to be less impressive. Moreover, PHBV itself is much tougher compared to PHB, but also more expensive, so that the need for an improvement in toughness via drawing is not crucial while the price may be prohibitive if the expected modest improvement in the properties of fibres is considered.

The procedure of preparation of fibres based on a copolymer of PHB with higher PHA is described by Fischer *et al.*[33] A copolymer PHB with hydroxyhexanoate (PHBH) shows high elongation but low tensile strength; cold drawing is believed to improve the strength behaviour. Solvent-cast films of PHB copolymer with 5 or 12% of hydroxyhexanoate were melted in a hot press and subsequently quenched in ice water. The more or less amorphous films were oriented by cold drawing to DR 2 to 5 and annealed at various temperatures (23–140 °C), then further drawing was applied at room temperature before annealing. PHBH films were easily drawn at low stress to DR 5. Similar to PHB drawn fibres, PHBH stretched films also showed an elastic behaviour after the sample was released from the clamps of the stretching equipment.[33] Therefore an annealing procedure was required for fixation of the extended polymer. It is interesting to note that no changes in the molecular weight were observed as a result of the drawing procedure: this may be attributed partially to rather mild thermal treatment as well as to higher thermal stability of PHBH as a result of steric hindrance of the chain scission due to the presence of the propyl side chains.[34]

Considering the mechanical properties of the copolymer PHBH with 5% of H, an increase of tensile strength was observed from 25 MPa for isotropic film up to 75 MPa for DR 5. In the same range of DR, modulus increased from 400 MPa up to about double value, while elongation at break decreased from an original 250% (isotropic sample) down to less than 100 % for DR 2 and then a monotonic increase back to the original 250% at DR 5 was observed. The increase in the H-copolymer content to 12% leads to a substantial increase in elongation of isotropic film and a certain decrease in tensile strength, while the decrease in modulus is substantial (from 400 down to 100 MPa). Monotonous dependences of all parameters with rising

Table 4.6 Mechanical properties (tensile strength (σ), elongation at break (ε) and modulus (E)) and crystallinity (C) of drawn PHBH films and dependence on the content of hydroxyhexanoate (H) and draw ratio (DR)

H content (%)	DR	σ (MPa)	ε (%)	E (GPa)	C (%)
5	1	32 ± 2	267 ± 30	480 ± 70	42 ± 5
	5	80 ± 1	258 ± 10	870 ± 40	47 ± 5
	10	140 ± 20	116 ± 10	1480 ± 150	65 ± 5
12	1	23 ± 4	871 ± 70	90 ± 3	31 ± 5
	5	53 ± 4	204 ± 18	30 ± 1	35 ± 5

DR were observed (a decrease in elongation at break and increase in tensile strength and modulus values). The two-step drawing resulted in continuing tendencies for all parameters, as seen in Table 4.6. Crystallinity changes (Table 4.6) correspond with the trends in mechanical properties of drawn films.

Oriented blends of poly(hydroxyalkanoates)

Besides drawing of PHB or its copolymers with other PHAs, several attempts were made to prepare fibres from blends of PHAs with other polymers. These attempts are mainly aimed at obtaining fibres with different properties or lower price; the latter being connected with either a less expensive second component of the blend or easier processing.

Park *et al.*[35] investigated the preparation and properties of fibres made from poly(L-lactic acid) (PLLA) and PHB of two different molecular weights. The two polymers are immiscible in the whole concentration range. The films of the blend after preparation by solvent casting were uniaxially drawn at either 2°C for PHB-rich blends (close to PHB's T_g) or 60°C for PLLA-rich blends (around PLLA's T_g). It is of interest to note that while blends based on the PHB matrix were impossible to draw above the T_g of the matrix due to rapid stress relaxation, PLLA-rich blends have to be drawn above the T_g of the matrix polymer. Although PLLA domains in normal molecular weight PHB matrix (600000) remained almost unstretched during cold drawing, good interfacial adhesion was suggested considering the good mechanical properties and the reinforcing effect of the PLLA presence. On the other hand, PLLA was found to be oriented if UHMW PHB (M_w almost 6000000) was used as the matrix. Interfacial entanglements, which are much more numerous in blends with UHMW PHB, are suggested to be responsible for the differences in orientation of the minor component. As a result, the mechanical properties of blends with UHMW PHB matrix improved considerably with increasing PLLA content, as seen in Table 4.7.

Table 4.7 Mechanical properties of the blends of PHB/PLLA depending on the M_w of PHB and the composition of the blends (data estimated from figures in Park *et al.*[35])

PHB portion	DR	σ (MPa)		E (%)		E (GPa)	
		NMW	UHMW	NMW	UHMW	NMW	UHMW
1	1	22	21	22	12	0.65	0.72
	5	25	41	60	45	0.88	1.18
0.9	1	22	23	30	13	0.65	0.73
	5	28	70	48	75	1.05	1.22
0.7	1	22	24	30	16	0.63	0.73
	5	34	102	39	102	1.12	1.55
0.5	1	24	30	18	33	0.70	0.81
	5	72	151	50	150	1.31	1.54
0.3	1	30	31	28	35	0.82	0.81
	5	165	158	92	158	1.80	2.05
0	1	44	44	22	22	1.12	1.12
	5	170	168	90	167	1.86	2.18

Oriented foils based on blends of copolymer PHB-co-hydroxyvalerate and polyalcohols are described by Cyras *et al.*[36] The blends were prepared by solvent casting, and castor oil or polypropylene glycol was used as the polyalcohol component. Dynamic mechanical behaviour indicates the formation of a two-phase immiscible blend. The addition of polyalcohols leads to an increase of crystallinity but lower storage modulus was observed due to an addition of the amorphous compound.

4.3   Poly(caprolactone)-based fibres

Poly(ε-caprolactone) (PCL) is a synthetic polymer that has many advantages: biodegradability; mechanical properties similar to polyolefins; hydrolysability similar to polyesters; compatibility with many other polymers; ease of melt processability; and high thermal stability. However, low melting point (around 60 °C) and slow rate of degradation *in vivo* (2–3 years) hinder its use as a homopolymer in many cases. Therefore PCL is used more frequently as a component in blends or as a comonomer if copolymers are to be formed.

Various methods have been described and used for PCL fibre production, some of them rather unconventional. Simple melt spinning is possible to apply.[37] Due to the low melting and crystallisation temperature of PCL, vertical direction of spinning, small distance between the die and cooling bath, and intensive cooling (ice water 5–10 °C) is recommended. The spinning temperature should be kept around 85–90 °C. At higher temperatures

(120°C), fibres with uniform diameter could still be obtained but signs of capillary instability were observed.[37] Fibre diameters were in the range 0.49–0.91 mm depending on the spinning conditions, e.g. ram speed, extrusion rate, take-up rate, and the ratio of take-up to extrusion rate. Melt spinning of PCL with incorporated additives was performed to produce PCL fibres containing N-(3,4-dimethoxycinnamoyl)-anthranilic acid, a drug suppressing fibroblast hyperplasia.[38]

An interesting method of PCL fibre preparation is described by Smith and Lemstra[14] as gravity spinning. The polymer is dissolved in a suitable solvent, in this case acetone, to produce solutions containing 6–20 wt % of PCL. The solution was transferred into a vessel and allowed to flow out through a spineret placed in the bottom of the vessel. The polymer solution was forced by gravity to flow into a non-solvent (methanol) forming a fibre. The 'as-spun' fibre was taken up on a mandrel using a variable speed. At concentrations of 5% and lower the fibre was not formed. Within a concentration range of 6–20% the production rate varied from 2.5 to 0.9 m min⁻¹ and fibres with diameters between 0.19 and 0.15 mm were formed. Both the production rate and diameter of the fibres decreased with the increase in the solution concentration. The fibres were round in diameter and exhibited a rough, porous surface.

This procedure can also be used for preparation of fibres containing various additives. Williamson et al.[39,40] prepared PCL fibres with the addition of ovalbumin as a hydrophilic macromolecule. The procedure consists of a preparation of 10% PCL solution in acetone and in situ formation of ovalbumin nanoparticles at a concentration of 1 or 5% re PCL content. Addition of poly(vinylpyrrolidone) is recommended to obtain better dispersion of nanoparticles. The fibres are then prepared by gravity spinning as described above. Progesterone, as a lipophilic steroid, was also shown to be incorporated into fibres at a concentration of 0.625 and 1.25% using the above-mentioned procedure.

Zeng et al.[41] investigated a preparation of ultrafine PCL fibres formed using electrospinning to achieve a biodegradable material with high surface area so that the rate of biodegradation could be substantially increased. The technique itself involves spinning the polymer from solution in a strong electrostatic field through a syringe with a capillary jet outlet. When the voltage is over a certain threshold value, the electrostatic forces are higher than surface tension. As the jet moves towards a collecting metal screen it acts as a counter-electrode, the droplets of the solution split into small charged fibres or fibrils and the solvent evaporates.[42] Nonwoven fabric is formed from the fibres produced. Chloroform solution of PCL was used; an addition of 1,2-dichloroethane resulted in an improvement of the process.[41] The voltage depends on the capillary diameter and was in the range of 29–36 kV. The diameter of fibres was strongly dependent on the capillary thickness which was between 0.1 and 0.4 mm, while the fibre diameter was in the range 300–900 nm. The effect of the driving pressure in the capillary as well as ambient temperature and air flow were also investigated in detail.

The mechanical strength of melt spun fibres was very low. The fibre could be easily stretched to DR values over 20 without breaking; it was suggested that the spinning process introduced very little if any orientation along the fibre axis.[37] After drawing to various DRs (between 5 and 25) the strength and modulus increased and elongation at break decreased substantially. Unfortunately the authors do not present exact data on mechanical properties; for the highest DR 25, the values 280 MPa and 450% for tensile strength and elongation at break, respectively, can be roughly estimated from the stress–strain curves presented.

The tensile properties of the gravity spun fibres were rather low, characterised by a tensile modulus between 10 and 100 MPa, tensile strength of 1.8–9.9 MPa and elongation at break of 175 up to 600 for 6 and 20% PCL concentrations, respectively. All values are substantially lower when compared to bulk PCL, obviously due to the porosity of the samples. Cold drawing of the fibres proceeded rather easily and resulted in an increase in strength properties and a decrease in elongation. Again, the properties depended on the concentration of the PCL solution during spinning. Fibres prepared from 6% solution could be drawn up to DR 2, while DR 5 could be achieved with the other fibres prepared from solutions containing 10–20% of the polymer. The highest values reached were 320 MPa for modulus, 39 MPa for tensile strength, and 136% for elongation at break. These values are not at all impressive and do not differ substantially from values for bulk PCL. Similarly, only a marginal effect was observed regarding the changes in melting temperature and crystallinity in response to changing DR.

Incorporation of additives (ovalbumin, progesterone) resulted in a decrease of tensile strength, modulus, and elongation at break compared to fibres without additive. A continuous decrease in the properties was observed with increasing concentration of the additive.[40] Similar concentration-dependent deterioration of mechanical properties was also observed if N-(3,4-dimethoxycinnamoyl)-anthranilic acid was mixed into PCL fibres during melt spinning.[38]

As indicated, caprolactone can also be used as a comonomer for preparation of copolymers, in this case transesterification reactions can also be considered. Poly(ethylene terephthalate) (PET) copolymers with PCL were prepared by reactive extrusion. In the presence of stannous octoate, ring opening polymerisation of caprolactone is initiated by hydroxyl end groups of molten PET. A block copolymer with a rather low proportion of transesterification was formed in a twin screw extruder as a result of fast distributive mixing of caprolactone into high melt viscosity PET and short reaction time. The copolymer was directly fed into a spin pot and extruded filaments were spun and drawn to DR 6.6. The interactions between PET and PCL segments may result in the formation of a miscible phase with a single T_g being around 45 °C.

4.4   Structure of drawn fibres

Isotropic PHB crystallises in an orthorhombic lattice crystalline structure (α-form) with the chains in the left-handed 2/1 helix, as reported by Yokouchi et al.[23] and Pazur et al.[43] X-ray diffraction patterns indicating an orthorhombic crystal structure (α modification, 2/1 helix) were also described by Yamamoto et al.[44] Changes in the crystalline structure may be expected as a result of drawing. Frequently, an additional crystal structure is observed, assigned to a zigzag conformation of a hexagonal β modification.[31] Thus, the appearance of both helical and planar conformations and their ratio depend on preparation conditions, mainly on DR. A paper by Yamane et al.[26] describes an appearance of higher crystalline orientation with increasing DR with c-axis parallel to the fibre axis, which seems to be the most preferential orientation direction. A reflection at $2\theta = 19.7°$ was assigned to a reflection of the pseudohexagonal phase as proposed by Furuhashi et al.[45] Using a high-speed spinning procedure, Schmack et al.[27] reported no signs of hexagonal modification for high-speed spinning up to DR 4.0 and proposed lack of stress-induced crystallisation at that degree of orientation. However, an increase to 4.5 leads to wide-angle X-ray scattering (WAXS) patterns indicating the presence of both crystal structures. The observed effects are generally in accordance with mechanical properties, but the effect of extrusion speed is not clear from this study, i.e. whether stress-induced crystallisation would be sufficient for the formation of hexagonal modification if the DR would stay low at higher extrusion speed.

A similar orthorhombic structure was also observed for drawn fibres based on a copolymer PHB-co-hydroxyhezanoate.[33] It was concluded that the higher hydroxyalkanoate units were excluded from the crystal structure.

The surface of melt spun fibres consists of many large spherulites, as indicated by scanning electron microscopy (SEM) observations.[46] After drawing to DR 6, the fibres have a fibrillar structure and their surface is fairly smooth. This fibrillar structure is formed mainly in surface areas of the fibres as indicated by the appearance of fibres observed by SEM after various periods of enzymatic degradation. The annealing results in a higher proportion of fibrillar structure which is also formed in core parts of the fibres. Annealing under tension leads to the formation of a more perfect fibrillar structure in the core and this effect seems to increase with the rising tension load during annealing.[46]

A similar structure was also observed for melt spun PCL fibres. SEM observations showed spherulites in the undrawn fibres which change to fibrillar stripes along the fibre axis in drawn fibres.[47] Differences in the crystal structure were also observed with rising DR.

4.5   Thermal properties

Generally, thermal properties of polymers change after drawing. The melt temperature of PHB determined by DSC depends to a certain extent on the material itself as well as on the measuring conditions; Yamane et al.[26] reported T_m = 171 °C (second run) and T_c = 91 °C at a cooling rate of 5 °C min^{-1}, while T_m of virgin powder was found to be 177 °C.[27] Melting temperature also depends on the molecular weight, as revealed by Park et al.[35] who determined T_m to be around 165 and 177 °C for PHBs with a molecular weight of 590 000 and 5 300 000, respectively. Equilibrium melting temperature was determined to be around 186 °C.[48] 'As-spun' fibres showed a melting temperature of 176 °C irrespective of DR,[26] drawing results in a significant increase in the T_m. An increase from 177 °C for virgin PHB powder up to 181.6 °C, and even to 188.9 °C, was reported,[27] depending on the DR being between 4.5 and 6.9. It is interesting that melt spinning/drawing at DR 4 does not lead to any increase in T_m; also mechanical properties are not improved due to drawing, although a little higher DR, namely 4.5, was reached. It has also to be mentioned that fibres drawn to a DR around 6 have a melting temperature higher than that reported for the equilibrium temperature of isotropic PHB.[27] Thorough drying itself leads to a substantial increase in the melting temperature, namely from 177.2 °C for virgin PHB powder up to 189.9 °C for pellets dried for 16 hours at 120 °C.[27] Thorough drying may have certain effects on the melting temperature, changes in the supramolecular structure occur after several hours treatment at 120 °C, even without the effect of losing moisture.

Annealing leads to changes in the thermal behaviour of drawn fibres. DSC melting peaks of annealed fibres are larger and tend to be sharper compared to unannealed fibres;[26] both effects are more pronounced with rising annealing temperature. Annealing under tension results in even sharper peaks, the results indicate[26] that annealing without tension leads to a recrystallisation of the material to a more perfect α-form, while the molecules between α-form lamellae crystallise into β-form if tension is applied during annealing; the effect is more pronounced at higher annealing temperatures.

4.6   Enzymatic and hydrolytic degradation

Yamane et al.[46] investigated the enzymatic degradation of melt spun fibres, PHA depolymerase from *Comomonas testosteroni* in potassium phosphate buffer was used at 37°C for this study. The enzymatic degradation of PHB occurs on the surface of the material and the rate of degradation strongly depends on the structure.[46] SEM observations of as-spun fibres indicate that enzymatic degradation begins on the surface, preferentially in less-ordered regions between spherulitic crystals. At the same time, the fibre diameter decreases as the degradation proceeds to the core of the fibre; similarly, degradation of drawn fibres started in less-ordered regions leaving the fibrillar structure to resist for a longer time. Clear differences in the morphology of fibres drawn to various DRs (4, 5 and 6) were not observed. After 4 days of enzymatic degradation, the drawn fibres changed to aggregates of small fibrous fragments while as-spun fibres retained their fibre shape with a spongy structure, although the diameter significantly decreased. This difference was attributed mainly to the original thickness of the as-spun fibres being thicker compared to drawn fibres while the degree of crystallinity was similar. It has to be noted that both the as-spun as well as the drawn fibres decomposed rather fast so that mechanical properties could not be measured after 24 hours of degradation. However, the annealed fibres (DR 6) kept their consistency longer, so that the mechanical tests could be performed even after 50 hours of degradation. The resistance of fibres against enzymatic degradation increases with increasing temperature of annealing; higher tension during annealing has a retarding effect on degradation.[46] WAXS study revealed that the disordered β-form is attacked by PHB depolymerase more rapidly than the more ordered α-form.

Similar tests were done with an oriented copolymer PHB-poly(hydroxyhexanoate) using PHB depolymerase purified from *Ralstonia pickettii* T1.[33] It was found that the rate of enzymatic degradation of two-step drawn films decreased with increasing DR, i.e. with increasing crystallinity, as expected. However, the decrease in the rate for one-step films was irrespective of DR 3 or 5 and much more pronounced compared to the two-step drawn films even though the DR was significantly higher in the latter case (DR up to 10). The authors admit having difficulties with offering a reasonable explanation for this behaviour.

Special centrifugally spun fibres as an alternative for gel spun fibres were tested regarding hydrolytic degradation (pH 10.6, 70°C).[32] Fibres degraded by gradual fragmentation and erosion to fibre fragments, particles, and eventually monomer. Mammalian and human epithelial cells were used to investigate the cellular interactions with the fibres. To achieve good cell adhesion, the surface of the fibres has to be treated by alkali or acids; the introduction of hydroxyls or carbonyls on the surface is suggested as an explanation of the effect. Neither cell line exhibited any cytotoxic response to the fibres.

Although no more detailed studies were found on the environmental degradation of PHB fibres, the above-mentioned effects may be considered to be generally valid for the anisotropic materials depending on the DR, structure parameters (content of α- and β-phase, etc.).

The rate of enzymatic degradation by lipase of PCL fibres drawn to various ratios was dependent on DR, suggesting that crystallinity and orientation degree are important parameters affecting the degradation kinetics.[49] The degradation behaviour of fibres differing in details of the preparation procedure revealed the changes in degradation kinetics on supramolecular structure of the material. From SEM observation it is observed that the enzyme preferentially attacks amorphous or less ordered regions.[47] Differences in crystal structure were revealed by SEM consisting mainly a portion of spherulites and fibrilles depending on the drawing conditions, these structural parameters affect the enzymatic degradation kinetics. However, in spite of different degradation rates in amorphous and crystalline regions, the crystalline part is also attacked by the enzymes and biodegrades.[47] It is interesting to note that when investigating enzymatic degradation of films made from butylene succinate-co-ethylene succinate copolymer it was found that the rate of degradation depends on the crystallinity rather than on the primary chemical structure. Thus, the degree of crystallinity seems to be the major rate-determining factor of the biodegradation of solid polymers, while the crystalline structure seems to be an additional parameter.

4.7   Other biodegradable and sustainable polyesters

Poly(glycolic acid) (PGA; Fig. 4.3) is an aliphatic polyester that has been widely used in biomedical applications since the early 1970s[50-54] and degrades via a simple hydrolysis mechanism (bulk degradation). However, the homopolymer has found little application as a fibre outside medicine, but such fibres do offer potential for the future and may be an area worthy of further research.

Poly(trimethylene terephthalate) (PTT; Fig. 4.4) was first synthesised in the 1940s by Whinfield and Dickson,[55] but has only recently received attention as a viable textile fibre as it can be synthesised via a more economical

4.3 Poly(glycolic acid) (PGA).

4.4 Poly(trimethylene terephthalate) (PTT).

process.[56] PTT is desirable because it has several unique properties, such as its force–elongation behaviour, resilience, and dyeing properties. It has outstanding elastic recovery; the fibre recovers 100% from approximately 120% strain.[57]

Although PTT is not a biodegradable fibre, it is worth mentioning as a fibre that offers a level of sustainability in the form of Sorona®. In 2004, DuPont and Tate & Lyle PLC announced a joint venture (DuPont Tate & Lyle BioProducts, LLC) to create products from renewable resources such as corn for numerous applications including clothing, interiors, engineered polymers, and textile fibres. The company uses a proprietary fermentation and purification process to produce 1,3-propanediol (PDO), one of the two base chemicals for producing PTT. Rather than using PDO from petrochemical sources, in this process it is derived from renewable sugar. The resultant fibre formed from polymerisation of renewable PDO and terephthalic acid is Sorona®.[57] However, the polymer is not fully sustainable due to the non-renewable sources of terephthalic acid, so cannot be compared to poly(lactic acid) in terms of its sustainability. Nevertheless, this fibre is a demonstration of positive moves by multinational companies to reduce demands on fossil fuels.

4.8   Application of polyester-based biodegradable fibres

PHA fibres are frequently aimed at medical applications; a combination of biodegradability, hydrophobicity, and biocompatibility seems to be of importance for many medical applications. PHB fibres are mainly used for the production of scaffolds.[58] From the point of view of medical applications, an interesting paper by Schmack et al.[59] deals with the effect of electron irradiation on the properties and degradation of PHB fibres with the aim of estimating the consequences of sterilisation of medical devices via electron beam irradiation. In this paper, melt spun PHB fibres were drawn to DR 7 and textiles with a mesh size of 0.5 mm were produced using embroidery technology.[60] Irradiation resulted in a decrease in tensile strength, while changes in modulus and elongation at break were negligible, as seen in Table 4.8; the changes can be attributed to a decrease in

Table 4.8 The effect of irradiation dose on changes in average number M_n, and molecular M_w weights, tensile properties (strength (σ), elongation (ε), modulus (E)) and relative change of M_w after *in vitro* degradation

Dose (kGy)	σ (MPa)	ε (%)	E (GPa)	M_n (kg mol^{-1})	M_w (kg mol^{-1})	M_n^{84}*	M_w^{84}*
0	307	41	4.5	77	157	0.87	0.84
5	295	41	4.3	65	145	–	–
10	248	39	4.3	61	126	–	–
15	257	39	4.1	53	111	–	–
25	236	39	4.4	44	92	0.82	0.74

*Relative molecular weight (M_t/M_o) after 84 days in a Sorensen buffer.

molecular weight, which is quite substantial. Hydrolytic degradation of irradiated fibres in a Sorensen buffer was found to be only a little more pronounced if the fibres were irradiated by a 25 kGy dose compared to non-irradiated fibres.[59]

Applications as scaffolds were also suggested for centrifugally spun fibres.[32] Although almost no cell adhesion was observed for unmodified fibres investigated by SEM, subsequent acid or alkali treatment resulted in a cell-adhesive material that may have a potential value as a wound scaffold.

In fact, most biodegradable polyester-type fibres are intended for medical purposes, although the applications are not limited to scaffolds. PCL gravity spun fibres were found to attach to fibroblasts and myoblasts. Due to high fibre compliance and a potential for controlling the fibre surface architecture, the fibres can be recommended for use as compliance-matched implants for soft tissue engineering.[39]

PCL gravity spun fibres were used also as a carrier for ovalbumin or progesterone.[40] Delayed drug release was observed; the rate depends on several factors, e.g. concentration and size of the protein particulates, which enables a programmed delivery of drugs or supporting agents for tissue engineering.

PCL was used as a matrix for a composite with poly(lactic acid) and PGA long fibres. The composite was prepared by *in situ* polymerisation of caprolactone with dispersed fibres, and a bioabsorbable composite material was obtained that was investigated for an application in craniofacial bone reconstruction.[61] PCL fibres containing *N*-(3,4-dimethoxycinnamoyl)-anthranilic acid can release the agent that suppresses the fibroblast hyperplasia.[38] The drug release rate was found to decrease with increasing DR, obviously as a result of increased crystallinity of the polymeric matrix.

An industrial application was suggested for copolymers of PCL and PET as a material for elastic seat belts. In this case the impact on the passenger was damped by the increased elasticity of the belt compared to the belt made of pure PET, decreasing the extent of injuries caused by contact with the seat belt.[62]

4.9   Future trends and concluding remarks

It can be said that the research on PHA fibre techniques is in quite an advanced stage so that a feasible technology could be designed if a request for a large quantity of fibres arose. The high volume production, however, depends on two factors, the quality of the fibres and the ultimate price of the product. At the moment, apparently, the quality improvement resulting from drawing does not compensate for the rather high price of the fibre, caused mainly by the price of the PHAs as the parent materials. The reason for the high price is partly related to the low volume of production of PHAs; it can hardly be expected that the production of fibres can influence the volume of production to such an extent that the price would fall substantially. Thus, in the near future, the PHA fibres can be expected to be applied mainly as low volume special materials. From this point of view, applications in medicine are the most obvious. The biodegradability of the fibres is certainly considered in many applications outside of medicine, but at the current material price it can hardly be expected to be a decisive factor regarding the high volume applications of PHA fibres.

Electrospinning may play a role in the spread of biodegradable fibres, especially if feasible ideas for applications of nonwoven biodegradable textiles appeared.[63] Electrospinning of PCL was successfully accomplished at laboratory scale as mentioned above; PHA electrospinning should be investigated as well, especially regarding PHB; electrospinning could bring some interesting, possibly surprising, results.

Considering oriented biodegradable bacterial polyesters, attention should be paid to foils. Manufacturing the oriented or even biaxially oriented biodegradable foils could result in a material with high potential for applications in packaging. In this case the volume of the biodegradable plastics could be relatively high so that the price of the polymer could also be affected. Thus, success in this direction may have a positive influence on the future of biodegradable polyesters.

The potential of oriented PCL seems to be lower compared to that of PHAs. Apparently it will be an important special polymer, but high volume applications are less probable. The main problem is related to its low melting temperature, although this can be improved to a certain extent via various modifications, especially transesterifications with poly(caprolactames)[64] or other polymers. However, even in this case the consumption will stay at low levels. The most promising way seems to be Novamont's (Italy) attitude of mixing PCL with starch or other biodegradable species to prepare composites. This material (MaterBi) is used routinely for production of biodegradable foils of good quality.[3] At the moment no data are available on drawing of the composites to produce fibres; neither does a need for such fibres seem to exist. PCL can be used in quite high volumes as a modifying component in blends with other biodegradable plastics, including PHAs and PLA. In this case, its role is related to an improvement of toughness and an increase in drawability. The significance of such blends for biodegradable fibre production is obvious.

In any case, biodegradable fibres based on polyesters while investigated and developed by researchers, should be considered seriously by industry and consumers. Knowledge concerning the properties and new techniques of production of the fibres should bring new ideas for applications of these materials followed by development of the production technology.

4.10   References

1 SUDESH, K., ABE, H. and DOI, Y., *Progr. Polymer Sci.*, 2000, **25**, 1503.

2 CHODAK, I., *Degradable Polymers Principles and Applications*, 2nd edn, editor: G. Scott, Kluwer Acad. Publ. Dordrecht, Boston, London, 2002, chapter 9.

3 BASTIOLI, C. and FACCO, S., *Biodegradable Plastics 2001 Conference*, Frankfurt, November 26–27, 2001.

4 HORI, Y., HONGO, H. and HAGIWARA, T., *Macromolecules*, 1999, **32**, 3537.

5 BRAUNEGG, G., *Degradable Polymers Principles and Applications*, 2nd edn, editor: G. Scott, Kluwer Acad. Publ. Dordrecht, Boston, London, 2002, chapter 8.

6 BRAUNEGG, G., *ICS UNIDO Workshop on Environmentally Degradable Plastics*, Seoul, September 19–22, 2000, 181–200.

7 KUNIOKA, M., NAKAMURA, Y. and DOI, Y., *Polymer Commun.*, 1988, **29**, 174.

8 FRITSCHE, K., LENZ, R.W. and FULLER, R.C., *Int. J. Biol. Macromol.*, 1990, **12**, 92.

9 BEAR, M.M., LEBOUCHERDURAND, M.A., LANGLOIS, V., LENZ, R.W., GOODWIN, S. and GUERIN, P., *React. Funct. Polymers*, 1997, **34**, 65.

10 KIM, Y.B., LENZ, R.W. and FULLER, R.C., *Macromolecules*, 1991, **24**, 5256.

11 DOI, Y. and ABE, C., *Macromolecules*, 1990, **23**, 3705.

12 FRITSCHE, K., LENZ, R.W. and FULLER, R.C., *Int. J. Biol. Macromol.*, 1990, **12**, 85.

13 CHODÁK, I., *Progr. Polymer Sci.*, 1998, **23**, 1409.

14 SMITH, P. and LEMSTRA, P.J., *Makromol. Chem.*, 1979, **180**, 2983.

15 KUSAKA, S., IWATA, T. and DOI, Y.J., *Macromol. Sci. Pure Appl. Chem.*, 1998, **A35**, 319.

16 LEE, S.Y., LEE, K.M., CHANG, H.N. and STEINBUECHEL, A., *Biotechnol. Bioeng*, 1994, **44**, 1337.

17 KUSAKA, S., ABE, H., LEE, S.Y. and DOI, Y., *Appl. Microbiol. Biotechnol.*, 1997, **47**, 140.

18 LUEPKE, T., RADUSCH, H.J. and METZNER, K., *Macromol. Symp.*, 1998, **127**, 227.

19 BASTIAANSEN, C.W.M., *Mater Sci. Technol.*, 1997, **18**, (Chapter 11) 1373, eds: Cahn, R.W., Haasen, P. and Kramer, E.J.

20 LIU, W.J., YANG, H.L., WANG, Z., DONG, L.S. and LIU, J.J., *J. Appl. Polymer Sci*, 2002, **86**, 2145.

21 BARHAM, P.J. and HOLMES, P.A., *US Patent* 4,427,614, 1984, Jan. 24.

22 GORDEYEV, S.A., NEKRASOV, YU. P. and WARD, I.M., *IVth Int Symposium on Polymer for Advanced Technologies*, Leipzig, 1977, PVII.10.

23 YOKOUCHI, M., CHATANI, Y., TADAKORO, K., TERANISHI, K. and TANI, H., *Polymer*, 1973, **14**, 267.

24 NICHOLSON, T.M., UNWIN, P.A. and WARD, I.M., *J. Chem. Soc. Faraday Trans.*, 1995, **91**, 2623.

25 GORDEYEV, S.A. and NEKRASOV, YU. P., *J. Mater. Sci. Letters*, 1999, **18**, 1691.

26 YAMANE, H., TERAO, K., HIKI, S. and KIMURA, Y., *Polymer*, 2001, **42**, 3241.

27 SCHMACK, G., JEHNICHEN, D., VOGEL, R. and TÄNDLER, B., *J. Polymer Sci., B Polymer Phys.*, 2000, **38**, 2841.

28 GORDEYEV, S.A., NEKRASOV, YU. P. and SHILTON, S.J., *J. Appl. Polymer Sci.*, 2001, **81**, 2260.

29 BARHAM, P.I. and KELLER, A., *J. Polymer Sci., Polymer Phys. Ed.*, 1986, **24**, 69.

30 DOI, Y., IWATA, T. and KUSAKA, S., *Eur. Pat. Appl.* EP 849311.

31 ORTS, W.J., MARCHESSAULT, R.H., ALLEGREZZA, JR A.E. and LENZ, R.W., *Macromolecules*, 1990, **23**, 5368.

32 FOSTER, L.J.R., DAVIES, S.M. and TIGHE, B.J., *J. Biomaterials Sci., Polymer Ed.*, 2001, **12**, 317.

33 FISCHER, J.J., AOYAGI, Y., ENOKI, M., DOI, Y. and IWATA, T., *Polymer Degrad. Stability*, 2004, **83**, 453.

34 ASRAR, J., VALENTIN, H.E., BERGER, P.A., TRAN, M., PADGETTE, S.R. and GARBOW, J.R., *Biomacromolecules*, 2002, **3**, 1006.

35 PARK, J.W., DOI, Y. and IWATA, T., *Biomacromolecules*, 2004, **5**, 1557.

36 CYRAS, V.P., FERNANDEZ, N.G. and VAZQUEZ, A., *Polymer Int.*, 1999, **48**, 705.

37 CHARUCHINDA, A., MOLLOY, R., SIRIPITAYANANON, J., MOLLOY, N. and SRIYAI, M., *Polymer Int.*, 2003, **52**, 1175.

38 YAMANE, H., INOUE, A., KOIKE, M., TAKAHASHI, M. and IGAKI, K., *Sen-I-Gakkaishi*, 1999, **55**, 261.

39 WILLIAMSON, M.R. and COOMBES, A.G.A., *Biomaterials*, 2004, **25**, 459.

40 WILLIAMSON, M.R., CHANG, H.-I. and COOMBES, A.G.A., *Biomaterials*, 2004, **25**, 5053.

41 ZENG, J., CHEN, X., XU, X., LIANG, Q., BIAN, X., YANG, L. and JING, X., *J. Appl. Polymer Sci.*, 2003, **89**, 1085.

42 RENEKER, D.H., *J. Appl. Phys.*, 2000, **87**, 4531.

43 PAZUR, R.J., HOCKING, P.J., RAYMOND, S. and MARCHESSAULT, R.H., *Macromolecules*, 1998, **31**, 6585.

44 YAMAMOTO, T., KIMIZU, M., KIKUTANI, T., FURUHASHI, Y. and CAKMAK, M., *Int. Polymer Process*, 1997, **12**, 29.

45 FURUHASHI, Y., KIKUTANI, T., YAMAMOTO, T. and KIMIZU, M., *Sen-i-Gakkaishi*, 1997, **53**, 356.

46 YAMANE, H., TERAO, K., HIKI, S., KAWAHARA, Y., KIMURA, Y. and SAITO, T., *Polymer*, 2001, **42**, 7873.

47 MOCHIZUKI, M., HIRANO, M., KANMURI, Y., KUDO, K. and TOKIWA, Y., *J. Appl. Polymer Sci.*, 1995, **55**, 289.

48 ORGAN, S.J. and BARHAM, P.J., *Polymer*, 1993, **34**, 2169.

49 MOCHIZUKI, M. and HIRAMI, M., *Polymers Adv. Technol.*, 1997, **8**, 203.

50 UHRICH, K., CANNIZZARO, S.M., LANGER, R. and SHAKESHEFF, K.M., *Chem. Rev.*, 1999, **99**, 3181.

51 LANGER, R. and VACANTI, J.P., *Science*, 1993, **260**, 920.

52 WONG, W.H. and MOONEY, D.J., in: Atala, A., Mooney, D.J., Vacanti, J.P. and Langer, R. (eds), *Synthetic Biodegradable Polymer Scaffolds*, Boston, Birkhäuser, 1997, 51–82.

53 FRAZZA, E.J. and SCHMITT, E.E., *J. Biomed. Mater. Res. Symp.*, 1971, **1**, 43.

54 KIMURA, Y., in: Tsuruta, T., Hayashi, T., Kataoka, K., Ishihara K. and Kimura, Y. (eds), *Biomedical Applications of Polymeric Materials*, Boca Raton, CRC Press, 1993, 163–190.

55 WHINFIELD, J.R. and DICKSON, J.T., *BP 578,079*, 1941; *USP 2,465,319*, 1949.

56 CHUAH, H.H., *Chem. Fibers Int.*, 1996, **46**, 424.

57 http://www.dupont.com/sorona/home.html

58 *Patent EP 0567 845 B1*, 1998.

59 SCHMACK, G., KRAMER, S., DORSCHNER, H. and GLIESCHE, K., *Polymer Degrad. Stability*, 2004, **83**, 467.

60 SCHMACK, G., GLIESCHE, K., NITSCHKE, M. and WERNER, C., *Biomaterialien*, 2002, **3**, 21.

61 CORDEN, T.J., JONES, I.A., HUDD, C.D., CHRISTIAN, P., DOWNES, S. and MCDOUGALL, K.E., *Biomaterials*, 2000, **21**, 713.

62 TANG, W., MURTHY, S., MARES, F., MCDONNELL, M.E. and CURRAN, S.A., *J. Appl. Polymer Sci.*, 1999, **74**, 1858.

63 DZENIS, Y., *Science*, 2004, **304**, 1917.

64 BERNÁŠKOVÁ, A., CHROMCOVÁ, D., BROŽEK, J. and RODA, J., *Polymer*, 2004, **45**, 2141.

5

Enzyme biotechnology for sustainable textiles

P. H. NIELSEN, H. KUILDERD, W. ZHOU and X. LU,
Novozymes A/S, Denmark

Abstract: Enzymes are used in a broad range of processes in the textile industry: scouring, bleach clean-up, desizing, denim abrasion and polishing. Enzymes are specific and fast in action and small amounts of enzyme often save large amounts of raw materials, chemicals, energy and/or water. This chapter describes enzyme use in the textile industry in the context of sustainable production and reports life cycle assessments (LCAs) on two enzyme applications: bioscouring and enzymatic bleach clean-up. The results show that resource use and impact on the environment can be reduced considerably when enzymes are implemented in the two processes.

Key words: enzyme, biotechnology, scouring, bleach clean-up, desizing, denim abrasion, life cycle assessment (LCA), cleaner production, water saving, energy saving, climate change.

5.1   Introduction

In many industries, enzymes are used as biological catalysts to replace harsh chemicals or perform reactions under milder conditions. The textile industry is no exception. Not only do enzymes make good economic sense by saving energy, water and chemicals or by improving quality, they also give valuable environmental benefits. These benefits are becoming more and more important at a time of increasing awareness about sustainable development and climate change.

Novozymes recently performed two life cycle assessment (LCA) studies at textile mills in China, one of the prime producing countries for cotton and textiles. One of the studies was of a process known as bioscouring for removing impurities from cotton yarn. This is an alternative to traditional scouring that involves a number of high-temperature steps with a large consumption of chemicals. The other process is known as bleach clean-up to remove excessive bleaching agent prior to dyeing. This enzymatic process required less water and less energy than the conventional process used in China.

On the basis of a qualitative assessment, it might seem obvious to assume that enzymes contribute to sustainable development. But what are the hard facts about the environmental impact of the use of enzymes in the textile industry? It should not be forgotten that the production of enzymes is also associated with environmental burdens (Nielsen *et al.*, 2007). The purpose of the LCA studies presented here is to assess and compare the environmental burdens created by enzyme production and distribution in comparison with the environmental burdens avoided in the processes at textile mills. Here are two concrete examples from China based on the specific conditions at two mills. It could be argued that these facts only apply to the two specific mills. Therefore sensitivity analyses have been performed to look at a variety of scenarios, such as the use of fuels other than coal to generate electricity or the use of further optimised processes. In all cases, enzymes gave clear environmental benefits and helped to reduce contributions to global warming.

Enzymes have been used increasingly in the textile industry since the late 1980s. Many of the enzymes developed in the last 20 years are able to replace chemicals used by mills. The first major breakthrough was when enzymes were introduced for stonewashing jeans in 1987. Within a few years, the majority of denim finishing laundries had switched from pumice stones to enzymes. More than one billion pairs of denim jeans require some form of pre-wash treatment every year. A brief summary of the main commercial applications is given below as well as future trends at the end of the chapter.

5.2   Enzyme applications in textile processing

Some of the important commercial applications of enzymes are reviewed here with special emphasis on their environmental performance.

5.2.1 Biopolishing

Cotton and other natural fabrics based on cellulose can be improved by an enzymatic treatment known as biopolishing. As the name suggests, the treatment gives the fabric a smoother and glossier appearance. Cellulases hydrolyse the microfibrils (hairs or fuzz) protruding from the surface of the yarn. A ball of fuzz is called a 'pill' and these pills can present a serious quality problem since they result in an unattractive knotty fabric appearance. After biopolishing, fabric shows a much lower pilling tendency. The other benefits of removing fuzz are a softer and smoother handle, and better colour brightness. An alternative way to carry out polishing is by singeing with a gas flame and therefore the use of enzymes saves gas and emissions from the combustion process.

5.2.2 Denim abrasion and finishing

Many denim garments are subjected to a wash treatment to give them a worn look. In the traditional stonewashing process, the blue denim is faded by the abrasive action of pumice stones. Cellulases were first introduced in the 1980s and nowadays more than 80% of denim finishers use cellulases or a combination of stones and cellulases to create the worn look on denim. Cellulases work by loosening the indigo dye on the denim in a process known as 'biostoning'.

A small dose of enzyme replaces several kilogrammes of pumice stones so the handling of enzymes is easier. The use of fewer stones results in less damage to the garments, less wear on machines and less pumice dust in the laundry environment. Approximately 1 kg of stones is required to stonewash 1 kg of jeans. In a 1-hour wash cycle, the pumice stones will lose up to 50% of their weight. The pumice grit can block drains so it needs to be filtered out from the wastewater. Large amounts of pumice sludge can be produced. For example, a denim finisher processing 100 000 garments a week with stones typically generates 18 tonnes of sludge (BioTimes, 1997). An environmental assessment was performed on jeans (OECD, 1998). The conclusions were that the biostoning process proved to be more environmentally sound than the traditional stoning process using pumice.

5.2.3 Desizing

In the case of fabrics made from cotton or blends of cotton and synthetic fibres, the warp (longitudinal) threads are coated with an adhesive substance known as a size. This is to prevent the threads breaking during weaving because these threads are stretched across the loom and are subject to large amounts of wear. The most common size is starch or starch derivatives. After weaving, the size must be removed again in order to prepare the fabric for finishing.

The desizing process may be carried out by treating the fabric with chemicals such as acids, bases or oxidising agents. However, starch-splitting enzymes (amylases) have been preferred for many decades. As a result of their high efficiency and specific action, amylases bring about complete removal of the size without any harmful effects on the fabric. Desizing can also be done at different pH values and temperatures to suit the specific processing equipment being used by textile mills. The latest amylases to be developed allow desizing to take place over a broader pH range (pH 5–10). This gives far greater flexibility for textile mills; traditional amylases work in a pH range of 5–7.

5.2.4   Bleach clean-up

Natural fabrics such as cotton are normally bleached with hydrogen peroxide before dyeing. Bleaches are highly reactive chemicals and any peroxide left on the fabric can interfere with the dyeing process. Therefore a thorough 'bleach clean-up' is necessary. Enzymatic bleach clean-up was first launched in the 1980s. The traditional method is to neutralise the bleach with a reducing agent or to rinse with hot water. Both the methods require copious amounts of water. Enzymes represent a convenient alternative because they are easier and quicker to use. A small dose of catalase is capable of decomposing hydrogen peroxide to water and oxygen.

Compared with the traditional methods of removing bleach, the enzymatic process results in less polluted wastewater and/or reduced energy and water consumption, depending on the process that enzymes replace. Most of the chemical agents used for the neutralisation of residual bleach are hazardous to handle and problematic to the environment. These reducing agents also need to be adequately removed prior to dyeing. Residual reducing agents can have a detrimental effect on shade reproducibility from batch to batch. In contrast, enzymes can be handled safely, have no effect on dye and can even be used in the same bath as dyeing. Furthermore, they are completely degradable in nature.

5.2.5   Bioscouring

Scouring is a cleaning process that removes pectins and thereby assists with the removal of impurities such as waxes, mineral salts, etc. from cotton yarns and fabric before dyeing. Traditionally, scouring involves a number of high-temperature steps with a large consumption of chemicals such as sodium hydroxide, sodium carbonate and hydrogen peroxide. Enzymes provide a biological alternative with high specificity towards difficult-to-remove pectin compounds on the raw cotton. The problem with chemicals is that they do not just remove the impurities, but they attack the cellulose too, resulting in weight losses. Enzymes enable a faster and gentler scouring process with lower energy and chemical consumption. This enzymatic process is known as bioscouring and was first launched in 1999.

Bleach clean-up and bioscouring are the subject of in-depth environmental assessments in the following sections.

5.3   Life cycle assessments of enzymes used in the textile industry

Section 5.2 indicates that there are considerable environmental advantages when using enzymes in the textile industry. It is, however, necessary to

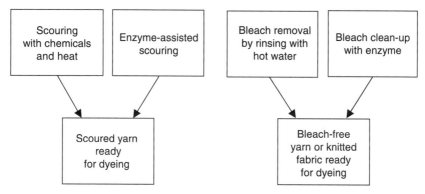

5.1 Two different scouring methods and two different bleach removal methods were compared from an environmental perspective.

investigate the advantages in more detail by including all processes in the product chain. Two concrete studies in China, from textile mills that have changed to enzymes recently, are presented below. The assessments are comparisons of two different scouring processes and two different bleach removal processes. In each case, one of the processes was enzymatic and one of the processes was non-enzymatic (see Fig. 5.1). LCA has been used as an analytical tool (see Box 5.1) and the two case studies presented here examined the changes that occur when enzymes are introduced in textile mills as an alternative to chemicals and hot water.

The industrial enzyme products used in the two case studies are from Novozymes and are produced by microorganisms in large fermentation tanks. Production relies to a large extent on inputs of agricultural products (sugar, protein, etc.), energy and water. The assessment of the environmental impact of enzyme production is based on data from Novozymes' production records in 2007. All processes in the production chain from the growing of agricultural products to the time the enzymes leave Novozymes' factory gate are included (see Nielsen *et al.*, 2007).

The environmental impacts avoided as a result of the reduced chemical, water, heat and electricity production are derived from the public LCA database Ecoinvent (2005). The environmental impacts avoided because of the reduced yarn production resulting from the use of enzymes in scouring are derived from the UMIPTEX database (Laursen *et al.*, 2006), except for the use of agricultural land, which is based on data on cotton yields in China (Petry and Xinping, 2006). The impact of the transport of chemicals and enzymes is included in the study based on data from Ecoinvent (2005).

Box 5.1 Principle of life cycle assessment (LCA)

LCA is a holistic environmental assessment tool that addresses raw material use and emissions in all processes in the product chain from raw material extraction through production to use and final disposal.

The environmental impact potential of a substance i emitted to the environment from a process is calculated as

$$EP(j)_i = Q_i \times EF(j)_i$$

where:

Q_i is the emitted quantity of substance i
j is the environmental impact (for example, global warming)
i is CO_2, CO, CH_4, N_2O, NO_3, PO_4, SO_2, etc.
EF is an effect factor

The environmental impact potential of a product

$$EP(j)_{product} = \Sigma (Q_i \times EF(j)_i) \text{ (Wenzel } et\ al.\ 1997)$$

Modelling has been facilitated in SimaPro 7.02 LCA software and EF is derived from Eco-indicator 95 (version 2.03). The study addresses changes induced by a switch from non-enzymatic processes to enzymatic processes. System boundaries and input data are determined in accordance with principles described by Ekvall and Weidema (2004).

5.4   Environmental assessment of the enzymatic scouring of package cotton yarn for dark-shade dyeing as an alternative to conventional chemical scouring

The assessment took place at the Rongxin Fibre Co. Ltd yarn dyeing mill located in Haining industrial park, 90 km from Shanghai in the Zhejiang province of eastern China. Here conventional scouring was replaced by enzymatic scouring with Scourzyme® 301 L in 2007 and enzymatic scouring is now running in full-scale production.

5.4.1 Scope of the study

Scouring is relevant for both package and hank yarns and for woven and knitted fabrics, but the present study focuses on package yarn (see Fig. 5.2). Yarns intended for light-coloured fabrics must be bleached prior to dyeing to remove residual cotton seeds and other impurities that can cause unsightly blemishes in the fabric if not removed. Bleaching and scouring are often

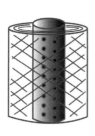   

Package yarn   Hank yarn

5.2 Package yarn and hank yarn. Package yarn is wound on to a
perforated cylinder so that water can be pumped through the yarn
during the scouring and dyeing processes. Hank yarn is not rolled but
simply hangs loose on a rod. The study focuses on dark-shade
package yarn.

performed in a combined scouring and bleaching process where sufficient
alkali and oxidant is added to avoid a two-step process. Yarns intended for
fabrics with a dark shade do not require the bleaching process and therefore
scouring on its own is sufficient.

Light-colour yarns where pre-bleaching is required constitute about 80%
of the market and dark shades constitute about 20%. The study addresses
the 20% part of the market because the potential of bioscouring is largest
for this product group. The reason for this is that pectate lyase does not
remove seeds and other seed fragments, which is a problem when dyeing
in lighter shades unless a natural look is preferred. Conventional scouring
and bioscouring can be performed on the same production line with the
same production equipment. Therefore there is no significant capital investment associated with a switch between the two scouring methods.

The study refers to production at a full-scale commercial yarn dyeing line
at Rongxin yarn dyeing factory with an output of 50 000 kg of dyed yarn per
day. Chemicals are delivered mainly by local producers. Water used in
production is extracted from a nearby river and treated by means of filtration and softening. Steam is supplied by the coal-fired Haining Dong Shan
combined heat and power plant located about 1 km from the factory, and
electricity is obtained from the national grid. Heat is not recovered in the
textile mill. Wastewater generated in the process is initially treated on site
and then fed to a central treatment plant before it is fed into a nearby river.

5.4.2   Process descriptions

Scouring is performed in a bath where a liquor of water and chemicals/
enzymes is pumped through the yarn. The liquor is heated with steam to
an appropriate temperature, and pumped through the yarn with an electric

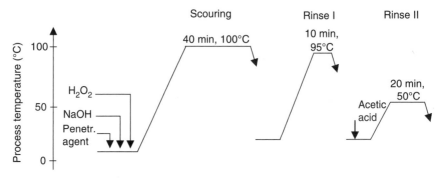

5.3 Process diagram for conventional scouring. Upward slopes indicate a rise in temperature. The final downward arrows indicate emptying of the bath.

pump. Water and chemicals/enzymes are mixed with each other in a separate tank next to the yarn bath.

Conventional scouring was performed in three steps: a scouring process followed by two rinsing processes (see Fig. 5.3). The scouring process was performed at 100 °C with penetration agent, sodium hydroxide (NaOH) and hydrogen peroxide (H_2O_2). The penetration agent is a surfactant that facilitated the water's entry into the yarn. Sodium hydroxide removed the unwanted impurities and hydrogen peroxide bleaches the yarn. Sodium hydroxide is not selective in the substrates that it attacks. In addition to the unwanted impurities, it also degrades other substances in the yarn, including the cellulose.

Removed impurities were washed out of the yarn in the two rinsing steps. The first rinsing step was performed at 95 °C. The second rinsing step was performed at 50 °C and acetic acid was added to reduce pH. The starting temperature of the two rinsing steps was higher than the starting temperature of the scouring process because the wet yarn and the dye machine contained a large amount of residual heat transferred from the high-temperature scouring step. The entire scouring process took 3 hours including filling and emptying the bath.

Bioscouring is performed in a single process at 60 °C with Scourzyme 301 L and penetration agent (see Fig. 5.4). Scourzyme 301 L is an enzyme product containing a pectate lyase that degrades pectins into soluble compounds. No rinsing steps or neutralisation steps are required prior to the yarn dyeing. The penetration agents used for bioscouring can be the same as used in the conventional scouring process. Bioscouring takes 90 minutes including the filling and emptying of the bath. Water from conventional scouring and bioscouring, as well as from any rinsing steps, is discharged directly for wastewater treatment after use.

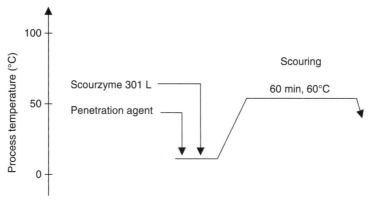

5.4 Process diagram for bioscouring.

Table 5.1 Savings when bioscouring replaces conventional scouring. All data are per tonne of scoured yarn (dry weight). Data are based on Rongxin production records unless otherwise stated

Material/utility	Unit	Conventional scouring	Bioscouring	Saving	Included in the LCA?
Scourzyme 301 L	kg	—	10	−10	Yes
H_2O_2 (27%)	kg	40	—	40	Yes
NaOH (100%)	kg	15	—	15	Yes
Penetration agent	kg	10	8.0	2.0	No
Sequestrant	kg	5.0	—	5.0	No
Acetic acid (98%)	kg	5.0	0	5.0	Yes
Yarn	kg	—	—	25*	Yes
Steam	tonnes	3	0.5	2.5	Yes
Electricity	kWh	300	150	150	Yes
Water	m^3	30	10	20	Yes

*Based on experiments performed by Novozymes (unpublished).

Bioscouring uses industrial enzymes but saves a range of chemicals. Furthermore, water, steam and electricity are saved because less water is needed for a single process step than for three process steps. In addition, less water needs to be heated and pumped through the yarn. Savings of water, electricity and steam are quite high because the liquor ratio (LR) is rather high (LR = ten tonnes of water per tonne of yarn) and because the amount of energy required to pump the liquor through the package yarn is also rather high. Quantities of saved and used materials and utilities when switching from conventional scouring to bioscouring are shown in Table 5.1.

Conventional scouring is a relatively harsh treatment and the yarn weight loss during the process can be 4% or more. Bioscouring is gentler and yarn weight loss is as little as 1.5% (based on experience from Novozymes). Scouring experiments with nearly the same conditions as at Rongxin have shown a yarn loss of 5.7% with conventional scouring and 3.3% with bioscouring. For this reason, a 2.5% yarn saving has been included in the assessment.

Other effects

Dye uptake is often higher with bioscouring than with conventional scouring owing to the resulting differences in the yarn and therefore dyestuff savings can be expected. Furthermore, bioscouring maintains more of the natural softness of the yarn than conventional scouring and less softener is needed at the end of the dyeing process. However, environmental data on the dyestuffs and softener production are not available and these savings are not included in the assessment. The system boundaries of the study are shown in Fig. 5.5.

Chemicals are delivered by local producers whereas Scourzyme 301 L is partly produced in Kalundborg, Denmark and partly in Hongda, China. Transport of enzymes and chemicals is included in the assessment.

5.4.3 Results of the study

The results of the environmental impact assessment are shown in Fig. 5.6. Figure 5.6 shows that resource consumption and environmental impacts induced by enzyme production generally are very small compared with the savings. The reason is that a small amount of enzyme saves large amounts of chemicals, energy and water in the scouring process. Transport of the enzyme from Denmark to China does not add significantly to the environmental impact of Scourzyme 301 L even though the transportation distance is rather long. The explanation is that the quantity of enzyme used is small and that ocean freight is energy efficient.

The main factors behind the saved contributions to global warming are shown in Fig. 5.7, which demonstrates that the heat saving in the bioscouring process is the main factor behind the reduced contribution to global warming, followed by electricity and yarn savings. Savings of water and chemicals in the process and the transport of chemicals from the manufacturers to Rongxin are less important. Reduced water consumption in the scouring process explains most of the water savings in Fig. 5.6. Water savings resulting from the reduced consumption of chemicals and yarn (primarily from the irrigation of cotton fields) are of minor importance. The saving of agricultural land observed in Fig. 5.6 can be entirely explained by

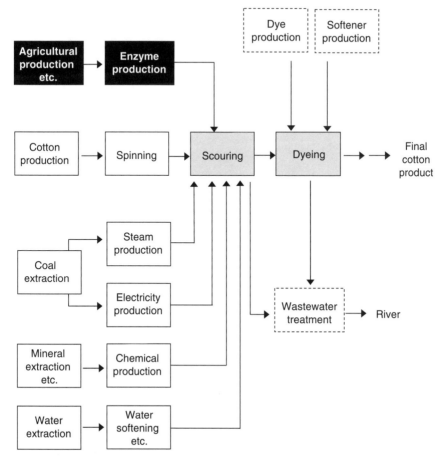

5.5 Main system boundaries of the environmental assessment. Black boxes indicate additional production, white boxes indicate saved production and grey boxes indicate unchanged production. All processes marked in the diagram are included in the study except for wastewater treatment and dye and softener production, which are marked with dashed boxes.

the reduced cotton production needed as a result of the use of enzymatic scouring.

Toxicity has not been included in the quantitative assessment because details of the emissions of toxic substances and the fate of these after emission are not known in large parts of the considered product chains. The contribution to toxicity is, however, likely to be linked to energy use (emissions of toxic substances with exhaust gases during combustion of fuels), use of agricultural land (pesticides in cotton production and raw

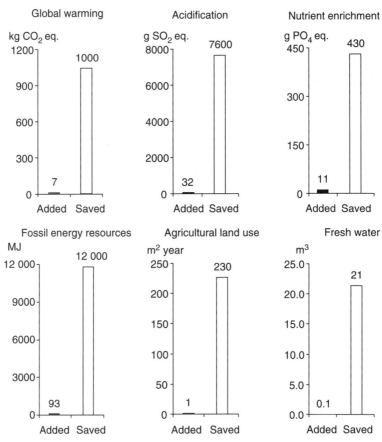

5.6 Added and saved resource consumption and environmental impacts when switching from conventional scouring to bioscouring. Added impacts are due to Scourzyme 301 L production and saved impacts are due to chemical, water, yarn, electricity and heat savings at Rongxin dyeing mill. All data are per tonne of yarn (dry weight).

material production for enzymes) and the release of toxic substances with wastewater from the scouring process. A small amount of enzyme with low toxicity replaces a large amount of aggressive chemicals (see Table 5.1) directly in the scouring process and large amounts of energy and agricultural land are saved by suppliers. Based on this, it is considered very likely that the toxicity impact reflects the other impact categories in Fig. 5.6 and decreases considerably when bioscouring replaces conventional scouring.

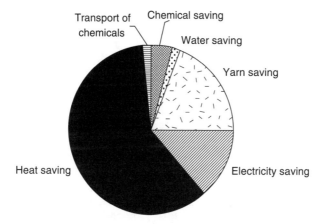

5.7 Main factors behind saved contributions to global warming.

5.4.4 Variation and uncertainty

Variation and uncertainty are important issues in environmental assessments. Therefore sensitivity analyses and uncertainty assessments have been made.

Steam supply

Steam used in the scouring process is delivered by a local coal-fired power station. It is interesting to see how results of the environmental assessment would change if the power station switched to other fuels such as wood pellets, natural gas, heavy fuel oil or lignite (brown coal). A sensitivity analysis has therefore been performed where coal (base case) is replaced by other fuels. Figure 5.8 shows the results for global warming and indicates that the advantages of using bioscouring as an alternative to conventional scouring increase if the power station switches from coal to lignite whereas they decreases if the power station switches to heavy fuel oil, natural gas and particularly wood pellets. The reason is that lignite generates more CO_2 emissions than coal per tonne of steam produced whereas other fuels generate less. The sensitivity analysis shows that the results of the study are very sensitive to the type of fuel used in steam production. It is, however, interesting to note that saved contributions to global warming exceed added contributions to global warming in all cases.

Heat recovery

Heat from the scouring process is not recovered at Rongxin and it is interesting to see how results of the assessment would change if heat were

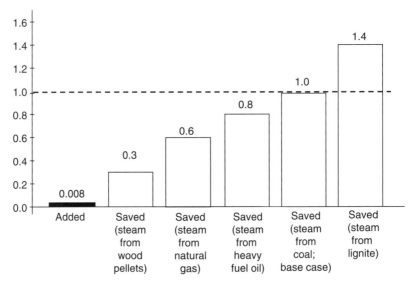

5.8 Added and saved contributions to global warming when switching from conventional scouring to bioscouring in the base case and in hypothetical cases where steam is produced from fuels other than coal. Added impacts are due to Scourzyme 301 L production and the saved impacts are due to fewer emissions resulting from chemical, water, yarn, electricity and heat savings at the mill. All data are given relative to the base case.

recovered. A sensitivity analysis where 50% of the heat is recovered in all processes has therefore been made. The results show that the advantages of bioscouring would be reduced considerably if heat recovery is introduced into production at the Rongxin mill (avoided contribution to global warming and avoided fossil fuel consumption are both reduced by around 40%) but nevertheless bioscouring still gives clear environmental advantages.

Liquor ratio

Rongxin uses 10 tonnes of water per tonne of yarn in the scouring process. The liquor ratio in textile mills usually ranges from 6 to 10 and it is interesting to see how the results would change if the liquor ratio was reduced to 6. A sensitivity analysis where water savings are reduced by 40% has therefore been made. The results show that the advantages of bioscouring are reduced considerably when the liquor ratio is reduced (avoided contribution to global warming and avoided fossil energy consumption are both reduced by around one-third) but nevertheless enzymatic scouring still gives clear environmental advantages.

Yarn saving

It is assumed in the base case that the gentler treatment of yarn with bio-scouring than with conventional scouring leads to reduced yarn weight loss and therefore reduced yarn production and cotton production (see Table 5.1). Reduced yarn weight loss may, however, not reduce the yarn use in practice and a sensitivity assessment has therefore been made where the saving in yarn has been ignored. The results show that disregarding the yarn saving reduces the environmental advantage of bioscouring by 20% in terms of the contribution to global warming and eliminates the saving of agricultural land.

Magnitude of steam saving

Data on steam consumption at Rongxin are given approximately and the actual steam saving could be somewhat higher or lower. The steam saving is important for the overall result of the assessment and a sensitivity assessment has therefore been made where the steam saving is 2 and 3 tonnes in comparison to the saving of 2.5 tonnes quoted in the original assessment. The results show that a lower steam saving reduces the advantage of bio-scouring considerably and vice versa. The environmental impacts generated by the production of enzymes are, however, small and bioscouring still gives clear environmental advantages.

Combined sensitivity analysis

The above sensitivity analyses show that the environmental advantages of enzyme use observed in Fig. 5.6 can change considerably if the production system or important assumptions are changed. A conservative sensitivity analysis that combined all the sensitivity analyses shows that enzymatic scouring still gives clear environmental advantages in any combination of cases.

Data quality assessment

The quality of key data on the scouring process at Rongxin and the produc-tion of enzymes at Novozymes is considered high and data are representa-tive. Quality of data on saved chemicals and energy, etc. is also considered high but poorly representative as it refers to European conditions (Ecoinvent 2005). European production is generally considered more envi-ronmentally efficient than Chinese production and the magnitude of the saved impacts given here should in most cases be seen as underestimates of actual saved impacts.

5.5   Environmental assessment of enzymatic bleach clean-up of light-coloured package yarn and knitted fabrics as an alternative to rinsing with hot water

Fabric and yarn need to be dyed evenly and reproducibly to meet the quality requirements of customers. For this reason, fabrics and yarns intended for final products in light colours are often bleached prior to the dyeing process. Bleaching is usually performed with hydrogen peroxide and the excess bleaching agent must be carefully removed prior to dyeing to avoid breakdown of the dye during the dyeing process. Bleach removal has traditionally been performed by rinsing in several baths with water or by adding a reducing agent such as sodium thiosulphate. Both of these traditional processes use considerable amounts of water and energy for heating the water. Enzymes are now used as an alternative by many textile mills that dye yarn and fabrics. Enzymatic bleach clean-up is based on a catalase, which catalyses the conversion of hydrogen peroxide into water and oxygen. The LCA study took place at a textile mill in the Esquel Group located in Guangdong in South East China where bleach removal based on rinsing with hot water was replaced by enzymatic bleach clean-up with Novozymes' Terminox Ultra® 50 L in 2002. The enzymatic process has been running in full-scale operation since then.

5.5.1 Scope of the study

Bleach clean-up is relevant for knitted fabrics and yarns intended for use in final products with light colours but not for dark shades (because bleaching prior to dyeing is not necessary) or white products (because no dyeing is applied and peroxide removal is not necessary). Woven fabrics are treated in a continuous process and hydrogen peroxide is removed together with caustics and other impurities during the pre-treatment process. Therefore enzymatic bleach clean-up or the use of a reducing agent is not applicable. In contrast, yarns and knitted fabrics are treated in batch processes and a specific bleach clean-up step is required. Material and utility consumption for bleach clean-up is nearly the same for package yarn and knitted fabric even though the machinery is different. The two products are treated as the same for the purposes of the study. Hank yarn and woven fabrics are not part of the study (see Fig. 5.9).

As the Esquel textile mill was using rinsing with hot water, this was used as the reference in the assessment. Bleach removal with a reducing agent has not been studied. The removal of bleach by rinsing with cold water and enzymatic bleach clean-up can be performed on the same production line with the same production equipment. Therefore there is no capital

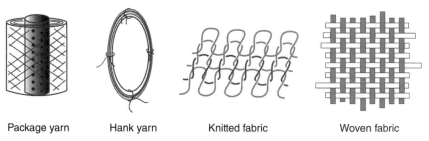

Package yarn   Hank yarn   Knitted fabric   Woven fabric

5.9 The study addresses package yarn and knitted fabrics but not hank yarn and woven fabrics.

investment associated with a switch from the reference process to enzymatic bleach clean-up. The study is based on full-scale commercial production at the Esquel Group's mill in Guangdong. Water are taken from a nearby river and treated by carbon filtration and ion exchange. Steam is produced at a combined heat and power plant owned by the mill. Heat from hot wastewater is recovered by a heat exchanger and condensation water and cooling water are recovered. Wastewater is treated in the mill's own wastewater treatment plant before it is discharged to the public sewer. Five hundred tonnes of knitted fabric and 1500 tonnes of yarn are dyed per month on the production lines that were part of the study.

5.5.2 Process descriptions

Bleach clean-up of package yarn is performed in a series of batch processes in water mixed with chemicals/enzymes at appropriate temperatures. Liquid is continuously pumped through the yarn rolls or the knitted fabrics to establish the necessary contact with the cotton. Baths are heated with steam, and the pumping of water is facilitated by electric pumps. The liquor ratio is 10 tonnes of water per tonne of yarn.

- *Reference for bleach removal.* The traditional way for the mill to remove bleach is a three-step process composed of one hot rinse and two cold rinses with pH adjustment in the second cold rinse (see Fig. 5.10). The output from the process is bleached yarn or fabric without the presence of hydrogen peroxide. The yarn or fabric is then dyed in a new bath.
- *Enzymatic bleach clean-up.* Enzymatic bleach clean-up is performed in a two-step process composed of one cold rinse and a cold hydrogen peroxide decomposition step (see Fig. 5.11) where the enzyme removes the hydrogen peroxide. Acetic acid is used for pH adjustment. The enzyme works at high speed and the bath is free of hydrogen peroxide after 25 minutes. Then dyeing is performed in the same bath with the same water.

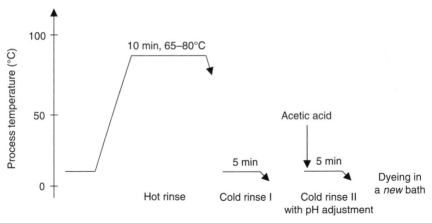

5.10 Process diagram for the traditional way of removing bleach by rinsing with hot water. Upward slopes indicate a rise in temperature. The final downward arrows indicate emptying of the bath.

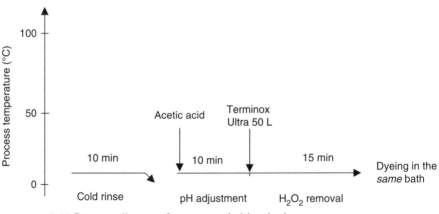

5.11 Process diagram for enzymatic bleach clean-up.

Enzymatic bleach clean-up uses enzymes and somewhat more acetic acid for pH adjustment than the process previously used by Esquel but saves water, heat and electricity because the process temperature is lower, the process time is shorter and the number of baths is two instead of four (see Table 5.2). The main system boundaries of the study are illustrated in Fig. 5.12. Acetic acid is delivered by local producers whereas Terminox Ultra 50 L is produced in Kalundborg, Denmark. The transport of acetic acid and enzymes is included in the study based on data from Ecoinvent (2005).

Table 5.2 Savings when enzymatic bleach clean-up replaces rinsing with hot water. All data are per tonne of package yarn or knitted fabric (dry weight)

Material/utility	Unit	Saving	Included in the LCA?
Terminox Ultra 50 L	Kg	−1.0	Yes
Acetic acid (98%)	Kg	−3.0	Yes
Steam	kg	1820	Yes
Electricity	kWh	33	Yes
Water	m³	20	Yes

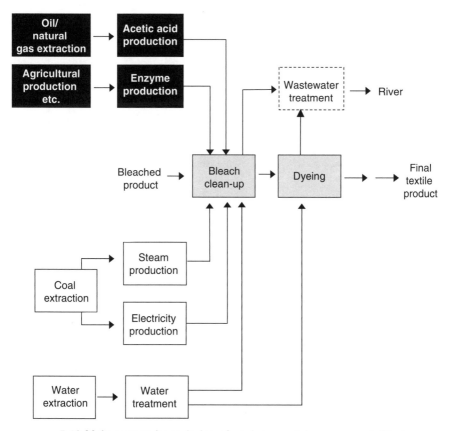

5.12 Main system boundaries of environmental assessment. Black boxes indicate additional production, white boxes indicate saved production and grey boxes indicate unchanged production. All processes marked with black or white boxes in the diagram are included in the study except wastewater treatment, which is marked with a dashed box.

5.5.3 Results of the study

The results of the environmental impact assessment are given in Fig. 5.13 which shows that the consumption of resources and environmental impacts caused by enzyme production are generally very small compared with the savings, except in the case of agricultural land use. The explanation is that

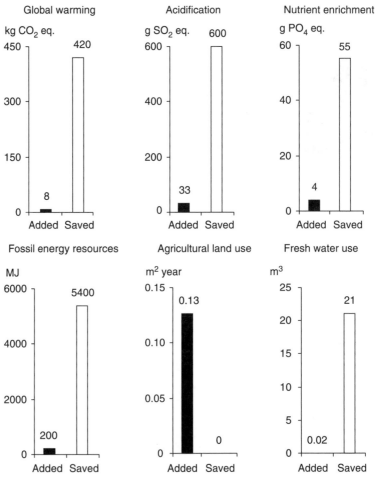

5.13 Added and saved resource consumptions and environmental impacts when switching from the traditional removal of bleach based on rinsing with hot water to enzymatic bleach clean-up. Added impacts are due to Terminox Ultra 50 L and acetic acid production whereas saved impacts are due to steam, electricity and process water savings. All data are per tonne of package yarn or knitted fabric (dry weight).

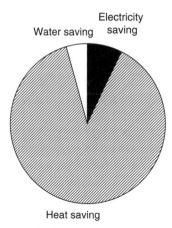

Electricity
Water saving   saving

Heat saving

5.14 Main factors behind saved contributions to global warming.

a small amount of enzyme saves large amounts of energy and water in the bleach clean-up process. Just as in the case of scouring with enzymes, transport of the enzymes from the enzyme factory in Denmark to China does not add significantly to the environmental impact of using enzymes. The main factors behind the saved contributions to global warming are indicated in Fig. 5.14 which shows that the heat saving is by far the most important factor behind a reduced contribution to global warming.

Toxicity has not been included in the quantitative assessment for the same reasons as explained for scouring. For the same reasons as before, it is considered very likely that the toxicity impact reflects the other impact categories in Fig. 5.13 and that the use of enzymes reduces the toxicity impact on the environment.

5.5.4 Variation and uncertainty

The advantages of enzymatic bleach clean-up are subject to variation depending on the conditions at a particular textile mill. Therefore a number of sensitivity analyses were carried out in the same way as for the scouring case study. In summary, the sensitivity analyses show that steam supply as well as water and heat management are very important for the final outcome of the study. Nevertheless, enzymatic bleach clean-up was advantageous from an environmental viewpoint in all of the cases considered. The assessments regarding data quality are the same as for the scouring case study. As before, it was concluded that the exact magnitude of the saved impacts given here should in most cases be seen as underestimates of actual saved impacts.

5.6   Conclusions on environmental assessments of enzymatic scouring and bleach clean-up

The use of enzymes in scouring and bleach clean-up as alternatives to chemical treatment and rinsing with hot water, respectively, led to considerable environmental improvements at the two production lines at textile mills in China. The explanation is that a small amount of enzyme saves considerable amounts of energy and water in both cases and also chemicals in the case of scouring. Sensitivity analyses indicate that the general conclusion of the assessment holds up under different energy supply scenarios although the sizes of the reductions in environmental impacts are subject to much variation and uncertainty. The impact of the transport of enzymes from the manufacturer to the final user is insignificant even though the transportation distance is long. The main findings of the study are therefore applicable to other textile mills with similar production systems elsewhere in the world.

The magnitude of the environmental improvements obtained by replacing the existing production methods with the enzymatic technologies are highly dependent on the type of fuel used and the actual production conditions. An estimation of environmental improvements at other factories must therefore rely on specific information on production processes and energy supply systems. The study has not addressed the removal of bleach with a reducing agent and further environmental assessments are required before any conclusions can be made about this method.

5.7   Perspectives

The pressure from the authorities on the textile industry to reduce the use of water and the impact of effluents has been growing in many parts of the world in recent years. For instance, in China, increased enforcement of the regulations has occurred during 2007 and 2008. Concerns about the pollution of rivers or large lakes used for drinking water and falling water tables are two of the drivers for the stricter enforcement of environmental regulations. Indeed, some large dyehouses have had to close down in the past few years because they did not have effective effluent treatment in place and were unable to meet the limits set by local regulations.

There is also a growing emphasis on reducing the carbon footprint (CO_2 emission in the entire product chain) of many industrial sectors in the world, and the textile industry is no exception. Glaciers are melting around the world at an alarming rate and there is growing pressure from consumers and governments to improve environmental performance. The link between the emission of greenhouse gases and global warming is now generally acknowledged. In combination with increasing energy prices, this has made many industries rethink their production processes.

5.7.1 Lower water consumption

It is estimated that it currently takes 100 kg or litres of water to process 1 kg of textiles. The more water used, the more wastewater is generated for treatment. A recent trend is to reduce the liquor ratio in each bath from 10 litres for every 1 kg of textiles to just 4 litres/kg in many cases. In addition, the number of separate baths required to process textiles has been reduced by 10–20% depending on the process. So, for example, instead of using 15 separate baths, only 12 or 13 are now required to process textiles.

Enzymes are seen to have a vital role in reducing water consumption and pollution. As the bleach clean-up case study shows, savings in water of 20 m^3/tonne yarn were made when using 1 kg of enzyme compared with the conventional method of rinsing to remove the hydrogen peroxide. The case study with bioscouring shows that a small dose of 10 kg of enzyme can save 20 m^3 of water/tonne yarn.

5.7.2 Reduced carbon footprint

Textile production is responsible for considerable CO_2 emissions, not only from the energy used in the industry itself but also from energy used for the production and delivery of raw materials (cotton, wool, etc.), chemicals and water. Enzymes of various types are able to save raw materials, replace chemicals and save water and energy throughout the production chain, and thereby help to reduce the carbon footprint of textiles.

5.8   Future trends and applications

In future, the combining of different enzymatic processes in the same bath will help to save more water and streamline processing to save time and ultimately costs. Some examples are given here.

5.8.1 Combined bleach clean-up and dyeing

Already the use of a catalase enzyme for bleach clean-up allows mills to reduce at least one process step by adding the catalase to the dyebath, thereby allowing the removal of hydrogen peroxide to take place in the same bath as dyeing. The hydrogen peroxide is first removed in a quick 5–10 minute step followed by dyeing in the same bath. This is made possible by the fact that the catalase only targets the hydrogen peroxide as a substrate to neutralise. It does not target the dyes and these are therefore unaffected.

5.8.2 Combined dyeing and biopolishing

In April 2008, Novozymes launched a new formulated product called Cellusoft Combi that allows mills to carry out combined biopolishing (see Section 5.2) and dyeing in the same bath. Traditionally, biopolishing is an additional process and this adds extra processing costs to a standard dyeing process. In traditional cases, the biopolishing process will be carried out before or after dyeing. This typically adds 90–120 minutes to the complete process. With Cellusoft Combi, this entire process is eliminated by carrying out the biopolishing process in the dyebath itself. Simply adjust the pH for dyeing after bleaching, add the Cellusoft Combi and continue dyeing. Carrying out biopolishing with Cellusoft Combi in the same bath as dyeing means savings in water, energy and process time.

5.8.3 Combined bioscouring and biopolishing

A similar concept is to combine the bioscouring and biopolishing processes in the same bath prior to dyeing. Here Scourzyme L and Cellusoft CR from Novozymes are added together at the start of the process. This is made possible by the fact that both these products have overlapping temperature and pH profiles. Once again, this saves process time, water, effluents and energy. There is also a synergistic effect when combining these two products resulting in even better biopolishing and bioscouring effects than if these products were used separately.

5.8.4 Combined desizing and bioscouring

Novozymes has launched the concept of CDB (combined desizing and bioscouring). This allows the enzymatic scouring and desizing of fabric to be performed in one step. Traditionally, desizing is carried out as a separate step where an amylase is applied to the fabric by means of pad batch or exhaust. In most cases woven fabrics that need to be desized are treated by means of one of the so-called pad batch methods where the fabric is allowed to dwell for several hours. Alternatively this is done on a continuous basis at high temperatures. With Scourzyme L and Aquazym SD-L, it is now possible to carry out desizing and bioscouring in one step. The pH and temperature profiles for these two products overlap and this allows for combining the processes of desizing and bioscouring in pad batch conditions.

The combination of these enzymes allows mills to omit the conventional strong alkaline scouring process. Instead of having a traditional three-step process (desize–scour–bleach), it is possible to reduce this to a two-step process (CDB–bleach). This saves large amounts of energy, water and effluent while increasing production capacity.

5.8.5 Longer term perspectives

There is a huge potential to reduce energy costs in the pre-treatment processes in a textile mill. For example, the process of bleaching needs to be addressed in future in order to reduce energy costs and enhance the quality of bleached fabric. Enzymes could provide the answer. It should be mentioned that the vast majority of enzyme applications today are used for the treatment of cotton. The textile industry is also looking for new ways to deal with current problems on fibres other than cotton. This includes finding ways to improve the quality of the so-called bast fibres (hemp, linen, etc.), wool and even synthetic fibres. The search is on for industrial enzymes that are commercially viable for use on these substrates. There is also a need to find alternative ways of dealing with an array of different sizing agents other than starch. Enzymes could even be used in future to break down dyestuffs in the effluents from dyehouses and denim finishing laundries. Dyestuffs remain one of the most difficult substances to remove at an effluent treatment plant. With all these opportunities and the need to move towards sustainable development, industrial enzymes could provide some of the solutions the textile industry is looking for in the future.

5.9   Sources of further information and advice

- *Encyclopaedias.* The industrial production of enzymes and the use of enzymes in a broad range of industrial processes (including textiles) are described in the *Kirk-Othmer Encyclopedia of Chemical Technology* (2004) and in *Ullmann's Encyclopedia of Industrial Chemistry* (2003).
- *Research papers.* Environmental studies of biotechnology and enzymes used in the textile industry have only received limited attention in the scientific literature. Some examples are, however, highlighted here: Ahuja *et al.* (2004) who reviewed the utilisation of enzymes from an environmental point of view; Aly *et al.* (2004) who studied the biotechnological treatment of cellulosic textiles; Chen *et al.* (2006) who studied the application of biotechnology in the Chinese textile industry; and Vankar *et al.* (2006) who studied the enzymatic dyeing of cotton and silk fabrics.
- Further information can also be obtained from Novozymes' publications. *Enzymes at Work* by Olsen (2004) is an informative free booklet published by Novozymes giving an overview of commercial enzyme applications industry by industry plus the history of enzymes and future prospects. *BioTimes* is a quarterly customer magazine published by Novozymes in English, Chinese, Spanish and Portuguese giving the latest developments and case studies about industrial enzymes. Free subscription and download are available. Advice on enzyme use in textile industry can be acquired from Novozymes.

- Finally, Novozymes has two fully equipped global application laboratories, one in Beijing (mill applications) and one in Kuala Lumpur (laundry applications) that are able to perform various applications typically found in bulk applications. Customers are assisted in these laboratories with queries related to their own processing situations, and new applications are developed and tested.

5.10   References

AHUJA SK, FERREIRA GM and MOREIRA AR (2004), 'Utilization of enzymes for environmental applications'. *Critical Reviews in Biotechnology*, **24** (2–3), 125–154.

ALY AS, MOUSTAFA AB and HEBEISH A (2004), 'Bio-technological treatment of cellulosic textiles'. *Journal of Cleaner Production*, **12** (7), 697–705.

BIOTIMES (1997). The end of the stone age. www.biotimes.com.

CHEN J, WANG Q, HUA Z and DU G (2006), 'Research and application of biotechnology in textile industries in China'. *Enzyme and Microbial Technology*, **40**, 1651–1655.

ECOINVENT (2005), *The life cycle inventory database*. Version 1.2. Swiss Centre for Life Cycle Inventories. www.ecoinvent.com.

EKVALL T and WEIDEMA BP (2004), 'System boundaries and input data in consequential life cycle inventory analysis'. *International Journal of LCA*, **9** (3), 161–171.

KIRK-OTHMER ENCYCLOPEDIA OF CHEMICAL TECHNOLOGY (2004), Enzyme Applications, Industrial. John Wiley & Sons, Inc.

LAURSEN SE, HANSEN J, KNUDSEN HH, WENZEL H, LARSEN HF and KRISTENSEN FM (2006), *UMIPTEX – Environmental assessment of textiles*. Danish Environmental Protection Agency, Working Report no. 4 (In Danish).

NIELSEN PH, OXENBØLL KM and WENZEL H (2007), 'Cradle-to-gate environmental assessment of enzyme products produced in Denmark by Novozymes A/S'. *International Journal of LCA*, **12** (6), 432–438.

OECD (1998), 'Evaluating the cleanliness of biotechnological industrial products and processes'. In *Biotechnology for Clean Industrial Products and Processes – Towards Industrial Sustainability*. OECD, Paris.

OLSEN HS (2004), *Enzymes at Work*. Novozymes A/S, Denmark.

PETRI M and XINPING W (2006), *China, People's Republic of; cotton and products, cotton situation update*. GAIN Report no. CH6085. US Department of Agriculture. Foreign Agricultural Service.

ULLMANN'S ENCYCLOPEDIA OF INDUSTRIAL CHEMISTRY (2003), Enzymes. John Wiley & Sons, Inc., New York.

VANKAR PS, SHANKER R and VERMA A (2006), 'Enzymatic natural dyeing of cotton and silk fabrics without metal mordants'. *Journal of Cleaner Production*, **15** (15), 1441–1450.

WENZEL H, HAUSCHILD M and ALTING L (1997), *Environmental Assessment of Products, Volume 1: Methodology, tools and case studies in product development*. Chapman and Hall, London.

6

Key sustainability issues in textile dyeing

J. R. EASTON, DyStar, UK

Abstract: This chapter looks at the main sustainability issues affecting the wet processing stage of textile manufacture and the strategies available to minimise the environmental impact of the coloration step. In particular the chapter highlights the critical role that accurate colour communication and intelligent dye selection can play in minimising resource consumption, reducing pollution loads, enhancing product durability and shortening time to market. With the increased emphasis on traceability and the requirement to demonstrate compliance with rapidly developing consumer product safety and chemical legislation, the chapter describes how retailers and brand owners are responding to the pressure to know more about the dyes and chemicals used in the manufacture of their products and to 'green' their supply chains.

Key words: textile coloration, resource efficiency, supply chain greening, traceability, product durability.

6.1   Introduction

Commercial dyeing can be described as a method for colouring a textile material in which a dye is applied to the substrate in as uniform a manner as possible to obtain an even shade (or level dyeing) with a performance and fastness appropriate to its final end-use.

The aims of a dyer are therefore to achieve the correct shade and fastness properties on a substrate in a level manner as efficiently and profitably as possible. If the shade is not correct and a shading addition is required this will add on average 20% to the cost of dyeing. Two shading additions will mean that the dyehouse is not making any profit on the work and is most likely losing money. The cost of a repeat dyeing increases the total cost for producing the finished fabric by 100–130%.[1] Not only are shading additions and re-dyes bad for profit, they are also bad for productivity, meaning delivery dates may be missed resulting in loss of business; they are also bad for the environment as they contribute disproportionately to water and energy consumption.

Note: Dianix, Imperon, Levafix, Realan, Remazol, Controlled Coloration, econfidence and Optidye are registered trademarks of DyStar Textilfarben GmbH & Co. Deutschland KG, Germany.

Textile dyeing involves the use of a number of different chemicals and auxiliaries to assist the dyeing process. Some auxiliaries (e.g. dispersing agents, buffers, dedusting agents) are already contained in the dyestuff formulation, but other auxiliary chemicals are also added during processing to aid the preparation, coloration and afterwashing processes. Since auxiliaries in general do not remain on the substrate after dyeing, they are ultimately found in the emissions from a dyehouse.

Most dyes remain substantially exhausted or fixed on to the fibre. However, in the case of reactive dyes, the most widely used class of dyes for dyeing cellulosic fibres, a significant proportion of the dye is hydrolysed during the application process due to the side-reaction with water and this hydrolysate can contribute to colour in the wastewater and to an increased chemical oxygen demand (COD).

Environmentally responsible dye application embraces the well-established '3R' principles of pollution prevention – i.e. reduce, re-use, recycle – and the most effective pollution prevention practice for textile wet processing is 'right-first-time' dyeing. Corrective measures such as shading additions or in the worst case strip and re-dye are all chemical, energy and water intensive and add significantly to the pollution load.[2]

The general theory and practice of the application of dyes to fibres has been extensively covered elsewhere and is not within the scope of the present chapter. A useful summary of the various environmental impacts of the dyeing process with different classes of dye was published by Cooper in 1992[3] (see Table 6.1) and in a previous publication in this series Bide looked at the key issues arising from the use of specific dye classes on a fibre by fibre basis.[4]

6.2   Key factors for improving sustainability in dyeing and finishing

6.2.1   Accurate colour communication

A dyer is required to match the colour of the client's 'standard' or reference shade on a particular quality of fabric and with the equipment available to him in his factory. The target he is aiming at may be electronic, in the form of reflectance data, or a real physical sample of coloured material or a combination of both. In the case of a physical target shade this can be anything from a scrap of coloured paper or fabric to a fully engineered colour standard. The retailer or buyer usually requires the dyer to match the 'standard' under several light sources to produce a non-metameric match. In order to avoid gross changes of colour, the colour standard should be as colour constant as possible under different illuminants.[5]

Table 6.1 Type of pollution associated with various coloration processes[3]

Fibre	Dye class	Type of pollution*
Cotton	Direct	1 – salt
		3 – unfixed dye (5–30%)
		5 – copper salts, cationic fixing agents
	Reactive	1 – salt, alkali
		3 – unfixed dye (10–40%)
	Vat	1 – alkali, oxidising agents
		2 – reducing agents
	Sulphur	1 – alkali, oxidising agents
		2 – reducing agents
		3 – unfixed dye (20–40%)
Wool	Chrome	2 – organic acids
		5 – heavy metal salts
	1:2 metal complex	2 – organic acids
	Acid	2 – organic acids
		3 – unfixed dye (5–20%)
Polyester	Disperse	2 – reducing agents, organic acids
		5 – carriers

*Pollution catgeories: 1, relatively harmless inorganic pollutants; 2, readily biodegradable, moderate–high biological oxygen demand (BOD); 3, dyes and polymers difficult to degrade; 4, difficult to biodegrade, moderate BOD; 5, unsuitable for conventional biological treatment, negligible BOD.

Apparel and textile retailers are facing dynamic challenges including global sourcing, shorter lead-times, cost pressures, higher quality demands and the need to maintain the trust of their customers through commitments to social and environmental responsibility. Brand integrity and corporate reputation is now a key differentiator for retailers in many Western markets. The inability to get the right colour at the right time is one of the biggest sources of delay, and consequently increased cost, in the apparel sector's current business processes. These additional costs which are often not properly accounted for at a product level arise from having to air freight goods to meet required delivery deadlines or from lost sales due to stock not being available in-store at the correct time.

Achieving a particular colour typically involves a mixture of three dyes, or trichromie, of yellow, red and blue. A dyer will often have a preferred set of primaries that have good dyeing behaviour and from which the widest range of shades can be economically achieved, along with additional dyes for specific requirements of shade, fastness or metamerism. A non-metameric match can be achieved if the dyes used have identical reflectance values to those used to create the colour standard. It is much more difficult to achieve a non-metameric match if dyes with different chromophores and reflectance values are used to make the match than are present in the original standard.[4]

By communicating the dye combination used to make the colour standard it is possible for the dyer to save time by having a 'flying start' for a laboratory formulation of the shade even if later recipe correction is required to suit the particular substrate being processed and the dyeing machinery available in the factory. The recipe used in bulk should consist of dyes that have compatible dyeing profiles so that level and reproducible dyeings are more likely to be achieved. If environmental factors limit the choice of dyes to be used (e.g. as with the restrictions on dyes and chemicals included in some organic textile standards) then it can become more difficult to find a dye combination that performs well under bulk application conditions and produces a non-metameric match with acceptable fastness performance.

Specific fastness requirements for particular articles, such as multiple wash fastness or perspiration light fastness, e.g. for golfwear, can also cause delays in shade matching and colour approval unless detailed guidance on suitable dye combinations to meet the desired performance criteria is available from the dye supplier. It is important that designers and colour specifiers understand the implications of their selection of seasonal palette shades further up the supply chain. Some shade and substrate combinations are more difficult to dye than others and the dyehouse may consume large quantities of dyes, chemicals, water and energy trying to hit a difficult shade at the limits of achievability when a slightly different shade may have presented a more easily achievable target.

Colour specification and communication tools such as those offered by Colour Solutions International (North Carolina, USA) can greatly assist in minimising time delays for lab-dip approvals and can also lead to savings in water, chemicals and energy by reducing the rejection rate.

6.2.2   Intelligent dye selection for product durability

Fastness is the resistance of a dyed textile to colour removal or modification of shade under the action of a range of agencies including light, water, washing, perspiration, environmental contaminants, physical abrasion, etc. Standard test methods for assessing the fastness of dyed textiles are available from the International Organisation for Standardisation (ISO) or from the American Association of Textile Chemists and Colorists (AATCC). Meeting the fastness requirements of the customer is mainly achieved by:

(a)   intelligent dye selection;

(b) efficient washing-off processes in which loose or unfixed dye is removed from the fibre after dyeing.

As with issues of shade and levelness, restriction of dye choice for environmental reasons can limit the fastness achievable. This is particularly an issue for high-performance textiles such as golfwear where the dye recipe selected to meet the demanding conditions of combined light and perspiration fastness testing has to take into account the very different responses of the different dye chromophores available.[6]

As noted above, colour fastness also depends on the removal of unfixed dye from the fibre at the end of the dyeing process. The use of rinsing processes that are efficient in water and energy use can significantly reduce the impact of these rinses (e.g. Variable Power Rinsing from Then (Germany) or Controlled Rinsing from Thies (Germany)). Intelligent selection of dyes with good wash-off performance, e.g. alkali-clearing disperse dyes for polyester fibres, can simplify processes leading to savings in chemicals, water and energy without compromising on fastness performance.[7]

Colour fastness is a key element of product durability and is sometimes at odds with the current ethos of the fashion industry which has increasingly promoted rapid change and lowered expectations of durability for fast fashion items particularly in womenswear. For example, a recent report by the House of Lords Science Committee[8] stated that between 2001 and 2005 – the last published calculations – consumption of clothing for every man, woman and child in Britain rose by more than 30%. The report attacked the culture of 'fast fashion' and urged the Government to do more to cut Britain's mountains of commercial and domestic waste. Peers warned that cheap, fast fashion 'encourages consumers to dispose of clothes which have only been worn a few times in favour of new, cheap garments which themselves will also go out of fashion and be discarded within a matter of months.'[8] However, there are emerging signs that greater product durability (or 'slow fashion') is being recognised as a key component of a more sustainable clothing industry.[9]

Most of the major dye manufacturing companies have ranges of dyes for all fibres that can deliver high fastness and enhanced product durability at relatively low cost compared with the retail price of the garment. The challenge is to get the designers and buyers to specify fastness standards for durable clothing and then stick to them in the face of pressure from vendors to accept something cheaper that does not quite meet the standard. With the pressure to get clothes into store as fast as possible, cutting corners on quality can have significant hidden environmental costs.

Examples of DyStar's contribution via product innovation in this area include examples of dyes with excellent wet fastness properties, for example:

Levafix™ CA dyes on cellulosic fibres; Dianix™ SF/XF dyes on polyester; Realan™ EHF dyes on wool; and Imperon™ HF pigments for highly durable textile prints.

6.2.3 Intelligent dye selection for chemical compliance

Factories producing fabrics use a wide range of chemicals, some of which have the potential to harm workers and cause irreversible damage if allowed to enter the environment untreated. Small quantities (residues) of some harmful chemicals on clothing can also pose a risk to consumers and reputational damage for the retailer or brand. However, it is workers in the dyeing factories, rather than Western consumers, who are most at risk. It should be just as important to reduce the exposure of dyehouse workers in China, India and other producer countries to carcinogenic dyes and chemicals as it is to reduce the exposure of cotton farmers in Africa to toxic pesticides.

Most responsible dye manufacturers abandoned the production of carcinogenic benzidine dyes many years ago in the light of evidence of increased levels of bladder cancer among their own workers. However, many of these dyes are still available today in major textile manufacturing locations. In situations where there is little or no regard for health and safety in the workplace, this can have tragic consequences for those involved in handling these chemicals.

New, more demanding, safety regulations covering all products containing chemicals manufactured and imported into the European Union will be introduced over the next few years under the REACH regulation, which came into force in 2007 but which will not be fully implemented until 2018 at the earliest. REACH creates a legal framework for the evaluation of the risks associated with the use of chemicals and requires greater transparency and communication on hazardous chemicals in the supply chain.[10]

Transparency is key. As Esty and Winston[11] have noted in their book *Green to Gold*: 'In a world of rising transparency and low-cost information, who is responsible for what is becoming increasingly clear. As pollution and toxic chemicals become easier to track back to their sources, we will know which companies created them, shipped them, used them, and disposed of them.' A number of leading clothing retailers and brands now publish their standards on chemicals and operate systems to assess the performance of their suppliers with regard to safe use of chemicals. In July 2004, the Apparel and Footwear International Restricted Substance Management Group (AFIRM) was formed by a number of leading US and European brands and retailers to share knowledge of chemical legislation requirements in different national and international jurisdictions and to co-operate in communicating Restricted Substance List (RSL) requirements to suppliers via seminars and publications. The AFIRM group has developed a supplier toolkit (www.afirm-group.com) to aid best practice in chemical compliance and held its most recent seminar in New Delhi, India in November 2008; having previously hosted supplier events in Shanghai and Hong Kong to communicate the importance of this issue to the supply base.

Other groups with active RSL programmes include:

- American Apparel & Footwear Association;
- Outdoor Industry Group;
- World Sporting Goods Federation;
- Board for Social Responsibility.

A host of individual companies operating in the textile and apparel industries also now have their own RSLs.

Leading dyestuff and chemical suppliers provide guidance to customers on the suitability of their products for the manufacture of textile goods required to comply with the chemical restrictions imposed by their customers. DyStar's econfidence™ programme is designed to ensure that its products are free from contamination by restricted or banned substances and are able to fulfil the requirements of the leading retailers, brands and ecolabels.[12]

6.2.4 Intelligent process selection for improved resource efficiency

Fresh water is an increasingly scarce resource as the demands of an ever-growing world population and the agricultural activity needed to support it consume a steadily rising proportion of global fresh water resources. The textile industry generally needs to find ways to reduce its water consumption and as a major user, and potential polluter, of water, the textile wet processing industry is under particular pressure to reduce water consumption on both environmental and economic grounds. Securing a reliable and economic supply of water is now a strategic imperative for textile operations.[13]

We have therefore witnessed a recent upsurge of interest in the so-called 'water footprint' of products, in particular cotton textiles with their associated issues of irrigation and pesticide use.[14] The preparation, coloration and finishing stages of fabric manufacture are significant contributors to the overall water footprint of textiles and clothing and so there is a renewed interest in optimised water use and investigation of the possibilities of water re-use in dyeing.

As so many textile processing steps require the use of hot water, minimising the water consumption per unit of production also has a concomitant benefit in energy consumption. Given the levels of public and corporate concern about global warming and climate change the textile industry can not afford to ignore the pressure to reduce the amount of energy embedded in its products, in other words to reduce their 'carbon footprint'. In the Best Available Techniques (BAT) Reference document for the Textile Sector[15] the EU Integrated Pollution Prevention and Control (IPPC) Bureau in Seville identified the following measures as BAT for water and energy management:

- monitor water and energy consumption in the various processes;
- install flow control devices and automatic stop valves on continuous machinery;
- install automatic controllers for control of fill volume and liquor temperature in batchwise dyeing machinery;
- establish standard operating procedures in order to avoid wastage of resources;
- optimise production scheduling;
- investigate possibilities for combining process steps, e.g. scour/dye, dye/ finish;
- install low-liquor-ratio machinery for batch processing;
- install low add-on equipment for continuous application;
- improve washing efficiency in batch and continuous processing;
- re-use cooling water as process water;
- install heat recovery systems to win back thermal energy from dropped dyebaths and wash baths.

The importance of right-first-time dyeing to mimimise waste during the dyeing process has long been emphasised by leading dye manufacturers and much of their innovation in terms of both new dyes and new application processes over the last 20 years has been directed towards reducing the demand for both water and energy.

Best available technology/good practice guides

Many publications have highlighted the need to consider resource use in textile processing (see Section 6.6). The BAT reference guide of the EU IPPC Bureau mentions the following as examples of BAT for textile processing:[15]

- general good management practices;
- staff education and training;
- well-documented procedures;

- improved knowledge of inputs and outputs of the process;
- improving the quality and quantity of chemicals used, including regular revision and assessment of the recipes, optimal production scheduling, use of high-quality water in wet processing (but not for example for machine cleaning);
- systems for automated control of process parameters – e.g. temperature, liquor ratio, chemical dosing, etc. – allow a tighter control of the process for improved right-first-time performance;
- optimising water consumption, for example by operating at reduced liquor ratio, and increasing washing efficiency, combining process steps (e.g. scour/dye) and re-using or recycling water (opportunities for water recycling in the textile industry are comprehensively reviewed in a multi-author book published by the Society of Dyers and Colourists in 2003[16]);
- Optimising energy consumption, for example by heat insulation of pipes and recovery of thermal energy from hot waste streams via heat exchangers to heat incoming water.

6.2.5 Waste minimisation and pollution control

Basically there are two approaches to reducing pollution arising from the textile wet processing sector:

(a) effluent treatment – or end-of-pipe solutions;
(b) waste minimisation – or source reduction solutions.

The first approach has no financial payback and is literally money down the drain but provides the dyeing facility with its 'licence to operate'. In contrast, the waste minimisation approach not only reduces environmental impact but also delivers reduced costs – a situation that has been referred to as the 'win–win scenario'. For a given batch dyeing process and given fabric using a common preparation and dyeing procedure, it has been shown by modelling studies that waste minimisation can best be achieved by operating at the lowest liquor ratio possible and maximising right-first-time performance.[17]

The Controlled Coloration™ concept originally developed by ICI and now promoted by DyStar describes textile coloration processes carried out in a way that minimises the impact on the environment by achieving high levels of right-first-time production.[18–20] The controls that the dyestuff manufacturer can exert are:

- control of dyeing behaviour;
- control of product quality;
- control of application processes;
- control of environmental impact.

For example, some of the factors that must be taken into account when designing reactive dyes for reduced environmental impact are:[21]

- careful choice of intermediates – no banned amines, minimum adsorbable organic halogen (AOX);
- high colour yield – high-fixation, multifunctional dyes leading to reduced levels of colour in effluent;
- suitability for ultra-low liquor ratio dyeing machinery – to minimise energy, water and chemicals consumption;
- right-first-time dyeing through dyestuff compatibility – to minimise wasteful shading additions or reprocessing.

The environmental problems facing the textile wet processing industry can not be solved using outdated products, processes or machinery. Innovation is required to address the environmental issues facing the supply chain and dyestuff manufacturers have a key role to play. As well as product innovation the other major contribution that the innovative dyestuff company can make to cleaner textile production is application process innovation. This combination of novel dyestuffs and optimised application processes leads to:

- minimised resource consumption – lower environmental impact;
- maximised productivity – higher throughput from available assets.

In order to deliver environmentally oriented process innovation the dyestuff companies work closely with leading textile machinery and equipment suppliers. Examples of such collaborations from DyStar include:

- Monforts and DyStar: Econtrol™ T-CA (ITMA Munich, 2007);
- Thies and DyStar: Luft-roto™ plus;
- Then and DyStar: SynergyG2/Remazol™ Ultra RGB dyes (ITMA Shanghai, 2008);
- Zimmer and DyStar: Chromojet™ Printing System.

DyStar has also developed optimisation programs (Optidye™) for dyeing each major textile substrate; these can be used to identify the most efficient application conditions for a given dye recipe. These programs can be incorporated into the machine controllers routinely fitted to modern dyeing machines by leading manufacturers.

Effluent treatment

There are many ways of treating dyehouse effluent in order to reduce its impact when it is eventually discharged to surface water. Some of these treatments are best carried out immediately the effluent is produced, i.e. before it is mixed with other types of effluent that may interfere with the efficiency of the chosen treatment technology. This is known as the partial (or segregated) stream approach.

Another alternative is to create a mixed or balanced effluent to smooth out the peaks and troughs in flow or composition typical of a dyehouse effluent and to then treat this in a multi-stage wastewater treatment plant with a large hydraulic capacity. Depending on the availability of public sewerage systems this can either be carried out onsite at the dyehouse or offsite in a centralised third-party treatment facility, be this privately or publicly owned and operated.

Either of these strategies may be acceptable options when properly evaluated and applied to the actual wastewater situation. Well-accepted general principles for wastewater management and treatment include:[17]

• characterisation of the different waste streams arising from the processes carried out;
• segregation of effluents at source according to their contaminant load and type;
• allocating contaminated wastewater streams to the most appropriate treatment;
• avoiding the introduction of wastewater components into biological treatment systems that could cause the system to malfunction;
• treating waste streams containing a relevant non-biological fraction by appropriate techniques before, or instead of, final biological treatment.

Technology options for all these treatment scenarios have been extensively reviewed in the literature.[22–26]

6.3   What are ecotextiles?

According to Bide[4] environmental acceptability in textile products generally falls into one of two categories. The first and simplest is to demonstrate that the product does not harm the user or harm the environment in use. A primary example is the Oeko-tex Standard 100 ecolabel scheme that certifies that a textile product is free from harmful chemical residues. The second category is based on the environmental impact of the manufacturing processes involved in making the product and can cover the whole life cycle of the product from cradle to grave. However the textile chain is long and complex and weighing the balance of all the possible alternatives makes life cycle analysis of textile products difficult if not impossible.

The term 'greening the supply chain' has emerged to describe a wide variety of actions that a growing number of companies are presently conducting to install greater performance rigor and operational control over their extended supply chains. Greening the supply chain initiatives are part of a process for implementing a sustainable development plan aimed at achieving improved environmental, health and safety performance; increasing efficiencies in the use of energy, water or other natural resources or raw materials; reducing the environmental and societal impact of business operations upon local communities and the global biosphere; and expanding economic and quality of life enhancing opportunities that result from the company's business activities.[27]

As Brian Walter, of the leading furniture supplier Herman Miller, has noted: 'You are only as green as your supply chain.'[28] He identifies three things that companies can do in working with their supply chains to go green:

(a)   design products with sustainability as a core principle;
(b)   define goals and hold yourself accountable;
(c)   embrace *transparency* and meaningful metrics.

Case studies on the different approaches taken by large and small retailers to the challenge of supply chain greening have also been reported by Goldbach[29] and Kogg.[30] Eco-innovation and green design also have a major role to play in greening the textile supply chain. In a previous publication in this series, Bhamra[31] reviewed the status of eco-design within the textile and clothing sector and concluded that 'the sector has a long way to go to catch up with the levels of eco-innovation being achieved elsewhere'. Since then, thought-leaders such as Fletcher[32] and Black[33] have taken up the eco-design challenge and courses on green and sustainable fashion design have been introduced in fashion and design colleges in the UK and elsewhere.

6.4   Future trends

The textile supply chain is a classic example of a buyer-driven global commodity chain in which it is the large retail groups and brand name marketers in North America and Western Europe who exercise the greatest influence on the specification of the goods they source, increasingly from the Asia Pacific region.[34] Owing to the globalisation of the textile supply chain, and consumer/non-governmental organisation (NGO) pressure on corporate responsibility in Western markets, environmental drivers have been communicated to all the major textile- and clothing-producing regions of the world. So even if an individual producer country does not have tough pollution control legislation or enlightened labour laws, minimum standards of working conditions and safety, health and environmental requirements may be imposed by the European or North American purchasing organisations.[21] As Messner has noted recently:[35]

'the number of global social and environmental standards is growing rapidly in sensitive sectors (e.g. labour-intensive industries, industries close to raw materials, food industries). These are the sectors in which social and ecological problems and health-relevant impacts frequently occur and are highly visible to the public, the consumers, and to NGOs in advanced countries that are the driving forces behind the proliferation of social and ecological standards. In other words, it is precisely in industries with low levels of technological complexity, which include industries in developing countries that have 'natural competitive advantages', that global standards and the high demands which they imply for the global governance capacities of local actors are assuming ever greater significance. Thus, building competitiveness is often no longer dependent only on compliance with the classic parameters of competition (time, price and quality of products and services) but also requires the capacity to orient products and production processes to global social and environmental standards. Even on the 'low roads' of the world economy knowledge-based competitive advantages and governance capacities of local actors are gaining in importance.'

Some emerging trends, as the industry strives towards a more sustainable business model for textile and clothing manufacture and consumption, are presented below:

- low-energy dyeing processes (possible enzyme activated, or biomimetic systems);
- low-water dyeing and printing processes (development of foam techniques, ink jet printing and dyeing, and re-examination of solvent-based processes with full solvent recovery);
- a renewed emphasis on right-first-time dyeing;
- greater emphasis on product durability and a move away from 'fast fashion';
- increasing prescriptive specification of dyes and chemicals by retailers/brands to ensure achievement of performance specification with regard to metamerism, fastness and RSL compliance;
- increasing transparency and traceability in supply chains;
- development of meaningful metrics to assess sustainability of supply chains, e.g. carbon footprinting, water footprinting and pollution load/unit of production;
- increased collaboration between actors along the whole supply chain from farm to store;
- closer co-operation between retailer/brands on supply chain standards, environmental requirements and chemical restrictions;
- greater awareness of sustainability in fashion and textile design.

6.5   Conclusions

Despite the current economic difficulties it is difficult to see how the textile industry could ever turn the clock back and ignore issues of pollution, product safety and resource conservation. The world has changed. The internet and 24-hour news channels combined with the activities of NGOs and lobby groups ensure that it is no longer possible for the industry to hide the less than desirable practices that may be taking place back up the supply chain.

There is a saying 'It takes a village to raise a child'. Only by all the actors in the textile supply chain working together can environmental improvement be promoted through the introduction of ecologically designed products and cleaner production techniques. By establishing such collaborative supply chain partnerships the industry can succeed in meeting the sustainability challenge of the twenty-first century.

6.6   Sources of further information and advice

- Clean Technology Programme of the United Nations Environment Programme (UNEP).
- IPPC Guidance for the Textile Sector, Environment Agency, July 2002.
- Forum for the Future, 'Fashioning Sustainablity', June 2007 at www.forumforthefuture.org.
- J.M. Allwood, S.E. Laursen, C.M. de Rodriguez and N.M.B. Bocken, Well Dressed?, University of Cambridge Institute for Manufacturing Report, 2006.
- DEFRA, Sustainable Clothing Roadmap pages at www.defra.gov.uk.
- UNEP, The Textile Industry & the Environment Technical Report No. 16, Paris, 1993.
- M. Thiry, 'Color craft', *AATCC Review*, 2008, p. 20.
- Water Footprints at www.waterfootprint.org.

6.7   References

1 B. GLOVER, P.S. COLLISHAW and R.F. HYDE, 'Creating wealth from textile coloration', *Journal of the Society of Dyers and Colourists*, 1991, **107**, 302–304.
2 US ENVIRONMENTAL PROTECTION AGENCY (USEPA), Best Management Practices for Pollution Prevention in the Textile Industry, 1996, Washington DC, USEPA.
3 P. COOPER, 'The consequences of new environmental legislation for the UK textile industry', *Textile Horizons International*, 1992, **12**(10), 30–38.
4 M. BIDE, 'Environmentally responsible application of textile dyes', in *Environmental Aspects of Textile Dyeing*, ed. R. Christie, Cambridge, UK, Woodhead, 2007.
5 P.S. COLLISHAW, S. WEIDE and M.J. BRADBURY, 'Colour (In) constancy – what is achievable on retailer standards?', *AATCC Review*, 2004, **4**(9), 16–18.
6 K. IMADA, N. HARADA and T. TAKAGISHI, 'Fading of azo reactive dyes by perspiration and light', *Journal of the Society of Dyers and Colourists*, 1994, **110**, 231–234.
7 P.W. LEADBETTER and A.T. LEAVER, 'Color wizards', *Review of Progress in Coloration*, 1989, **19**, 33.

8 House of Lords Science & Technology Committee Report, August 2008, London, HMSO, p. 51.
9 *The Independent*, 11 August 2008.
10 Regulation No. 1907/2006 of the European Parliament and of the Council, OJEC L396, 31.12.06, pp. 1–849.
11 D.C. ESTY and A.S. WINSTON, *Green to Gold*, New Haven, Connecticut, Yale University Press, 2006, p. 18.
12 J.R. EASTON, 'Econfidence in the textile supply chain', in Proceedings of the Tex Summit International Conference, Mumbai, 2007, p. 61.
13 J.R. EASTON, 'General considerations in reuse of water: reuse from coloration processes', in *Water Recycling in Wet Processing*, ed. J.K. Skelly, Bradford, UK, SDC, 2005, pp. 3–15.
14 A.K. CHAPAGAIN, A.Y. HOEKSTRA, H.H.G. SAVENIJE and R. GAUTAM, 'The water footprint of cotton consumption: An assessment of the impact of worldwide consumption of cotton products on the water resources in the cotton producing countries', *Ecological Economics*, 2006, **60**(1), 186–203.
15 European Commission, IPPC BATREF Guide for the Textile Sector, Brussels, 2003.
16 J.K. SKELLY (ed.), *Water Recycling in Textile Wet Processing*, Bradford, UK, SDC, 2003.
17 B. GLOVER and L. HILL, 'Waste minimization in the dyehouse', *Textile Chemist & Colorist*, 1993, **25**(6), 15–20.
18 P.S. COLLISHAW, B. GLOVER and M.J. BRADBURY, 'Achieving right-first-time production through control', *Journal of the Society of Dyers and Colourists*, 1992, **108**, 13–17.
19 P.S. COLLISHAW, D.A.S. PHILLIPS and M.J. BRADBURY, 'Controlled coloration: a success strategy for the dyeing of cellulosic fibres with reactive dyes', *Journal of the Society of Dyers and Colourists*, 1993, **109**, 284–292.
20 A.D. CUNNINGHAM, 'The controlled coloration approach for right first time dyeing of polyester', *AATCC International Conference Book of Papers*, 1995, pp. 424–436.
21 J.R. EASTON, 'Supply chain partnerships for sustainable textile production', Ecotextiles04 Conference, Manchester, 2004, pp. 50–58.
22 *Choosing Cost Effective Pollution Control*, Environmental Technology Best Practice Programme, GG109, 1998, www.p2pays.org/ref/24/23103.pdf.
23 P. COOPER (ed.), *Colour in Dyehouse Effluent*, Bradford, UK, SDC, 1995.
24 A. REIFE and H.S. FREEMAN (eds), *Environmental Chemistry of Dyes and Pigments*, New York, John Wiley & Sons Inc, 1996.
25 A.B. DOS SANTOS, F.J. CERVANTES and J.B. VAN LIER, 'Review paper on current technologies for decolourisation of textile wastewaters: Perspectives for anaerobic biotechnology', *Bioresource Technology*, 2007, **98**, 2369–2385.
26 S.R. SHUKLA, 'Pollution abatement and waste minimisation in textile dyeing', in *Environmental Aspects of Textile Dyeing*, ed. R. Christie, Cambridge, UK, Woodhead, 2007, pp. 116–148.
27 T. YOSIE, 'Greening the supply chain in emerging markets: some lessons from the field', November 2008. GreenBiz Reports, www.greenbiz.com.
28 B. WALTER, 'You are only as green as your supply chain', *HBR Green*, February 2008, www.hbrgreen.org.

29 M. GOLDBACH, 'Coordinating interaction in supply chains – the example of green-ing textile chains', in *Strategy and Organization in Supply Chains*, eds S. Seuring, M. Müller, M. Goldbach and U. Schneidewind, Heidelberg, Physica Verlag, 2003, pp. 47–63.

30 B. KOGG, 'Power and incentives in environmental supply chain management', in *Strategy and Organization in Supply Chains*, eds S. Seuring, M. Müller, M. Goldbach and U. Schneidewind, Heidelberg, Physica Verlag, 2003, pp. 65–81.

31 T. BHAMRA, 'Building ecodesign throughout the supply chain: a new imperative for the textile and clothing industry', Ecotextiles04 Conference, Manchester, 2004, pp. 41–49.

32 K. FLETCHER, *Sustainable Fashion and Textiles: Design Journeys*, London, Earthscan, 2008.

33 S. BLACK, *Eco-chic: The Fashion Paradox*, London, Black Dog Publishing, 2008.

34 G. GEREFFI, 'The organisation of buyer-driven global commodity chains. How US retailers shape overseas production networks', in *Commodity Chains and Global Capitalism*, eds G. Gereffi and M. Korzeniewicz, Westport, Greenwood Press, 1993, pp. 95–122.

35 D. MESSNER, 'The concept of the world economic triangle. Governance patterns and options for regions', IDS Working paper 173, p. 89, Brighton, Institute of Development Studies, 2002.

7
Environmentally friendly plasma technologies for textiles

T. STEGMAIER, M. LINKE, A. DINKELMANN, V. VON ARNIM, and H. PLANCK, Institute for Textile Technology and Process Engineering (ITV) Denkendorf, Germany

Abstract: Plasma technologies present an environmentally friendly and versatile way of treating textile materials in order to enhance a variety of properties. This chapter begins by presenting the technologies of atmospheric pressure plasma. It then gives some examples of applications, before describing the environmental benefits of plasma treatment.

Key words: plasma technology, textile, surface treatment, environmentally friendly.

7.1   Introduction

Growing demands regarding the environmental friendliness of finishing processes as well as the functionality of textiles have increased the interest in physically induced techniques for surface modification and coating of textiles. In general, after the application of water-based finishing systems, the textile needs to be dried. The removing of water is energy intensive and therefore environmentally harmful and expensive. Plasma treatment, being a dry process, represents an economical alternative.

A plasma can be described as a mixture of partially ionised gases, where the constituents are activated by external energy addition. Thereby, atoms, radicals and electrons are present in a dynamic equilibrium. If the energy input is steered in such a way that the gas temperature remains in the range of room temperature, the process is referred to as 'cold' or 'low-temperature' plasma.[1] The main advantages of such plasma treatments are:

- the electrons in low-temperature plasmas are able to cleave covalent chemical bonds, thereby producing physical and chemical modifications of the surface of the treated substrate without changing the fibre properties;
- there is a minimal consumption of chemicals and no drying process is required;

- high environmental compatibility of the processes;
- the processes can be applied to almost all kinds of fibres.

In principle, all polymeric and natural fibres can be plasma treated to achieve the following effects:

- wool degreasing;
- desizing;
- change in fibre wettability (hydrophilic, hydrophobic properties);
- increase in dyestuff affinity;
- improved dye levelling properties;
- anti-felt finishing in wool;
- sterilisation (bactericidal treatment) etc.

During the treatment the textile stays dry and, accordingly, drying processes can be avoided, no wastewater is generated and no (or less) chemicals are required. Further advantages of plasma technology are the extremely short treatment time and the low application temperature. Therefore, plasma treatment represents an energy-efficient and economic alternative to classical textile finishing processes.

For plasma treatment, different types of power supply are used to generate the plasma:[2]

- Low frequency (LF), 50–450 kHz;
- Radio-frequency (RF), 13.56 or 27.12 MHz;
- Microwave (MW), 915 MHz or 2.45 GHz.

The power required ranges from 10 to 5000 watts, depending on the size of the reactor and desired treatment.

Gases commonly used for plasma treatment are:

- chemically inert (e.g. helium and argon);
- reactive and non-polymerisable (e.g. ammonia, air and nitrogen);
- reactive and polymerisable (e.g. tetrafluoroethylene, hexamethyldisiloxane).

For the surface treatment of textiles with low-temperature plasma, there are two distinct types of process: atmospheric pressure processes and low pressure processes (Table 7.1 and Fig. 7.1). For many years, mainly low pressure plasma processes have been developed for textile plasma treatment. However, the integration of these processes into continuously and often fast-running textile production and finishing lines is complex or even impossible. In addition, due to the need for vacuum technology, low pressure processes are expensive.

Plasma processes that operate at atmospheric pressure are therefore advantageous for the textile industry for the following reasons.

Table 7.1 Comparison between atmospheric pressure and low pressure
processes[3]

	Atmospheric pressure plasma	Low pressure plasma
Plasma generation	High tension (e.g. 1–100 kV; 20–40 kHz) between dielectric, coated electrodes	Direct current (DC) or high frequency (up to the microwave range)
Pressure	Atmospheric pressure	Low pressure (0.1–1 mbar)
Extension of the plasma	30 µm (micro-discharge)	electrode diameter
Life time	1–10 ns	Permanent
Continuous process control	Simple	Complex

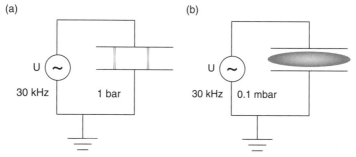

7.1 Types of plasma treatment: (a) atmospheric; (b) low pressure.
U, voltage (power source).

- The typical working width of textile machines is between 0.2 and 10 metres. Plasma modules suitable for use in textiles need to be able to be scaled up to these dimensions; this is easier for atmospheric pressure techniques.
- Textiles have large specific surfaces compared with foils, piece goods or bulk solids. Even with strong pumps the reduced pressure that is necessary for low pressure plasma will only be reached slowly owing to the time taken for desorption of adsorbed gases.
- Depending on the fibre material, textiles can adsorb and absorb relatively large amounts of water. To reduce the chamber pressure below the vapour pressure of water (23.4 mbar at 20 °C) a drying process is necessary. This is time consuming because diffusion processes within the fibre and the cooling due to evaporation of the water slow down the

drying (vapour pressure of ice: 4.02 mbar at −5°C). The evaporation time can be shortened by heating the substrate.

• Atmospheric pressure processes can be combined with spraying or aqueous aerosols.
• Atmospheric pressure processes are in-line capable in contrast with batch low pressure plasma processes.
• Investment and maintenance costs for atmospheric pressure plasma modules are moderate.

New markets can be opened if the plasma processes that have been developed for low pressure conditions to date are transferred to normal pressure conditions.[4] The dielectric barrier discharge (DBD) method is one plasma process that can also be applied at atmospheric pressure. Therefore, for many years, have been expended research efforts within cooperations between partners from textile and plasma research institutes and industry in order to make DBD technology a useful and practicable method for the textile industry, aiming at the development of new products, improved textile functions and sustainable processes.[1,5]

7.2   Atmospheric pressure plasma processes

A focus of ITV Denkendorf within its research and development projects is the development of new technologies for continuous plasma coating of wide fabrics under atmospheric pressure and the improvement of the profitability of the plasma coating processes. For example, oil- and water-repellent layers are deposited on the substrate. New processes open up by combining plasma treatments with sprayed aerosols.

Only low-temperature discharges can be used to functionalise textiles by plasma treatment.[6] In low-temperature plasmas, the thermal energy of the neutral atoms and molecules roughly stay at room temperature; however, the free electrons can transmit high energy. Therefore the electrons have to be heated to a minimum of 10 000 Kelvin while the gas temperature which is controlled by the thermal energy of atoms and molecules remains at moderate temperatures. These plasmas can interact with the polymer surface for some milliseconds without thermally changing the substrate. In thermal plasmas, however, the neutral particles are in thermal equilibrium with the charged particles and transmit too much energy to the polymer surface which instantly destroys the textile. Thermal plasma flame treatment is common in textile processing; however, this technology is mainly used for the removal of hairs on textiles – not for functionalisation.

Functionalising is a chemical process. Thereby, the reaction time of the reaction partner is an important process parameter. For a plasma treatment, this means that the longer the polymer surface is exposed to the activated reactive gas the more functional groups are formed, and within limits the better is the desired effect. In the following sections the most common forms of atmospheric pressure plasmas – corona discharge and DBD – are described.

7.2.1 Corona discharge

The best-known atmospheric plasma technology is the so-called 'corona discharge'. In material processing practice, the expression 'corona discharge' is often differently understood in physics and electronics. Strictly speaking, corona discharge occurs if a rather asymmetrical electric field has an effect in a gas, thus causing electric current to flow in the microamps to milliamps. The range considerable inhomogeneous field required is typically reached by an asymmetrical shaping of the electrodes. The simplest electrode configuration consists of a metallic needle opposite a conductive plate or a thin wire within a metallic cylinder. If the wire or the metallic needle, respectively, is connected with the negative pole of the power supply and the large, extensive electrode (cylinder and plate, respectively) is connected with the positive pole, a negative corona discharge occurs. In the reverse polarity case, a positive corona discharge results.[7] Corona discharge differs from glow discharge (weak current) and arc discharge (high current) with regard to the flow of current. Within a corona discharge the current varies within wide ranges and is mainly performed by charge carriers of only one polarity. The electric field mainly concentrates at the highly curved electrodes. Therefore, light emissions occur in this cathode region. The majority of the gap between the electrodes remains dark, however; this so-called 'drift zone' is occupied by negatively charged particles. In this area the charge carriers are decelerated by permanent collision between the gas particles and slowly drift towards the electrode. In the glowing region of high electric field strength the charge carriers can receive enough kinetic energy between two collisions with gas particles to transmit enough energy to fragmentise, ionise or set the collision partners in strong molecular vibration. If electrons are excited by the collision to energetically higher states, light quantums can be emitted when the electrons reoccupy the lower energy level.

In the case of a negative corona, electrons are generated in the strong electrical field at the curved electrode. These are accelerated within the field, however they soon transfer their energy to surrounding gas particles. Subsequently, they are taken up by neutral particles in the dark drifting zone and form negative ions which slowly drift towards the large and positively charged anode.

In the case of the positive corona, no electrons can be set free at the cathode. In the drifting zone the electric current is in practice carried only by positive charge carriers. When they impinge on the cathode they can not – in contrast to the negative corona – eject new electrons from the electrode. Negative charge carriers in the drifting zone can only be generated by photoionisation or natural radioactivity. These can only cumulatively multiply in the area of a very high field strength near to the positively charged needle-like electrode. Under similar external conditions, the positive corona discharge is inhibited, in contrast with the negative corona discharge. The discharge current is halved.

The typical field strengths for corona discharges are roughly around 40–50 kV/cm at radii between 2 and 10 mm. Discharge currents range from microamps to milliamps. A typical corona discharge current is below 200 µA.[8] Such 'real' corona discharges are used for textile treatment in order to either remove, or to specifically apply, electrostatic surface charges from or to fabric layers, but not for surface functionalisation or activation. The electrostatic discharge of fabric layers is, for example, advantageous in the case of hydrophobic fibres like polypropylene (PP) nonwovens or textiles finished with fluorocarbons. Owing to friction and contact with non-grounded surfaces, charges can occur that exert strong far-reaching coulomb forces on fabrics, thus, for example, avoiding proper plaiting. The charged fabric, moreover, tends to attract dust particles. Normally, dust is not quite neutral but slightly negatively charged so that dust particles whirled up by air flows are directed to the textile surface by positive charges and adhere owing to electrostatic forces. If the fabric layer is discharged, the dust particles can be sucked off. In the textile industry, re-damping by aerosols is improved by electrostatic charge.[9] Textile webs and yarns, respectively, with like electric charges repel each other, thus avoiding blocking.

In the case of fabrics and webs processed at high velocities, a gas boundary layer is coupled to the surface. Corona discharge helps to convert this layer from a laminar to a turbulent flow. This facilitates the drying process of the web as the steam is transported along with the fabric in the laminar boundary layer. As a turbulent flow, the humid layer can more effectively be sucked off from the fabric.[10,11] The prerequisite for switching a flow from laminar to turbulent, however, is that the flow is already relatively close to instability. This means that the discharge process can not cause any turbulence over slowly moving surfaces. This technology is distributed by the company Eltex (EFD-technology). Corona discharge can further be used in order to effectively exchange the air entrapped within a textile and to substitute it with inert or reactive gas. It has been shown that the efficiency of the discharge-supported exchange of oxygen and nitrogen within a fast-running web increased with web velocity; 63% of the entrapped oxygen was washed out by nitrogen at 55 m/min web speed, whereas at 150 m/min web velocity, 86% of the entrapped oxygen was exchanged with nitrogen.[12]

7.2.2   Dielectric barrier discharge

The set-up for DBD processes, although similar to the corona discharge, has characteristic deviations from the corona set-up. The most important characteristic is that at least one electrode is coated completely by an insulator, the dielectric barrier for electric currents. DBD belongs to the class of non-equilibrium, low-temperature plasmas. For a given energy input it is possible to calculate the temperature for all individual kinds of particles. The DBD is characterised by high energy of the free electrons while the ions and neutral particles gain only little additional kinetic energy. Kinetic energy and momentum are transferred to the particles upon collision with electrons that have been accelerated along the electric field lines. Because of their low weight and small collision cross-section, the electrons gain considerably more thermal energy compared with the heavy and large gas particles. Upon collision between the fast electrons and gas molecules or atoms, the molecules are fragmented or transferred into electronically excited highly reactive states. The formation of a DBD process is complex and will only be described in basic detail here. First an electric field is created. Free charge carriers are accelerated by coulomb forces in this field. Even a very few free electrons of natural origin, such as cosmic radiation or natural radioactivity, are sufficient to initiate a discharge process. These charge carriers are accelerated within the field and generate new ion-electron-pairs upon collision with neutral gas particles. It is mainly the light electrons that are accelerated and absorb thermal energy. Additional electrons are emitted into the gas gap upon impact of ions on the electrode. The avalanche-like cloud of charge carriers that is generated moves towards the positively charged electrode. The characteristic light emissions within the visible and ultraviolet (UV) range of plasma discharges result from the collision-induced electronically excited states of the gas atoms and molecules. The electric field between the counter electrode and the charged cloud becomes stronger while the field between the negatively charged electrode and the cloud breaks down. At the time the electron cloud reaches the positively charged electrode, high voltage occurs at the dielectric electrode coating and the gas gap is field-free. Currents can no longer flow and the discharge is extinguished although the whole gas gap is still electrically conductive. This breakdown process of initiation, avalanche-like charge carrier generation and extinguishing occurs within a few nanoseconds. However, the process varies according to the type of gas, gas gap size, gas pressure and the nature of the dielectric coating.

The DBD is described as 'filamented'. At normal pressure many single discharge filaments, also called streamers, are generated within the gap between the electrodes in random spatial and temporal distribution over the length of the gap. The individual discharge filaments do not interact with each other apart from the initial electron generation by UV radiation. One filament has a radius of around 0.1 mm and transports an average electric current of 0.1 A. The average electron energy ranges between 1 and 10 eV. The amount of energy dissipated per filament is in the microjoule range. A discharge that hits the dielectric layer of the counter electrode generates a so-called 'footprint' – an area where the charge is distributed on the dielectric surface by sliding discharges. The footprint is a few square millimetres in size and is considerably larger than the discharge filament diameter.[13–15] These surface discharges also occur on the surfaces of textile fibres in the discharge gap. These surface discharges enhance the effect of the DBD with respect to fibre surface functionalisation because reactive species are generated at the fibre surface exactly where they are needed for the reaction process.

DBD is the most commonly applied technology to achieve an easily scaled-up, expanded, non-equilibrium plasma for the treatment of wide webs and fabrics. The free electrons in the plasma discharge are heated up to 10 000–100 000 Kelvin while the gas itself can be kept at moderate temperatures, typically between room temperature and 100 °C. This allows the treatment of textiles without thermal damage to the fibrous substrate. Ambient air is often used as a process gas in DBD because of its easy availability and the reactivity of the oxygen. The molecular oxygen in the air can be converted into reactive atomic oxygen radicals or excited ozone molecules.

The set-up for DBD processes is typically characterised by the combination of two opposing electrodes with an intermediate millimetre-sized air gap. At least one electrode is coated by an insulator. The dielectric barrier covers the whole surface of the electrode and acts as an isolating barrier for electric currents. The dielectric layers need to be resistant to high voltages of a few kilovolts; they have to suppress electric currents even when the high voltage is applied. Only negligible currents are allowed through the barrier. Local avalanche breakdowns through the barrier may not occur The electrode arrangement of the barrier discharge acts like several condensers in series with different capacities and dielectric strengths within an electric circuit. The gas gap forms the condenser which leads to the avalanche breakdown within the gap when a sufficiently high voltage is applied. A spark develops and the condenser is temporarily discharged, caused by the electrically conductive plasma. The electric current that flows during the discharge is obstructed by collisions with and between gas particles. The gas forms a resistor that transfers energy into the gas. The dielectric coatings of the electrodes inhibit an uncontrolled current like an arc discharge between the electrodes. Only alternating currents can pass the condenser while direct currents are blocked. In order to repeatedly or almost continuously couple power to the discharge, a DBD must be run with alternating currents. A second discharge at the same site can only take place when either the voltage has been doubled or when the polarity sign has been inverted. The common frequencies range between 1 and 100 kHz. The excitation frequency depends on the resonance frequency of the oscillating circuit that is formed by the transformer coil, electrode gap and dielectrics. Relatively high power can be coupled to the plasma if the excited frequencies are close to the resonance frequency. Frequencies that deviate significantly from the resonance frequency can couple only a small amount of power into the system.

At the minimum ignition voltage the discharge ignites only at an especially favourable point. If the electrodes, for example, are not quite plane-parallel to each other the site with the smallest gap is the preferred site for the first discharge. If the voltage is increased more discharges are generated side by side at other sites which fill the whole gap between the electrodes with a discontinuous plasma.

The technical term 'dielectric barrier discharge' has been established in the textile and foil processing industries for the above-described processes. However, in practice, processes with discharge by commonly only one dielectric layer and preferably in air are also often called corona discharge. This is technically speaking incorrect, because DBD is fundamentally different from corona discharge. Both corona discharge and DBD are so-called 'silent discharges' but in contrast to the DBD, corona discharges have an extensive dark field and are generated between two highly asymmetrically formed metallic electrodes. The coating of at least one of the two electrodes by a high-resistance dielectric barrier in practice prevents operation under conditions of direct current. In corona arrangements, low-resistance barriers can prevent an arc discharge and distribute direct current equally to many point electrodes. The electrical current at each electrode is quite often limited by an inserted protective resistor. In DBD, however, high-resistance coatings such as glass, ceramics or silicone rubber (polydimethylsiloxane, PDMS) are used whose electrical conductivity can in practice be neglected. The total high voltage must be applicable to the barrier layer without leading to avalanche breakdown events.

One reason for the misuse of the 'corona' term is that in many cases a considerably asymmetrical electrode configuration is chosen for barrier discharges of fabrics or foils. The fabric is directed over a broad roller which forms one, often grounded, electrode. The high-voltage, bar-shaped electrodes are arranged around the roller parallel to the roller axis, however they are small compared with the radius of the roller.

This results in the following definition of DBD:

A dielectric barrier discharge (DBD) is a filamented discharge between two electrodes with at least one dielectric with a high-frequency excitation, mostly in the range between 1 and 100 kHz (Fig. 7.2).

(a)

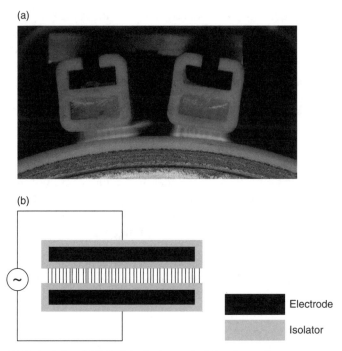

(b)

Electrode

Isolator

7.2 Principle of DBD, (a) photograph, (b) schematic.

The traditional and most common application of corona and DBD technology, respectively, is the improvement of wetting of non-polar surfaces. Such low-energy surfaces which are typical for polymeric materials have poor wetting properties with regard to polar high surface tension liquids. A high surface tension liquid like water or a water-based dispersion only spreads and forms a stable film on surfaces with comparable high surface tension. On non-polar polymeric surfaces, aqueous films will break and de-wet forming single isolated droplets. A stable film formation is important for printing, varnishing, coating and several other applications. DBD treatment often helps to avoid these de-wetting phenomenons by increasing the surface energy. The advantages of a DBD treatment are a better wetting and also a better adhesion of printed dyestuff or lacquer to the treated substrate. The positive effect of a DBD treatment is also used in combination with glue. Hot-melt, reactive adhesives and water-based adhesives layers have an improved peel adhesion on plasma functionalised polymers.

The DBD process is especially useful for applications with roll-to-roll systems for thermally sensitive materials like foils, textiles, or photographic print paper. However, DBD is also used to alter other materials such as wood, insulated cable or even fingernails. The benefits of DBD with regard to technical textiles is described in more detail in Section 7.3.

7.2.3 Adaptations of atmospheric pressure plasma technology to textiles

The potential of surface modification of roll goods using corona discharge technology has been known for a long time and has been established in areas such as the foil industry.[16] However, attempts to use the same technology for textile treatment usually failed.[17] This is because the corona treatment of foils has usually been performed with a set-up of metallic electrodes. A polymeric foil, although usually a non-porous electrical insulator, then acts as a dielectric barrier between the electrodes. The foil forms either a dielectric barrier on one electrode if it is in tight contact with the electrode and the discharge filament end on the surface of the foil, or two separate discharges occur, if there are gaps between the foil and both electrodes. Thus, corona treatment of foils has in effect many characteristics of a DBD. Problems with the corona treatment of textiles occur because textiles differ from foils in many aspects, as described in the next paragraphs.

Textiles are commonly thicker than foils and have an inhomogeneous structure. They are porous and gas permeable, and typically very rough, even hairy. Their specific surface area is much larger than those of foils. Therefore, textiles can not inhibit the penetration of the discharge in the way that foils do. Discharge streamers can pass through the textile along the electric field direction at least locally. These filaments are high-current, hot discharge streamers because the gap between the electrode has to be wide according to the thickness of the textile and no dielectric barrier on the electrodes limits the current in the corona set-up. The high current flow along the yarn surface through the textile can induce carbonisation of the polymer. Even minimal carbonisation increases the local electrical conductivity of the substrate permanently. This, however, results in a subsequent concentration and localisation of further discharges at this point, increasingly damaging the textile. The typical results are micro-fine holes within the substrate.

Since textile perforation is commonly an undesired result, the aim of textile plasma treatment is to achieve a discharge that is as homogeneous as possible. The surface functionalisation should take place as effectively as possible and as gently as possible for the substrate. This aim could be reached by different modifications to the established corona technology. Several combinations of metal, silicone-, glass- and ceramic-coated electrodes have already been tested for the treatment of textiles. It has been shown that the use of only one dielectric – as is common for foil treatment with a dielectric-coated roller – is not sufficient for the treatment of textiles. Textiles can be better treated by means of two dielectric-coated electrodes. This is because via two dielectric barriers the electric current distributes into more single filament discharges thus making the discharge pattern more homogeneous (Fig. 7.1).[11]

It is further recommended that the discharge gap selected is as narrow as possible. This is because with increasing distance the breakdown voltage raises and the size of the footprints as well as the current of a single discharge increases. High local currents, however, can damage the textile. Accordingly, voltage must be adapted to the dimensions of the discharge gap and the textile to be treated, i.e. the shorter the electrode distance, the lower the breakdown voltage necessary and the lower the current in the single filament. A narrow gap facilitates a homogeneous coupling of power to the discharge.

It was further found that textiles usually need to be treated at much lower substrate velocity than foils to gain comparable and satisfactory results. This is because textiles have a much larger specific surface than foils owing to their fibrous structure. In order to generate a surface density of functional groups comparable with foils, it is necessary to generate considerably more reactive sites. Furthermore, in textiles the filaments of a yarn are closely packed, thus obstructing the exchange of gases. Thus, textiles must be exposed to a plasma for a longer time compared with foils, in order to compensate both disadvantages – the large surface area and the reduced accessibility of the activated plasma gases to the fibre surface. This can be best achieved by reducing the fabric speed or by scaling up and arranging more electrodes in sequence.

Modifications of the electrode configuration were also helpful. A more symmetric electrode assembly proved to be advantageous for textile treatment. This is because an equal number of discharge filaments is generated during each half wave of the alternating potential of the electrodes when the electrodes are symmetrically assembled. Each discharge in each direction during each half wave carries approximately the same current, enhancing the overall homogeneity of the discharge. Creating an intermittant excitation frequency is another suitable tool to make the DBD more homogeneous. An intermittant excitation means that the generator is electronically switched on/off according to a given pulse-break ratio. This enables the coupling of power to be controlled independently of the applied voltage. An intermitting mode can be of further advantage for rather slow plasma-activated reactions. Reactive gases, e.g. for plasma polymerisation, can be fed into the plasma zone during a break. Chemical processes can continue during the break thus saving unnecessary energy expenditure. The results of DBD treatments of textiles from natural fibres, e.g. wool or cotton, but also man-made fibres, have been published frequently. Significant changes in textile properties (such as friction coefficient and surface energy) or improved antistatic behaviour have been reported.[7–9]

Principle of the chemical processes and equipment

The chemical processes that are initiated in the plasma are usually very complex and depend on the chemical species in the gas gap. In synthetic air alone more than 100 reaction paths and some 30 different species must be considered in order to understand the processes within a DBD. The gaseous molecules are excited and partially fragmented ready to react with a partner in a chemical reaction. Positive ions are generated upon collision with high-energy electrons, whereas the capture of slow electrons leads to the formation of negatively charged ions. These charge carriers, however, are only short-lived and disappear within approximately 50 ns after the initialisation of the discharge process.

Nitrogen and oxygen, as the main components of air, are the most important collision partners of the accelerated electrons, if the DBD is performed with air as plasma gas. Neutral radicals generated, such as $O(^3P)$ and $O(^1D)$ as well as $N(^4S)$ and $N(D)$, react with surface molecules or with molecular oxygen in different electronic states (O_2, $O_2(A)$, $O_2(a)$, $O_2(b)$) and nitrogen (N_2, $N_2(A)$) to form more stable intermediate products such as O_3^* or NO or NO_2. The excited ozone disappears after approximately 1 while stable ozone remains. Nitrogen monoxide is converted after the discharge into different nitrogen oxides – N_2O, N_2O_5, N_2O_3 and NO_3 – within a tenth of a second.[11] Hazardous gases like ozone which are typically generated within a DBD in air need to be exhausted, disposed of or destroyed, e.g. with activated carbon filters, in order to guarantee workplace safety. In order to generate a streamer discharge in air, the breakdown field strength of 27 kV/cm must be exceeded. The actual voltage to be applied also depends on the nature of the dielectric barrier.

Apart from oxygen and nitrogen, real air also contains water, CO_2, hydrogen and rare gases, which all are involved in the reaction mechanisms. Rare gases like Argon, however, are not truly inert gases in a plasma, because they can transfer energy and even be converted into short-lived energy-rich molecules, so-called 'excimers'. Although the DBD is a so-called 'cold discharge', the electrodes will become hot with time. Control of the electrode temperature, by cooling of the electrodes, may be necessary for several reasons. First, after a typical DBD treatment over several hours the temperature of the electrodes will rise to temperatures between 100 and 200 °C. The textile properties are likely to be changed by thermal stress. Textiles that remove heat insufficiently, e.g. owing to low web speed, can display increased brittleness or changed colour. Second, plasma processes are temperature dependent. If, for example, a polymer layer is to be deposited by the plasma, heat might impede the polymerisation and deposition. On the other hand, moderately heated electrodes can support hydrophilic treatment of textiles in air. A plasma regime change can be observed in air when the temperature approaches 80–100 °C. In the case of hot electrodes the discharge is more homogeneous and less filamented than with cold electrodes. The coupled power decreases measurably when switching from cold to hot electrodes, while the voltage remains constant. It is supposed that the change in the plasma regime is related to the stability of ozone. The plasma-generated ozone becomes unstable when the temperatures exceed 80 °C. It is assumed that it is mainly the enforced decay of ozone that causes the observed regime change. More oxygen atoms are generated which are available for the surface modification, thus accelerating the process to impart hydrophilic properties to the fabric.

The aim, therefore, is an efficient adjustment of temperature and cooling of the system to achieve optimal treatment results for continuous DBD operation. This can be reached efficiently by cooling the electrodes.

Technical possibilities

Fabrics such as woven fabrics, knitted and warp knits, nonwovens, laid fabrics, papers and foils as well as mono- and multifilaments, staple fibre yarns, roving yarn, carded nonwovens, and also single fibres and powders, can be treated with the DBD method described. Limitations or specific configurations are necessary, for example, for electrically conductive textiles. In a common DBD set-up the endless substrate should be an electric insulator. Otherwise it is likely that electric currents will flow along the substrate to the grounded parts of the machine it is in contact with, like grounded deflection rollers. High current discharges between textile and grounded element might occur which might damage the substrate. A specific tandem-DBD set-up has been described to treat potential-free electrically conductive substrates. The set-up consists of two symmetrically arranged pairs of electrodes. They are electrically controlled in such a way that a filament-like discharge occurs between the first electrode and the substrate. The discharge current flows along the substrate to the second discharge zone. The filamented discharge at this plasma zone closes the electric circuit.[18]

7.3   Examples of applications

7.3.1   Activation of fibre surface

For many applications, particularly technical applications of textiles, strong adhesive bonds between textiles and other substrates are needed. Thus, bonding, coating or lamination are important working steps. They are responsible for efficient processing and sustainable functionality of the product. Problems in ensuring durable adhesion to textile substrates occur for many reasons. The flexibility of a textile poses a specific challenge for adhesive bonds. On the fibres, residues from processing agents like spin finishings or sizing agents, which originate from the previous manufacturing steps, prevent good adhesion to fibre surfaces. However, even the untreated surface of a synthetic fibre such as polyester (PET), polyacrylonitrile (PAN) or highly drawn polyolefine has a crystalline and hydrophobic character. Wetting ability and chemical reactivity of synthetic fibres are therefore usually poor and adhesion forces are weak.

In the following, the potential of DBD treatments for improving the adhesion of coatings and bondings to other substrates will be demonstrated by practical examples. In all cases, the subsequent processing of the plasma-treated textiles took place after a period of at least a few hours after the DBD treatment. This will indicate the practicability of the process because it has to be tolerant to short storage times and typical delays in the manufacturing process. As a result of this waiting period, it is assumed that at the time of processing of the plasma-treated textiles, short-lived reactive functions such as unsaturated radicals no longer exist on the fibre, but polar oxygen-containing groups are the basis for the observed adhesion effects. For the practical tests, mainly polyurethane coatings were applied to different textile substrates.

Owing to their elastic and film-forming properties, polyurethanes are an important material class for coating and bonding of textiles. Although solvent-based polyurethanes are still processed in a few cases, in most cases water-based polyurethane dispersions are used. Improved wetting of the textiles by these dispersions should result in improved adhesion. For example, DBD-treated and non-treated polyester fabrics had been poly-urethane coated using a water-based polyurethane dispersion. After drying and curing, the delamination force, measured by peel tests, was found to be 50–200% higher on the plasma pre-treated fabric than on the non-activated fabric, depending on the dose. As a second example, two polyester fabrics had been bonded to each other by a direct-processed polyurethane film. Peel tests proved that the adhesion force could be enhanced many times over a non-activated sample by plasma pre-treatment, depending on plasma dose.

A DBD pre-treatment improved the adhesion of a transfer-coating-applied polyurethane coating significantly, even on glass fabrics. For both, warp and weft directions the delamination forces of the coating increased by a few hundred percent (Fig. 7.3). In addition to the better adhesion performance, the plasma pre-treated material was much more resistant to hydrolysis than the non-activated fabric and even after 90

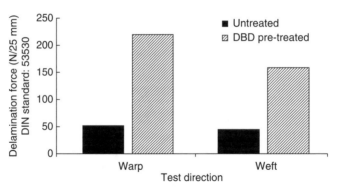

7.3 Enhanced adhesion performance of coatings after plasma activation of the fabric.

days of immersion in water only a few signs of delamination could be found.

A DBD pre-treatment is not only effective with regard to water-based coatings and glues. For example, copolyamide hot-melts were successfully coated on a polyester multifilament fabric at a critical low processing temperature with usually poor adhesion performance. The adhesion force of the hot-melt coating was measured by a peel test according to DIN 53357. As another example, to improve the adhesion of a seam sealing tape the top charmeuse layer of a three-layer laminate for functional clothing was DBD pre-treated. The adhesion force of the tape was measured by a peel test according to DIN 53530. In both cases, the hot-melt and the tape, the adhesion force was found to be doubled compared with non-pre-treated fabrics. Other applications of plasma activation of textiles are based on the enhanced water pick-up and intensified and improved dying results achieved with pre-treated textiles.

7.3.2 Stability of plasma-activated surfaces

It is a common observation that the chemical and physical-activity of plasma-activated polymeric surfaces diminishes with time. Although the chemical bonds that are generated by the plasma treatment are stable, the effect of a raised surface energy or chemical reactivity is not endlessly durable. This ageing of the surfaces is of specific relevance for polymers with low glass transition temperatures. Materials such as polypropylene or silicones have very high polymer chain mobilities. They age rather fast and the effects of a plasma activation are often lost after a day of storage under normal conditions.[19–21]

The explanation for this phenomenon is seen in the fact that the high surface energy of a plasma-treated polymer is an energetically unfavoured metastable state. A reorientation of polar molecules to each other and a migration of non-polar groups to the surface results in an energetically favoured state where the polar groups are buried within the bulk and therefore ineffective. In particular, if the polymer has large amorphous fractions at the surface or the glass transition temperature is low, such reorientations can take place soon after the surface modification. As a result of this mobility-based effect, high diffusion rates and storage at elevated temperature will accelerate this ageing process.

Long-term X-ray photoelectron spectroscopy (XPS) analysis of air-plasma-treated surfaces demonstrated the loss of even chemically stable functional plasma-induced groups with time. The oxygen/carbon ratio at the surface slowly decreases without the presence of potential reaction partners and under normal conditions that exclude a chemical reduction of the oxygen-containing groups.[22]

The ageing of plasma-induced surface activation effects is strongly dependent on the material. It can be diminished by cross-linking the activated surface which can result from the plasma treatment as well. If the surface becomes cross-linked by breaking and recombining of chemical bonds the chain mobility gets decreased and the stability of the surface activation effect increases. Very stable plasma activation effects can on the other hand be observed with polar or crystalline polymers such as cotton or polyester (PET). Air-DBD-treated cotton warp yarns remained hydrophilic over at least three weeks and activated polyester felts kept their plasm-induced hydrophilicity over two years of storage.

In addition to the influence of the pure fibre polymer the durability of a plasma treatment also depends on usage conditions and other chemical parameters. Because the plasma modification only modifies a very thin monomolecular top layer the effect is always affected by friction and abrasion. Polymeric additives might influence the effect and stability of plasma treatments, as will spin finishings, sizing agents and other functional finishing agents. If they cover the whole fibre, only the finishing agents will be modified by the plasma treatment.

7.3.3 Deposition of nano-layers by gas polymerisation

Plasma treatments of textiles are not limited to relatively unspecific activations of the textile surface. The gas within the discharge controls the discharge behaviour as well as the formation of reactive species. If reactive gases that are able to polymerise after excitation in the discharge are fed into the plasma zone, thin coatings can be deposited on the substrate from a non-equilibrium plasma by, for example, radical polymerisation.

The morphology of the coating and the deposition rate are controlled by the reaction mechanism and reaction rate. Readily polymerising systems form particles within the discharge that can be deposited on the substrate.

However, for textile treatments, these dust-forming plasmas are less relevant because the particles typically lay loosely on the surface. Of much greater interest are polymerisation processes that predominantly take place at the substrate surface by forming a permanent functional surface coating. For example, highly cross-linked layers with varying surface energies, depending on the chemical composition, can be deposited from non-equivalent plasmas.

- C–H–O-containing layers with compositions $C_{0.6-0.75}O_{0.01-0.05}H_{0.2-0.35}$ formed in carbon hydrogen atmospheres generate surfaces with surface energies between 45 and 56 mN/m.[23]
- From organosilicons-containing atmospheres, SiO_2-like coatings with surface energies up to 66 mN/m, as well as low-energy silicone-like layers with surface energies of 30 mN/m can be created.[22]
- Coatings with very low surface energies can be generated with gaseous fluorocarbon precursors. Polytetrafluoroethylene (PTFE)-like surfaces are very hydrophobic because their surface energy lies below 20 mN/m. The bi-radical and reactive fragment CF_2 is responsible for the plasma polymerisation of fluorocarbons and the efficient formation of PTFE-like layers. This polymerisable molecule results from the discharge-initiated fragmentation of fluorocarbons. Together with other fragments like CF_3 or CF, a highly cross-linked polymeric coating can be deposited via surface reactions on the substrate surface.[9,22]

For plasma-coating processes the plant design needs to be more complex than that of the relatively simply constructed surface activation modules. Compared with the one-step surface activation process, plasma polymerisation is a slow multi-step process. The polymeric chain growth is initiated by fragmentation in the plasma and completed by polymer and network formation at the surface of the substrate. Therefore, to create a functional polymeric surface layer of adequate thickness from gaseous monomers, a sufficient residence time of the precursors and the substrate within the plasma zone is needed in order to form the activated monomers, transport them to the surface and carry out the surface polymerisation. In addition to plasma residence time and plasma energy dose, the layer-forming plasma processes also depend on reaction processes without or outside the plasma discharge. The plasma-induced polymerisation process can carry on undisturbed without or outside the discharge, making a pulsed plasma source an advantage for plasma polymerising processes.[24]

Open plasma devices, as typically used for continuous surface activation of webs, are therefore not suited for plasma polymerisation. In order to define and control the necessary gas composition in the reaction zone accurately and even be able to run the plasma process under exclusion of air components (oxygen or nitrogen free) an encapsulation is necessary. In

(a)

(b)

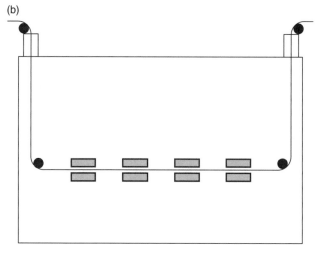

7.4 Encapsulated plasma unit for 1 m textile width , (a) photograph, (b) schematic.

Fig. 7.4 a device for plasma coating of textile webs is shown. The prototype has been designed to fulfil two sets of requirements: the demands of the academic and the industrial developer for efficient process development, and those of the textile industry for industrial use. Features worth noting are easy loading and unloading, integration into an in-line coating line, treatment widths of 1 m and upscale possibilities, various modules for monitoring and varying the gas balance, and separate control of all electrodes. A simply designed inlet and outlet module for the textile, acting as a gas lock to remove air fractions from the textile before entering the plasma chamber, even allows continuous in-line roll-to-roll processing of textiles.

Table 7.2 Plasma treatment versus traditional textile wet processing with regard to environmental and energetic aspects[2]

	Plasma process	Traditional wet chemistry
Medium	No wet chemistry involved. Treatment by excited gas phase.	Water-based
Energy	Electricity (only free electrons heated)	Heat (temperature of entire system must be raised)
Energy consumption	Limited for atmospheric pressure plasma. Greater for vacuum plasma.	Evaporation of water results in high energy consumption
Temperature	Room temperature	High temperature
Water consumption	Negligible	High

7.4   Environmental benefits of plasma technology

As described in the previous sections, plasma technology offers various opportunities for reducing the energy consumption and environmental impact of textile finishing processes. Table 7.2 compares plasma technology and a traditional wet process for textile finishing from energetic and environmental viewpoints. The table indicates that, compared with plasma technology, the traditional wet finishing processes require higher quantities of energy. In order to emphasise this even more, the following example shows not only the economic advantages but also the ecological advantages of a plasma treatment designed to impart oil repellence to a textile. We compared the energy consumption required for the generation of an oil-repellent functional layer on a polyester textile (monofilament woven fabric) by fluorocarbons. The traditional wet chemistry process (application in foulard, drying in a stentering frame) was compared with a plasma polymerisation at atmospheric pressure under continuous in-line conditions, as described in Section 7.3.3.

Fluorocarbon finishing using traditional wet chemistry (foulard, drying)

The calculation was carried out in consultation with the textile machinery manufacturer Brueckner (Leonberg, Germany) for a stentering frame with 9 dry zones, a 3 m length and a working width of 1.8 m.

Fabric velocity:	35.3 m/min
Fabric throughput:	3812.4 m²/h
Expected weight:	165.0 g/m²

From this follows the fabric throughput of   629.0 kg/h

Energy consumption of the heater:	730.0 kW
Energy consumption of the control system (air circulation, exhaust air, transport, etc.):	58.0 kW
Total energy consumption (heater + control system):	788.0 kW
Energy consumption per m² fabric:	0.21 kW/h/m²
Energy consumption per kg fabric:	1.25 kW/h/kg

In this case, the energy consumption for the application of a fluorocarbon finishing by wet processing is about 1.25 kW/h/kg fabric.

Oil-repellent fluorocarbon surface using atmospheric plasma polymerisation

The calculation is based on experiments at ITV Denkendorf with an Ahlbrandt System (Lauterbach, Germany) plant.

Mass per unit area:	110 g/m²
Fibre diameter:	100 μm
Polymer density:	1.38 g/cm³
From this follows the following fibre surface:	3.2 m²/m²
Corona dose:	13 W/h/m²
Corona dose relating to fibre surface:	4.06 W/h/m²
Energy consumption per kg fabric:	0.12 kW/h/kg

The energy consumption for the application of an oil-repellent layer (oil-repellent Grade 7 according to the AATCC) lies thus at 0.12 kW/h/kg fabric.

This simplified calculation makes clear that at least in some cases the energy consumption and thus the environmental impact can be reduced to a great extent by an alternative plasma treatment. In some cases the increase of the adhesion between the textile and a layer such as a coating or a glue as a result of plasma pre-treatment can avoid a pre-process such as applying a chemical primer or a solvent process. Regarding the water pollution, it has to be taken into account that the fluorocarbon resins of the wet processes are all characterised by poor biodegradability and bio-eliminability. Therefore, using the plasma technology, substantial environmental benefits in terms of reduced discharge can be achieved.

7.5   Future trends

Plasma systems that allow the surface treatment of substrates independent of substrate porosity, area size and topography will be of specific interest for the processing of textiles. Such systems are independent of the position of a counter electrode. In conventional systems, for example, the maximum substrate thickness is limited by the maximum discharge gap between electrode and counter electrode. Systems with so-called 'indirect' or 'remote' plasma sources will be able to penetrate into thick textile substrates and homogeneously modify three-dimensional structures. A special challenge is the design of an adequate plasma system for the typical wide working width of textiles.

7.5.1   Developments in discharge technologies

Low-pressure plasma treatments for surface functionalisation and coating have a broad field of application in the processing of semi-conductors, optical devices or medicine technologies and machinery manufacturing. Non-thermal atmospheric pressure plasma technologies have lead a shadowy existence beside the corona treatment of polymeric surfaces. However, in recent years it has been shown that a number of non-thermal processes, even chemical vapour deposition (CVD) processes, which were limited to low pressure plasmas could be successfully transferred to the atmospheric pressure range. The starting point was technical developments of individual plasma sources such as DBD, arc-jet sources, microwave sources or hybrid plasma sources. The new plasma sources are generally usable for surface modification, substrate pre-treatment, surface cleaning, etching and plasma chemical coating.

Compared with plasma spraying and other thermal plasma treatment techniques the advantage of non-thermal atmospheric pressure plasma techniques is the ability to coat large surfaces of heat-sensitive substrates like textiles by nano-scaled functional layers. The treatment width of such plasma systems varies widely, starting from a few millimetres (coplanar discharges, jets). These plasma sources can be scaled up to larger treatment widths by arranging arrays of single sources. Surface patterning can be realised by structuring dielectrics via print-like methods, generating hollow discharges.

Two systems can be differentiated in the field of atmospheric pressure plasma deposition processes.

1   Precursors, either gaseous or as aerosols, are applied directly into the discharge gap where the plasma is generated. Plasma chemical deposition of fluorocarbons and silicon oxide layers have been realised on a pilot plant scale.

2   By means of remote processes the plasma generation and the activation zone of the precursors are separated. Deposition rates of 100 nm/s are possible.[25]

7.5.2   Developments in computer simulation

The ongoing expansion of fundamental knowledge about the generation and effect of plasmas in combination with further evolving software and hardware for numerical simulations are the basis for simulating the chemical/physical processes of plasma treatments. Numerical simulations can support the optimised design of plasma devices with regard to gas flow, electrical modules and effect.

7.6   Source of further information and advice

Plasma Technologies for Textiles (Woodhead Publishing, 2007), edited by R. Shishoo (Shishoo Consulting AB, Sweden), gives an excellent overview of this topic for readers who wish to obtain more detailed information.

7.7   References

1   STEGMAIER, T., VON ARNIM, V., DINKELMANN, A., SCHNEIDER, P., Behandlung von laufenden Textilbahnen im Atmosphärendruckplasma, *Technische Textilien*, **4**, 2001, 147–149. *Melliand Textilberichte*, **6**, 2004, 476–481.

2   SHISHOO, R. (ed.), *Plasma Technologies for Textiles*, Woodhead Publishing, Cambridge, UK, 2007.

3   SCHÖNBERGER, H., SCHÄFER, T., *Beste verfügbare Techniken in Anlagen der Textilindustrie*, Umweltbundesamt, Berlin, 2002.

4   INAGAKI, N., *Plasma Surface Modification and Plasma Polymerization*, Technomic Publishing, Lancester, Pennsylvania, 1996.

5   BAHNERS, T., BEST, W., EHRLER, P., KIRAY, Y., LUNK, A., WEBER, N., Barriereentladung bei Atmosphärendruck als In-Line Veredlungswerkzeug zur Hydrophilierung über Hydro- bis Oleophobierung, Proceedings of the 26th Aachen Textiles Conference, November 1999, pp. 7–15, Aachener Textiltagung, 1999.

6   D'AGOSTINO, R., FAVIA, P., OEHR, C., WERTHEIMER, M.R., Low-temperature plasma processing of materials: past, present, and future, *Plasma Processes and Polymers*, **2**(1), 2005, 7–15.

7   JANZEN, G., *Plasmatechnik: Grundlagen, Anwendungen, Diagnostik*, Hüthig Verlag, Heidelberg, Germany, 1992.

8   AKISHEV, YU., GRUSHIN, M., KOCHETOV, I., KARAL'NIK, V., NAPARTOVICH, A., TRUSHKIN, N., Negative corona, glow and spark discharges in ambient air and transitions between them, *Plasma Sources Science and Technology*, **14**, 2005, 18–25.

9   HAHNE, E.A., KNOPF, F., Device for humidifying a material web, United States Patent 6,827,781, 2004.

10 SUMOREK, A., PIETRZYK, W., The influence of electric field on the energy consumption of convective drying processes, *Agricultural Engineering International: the CIGR Journal of Scientific Research and Development*, Manuscript FP 00 017, Vol. III, 2001.

11 STEGMAIER, T., ABELE, H., PLANCK, H., Energieeinsparung und Reduzierung der Trocknerlänge bei IR- und Konvektionstrocknern, *Melliand Textilberichte*, **5**, 2005, 348–353.

12 VINOGRADOV, G.K., TIMOKHOV, A.G., ZIMENOK, A.I., Comprehensive study of reaction kinetics of simplest unsaturated fluorocarbons in non-equilibrium plasma. Mechanisms of gas phase transformations. In Proceedings of 12th International Symposium on *Plasma Chemistry* (ISPC), Vol. IV, 21–25 August 2000, Minneapolis, Minnesota, USA, pp. 1867–1872.

13 KOGELSCHATZ, U., dielectric-barrier discharges: their history, discharge physics and industrial applications, *Plasma Chemistry and Plasma Processing*, **23**(1), 2003, 1–46.

14 ELIASSON, B., KOGELSCHATZ, U., Modeling and applications of silent discharge plasmas, *IEEE Transactions on Plasma Science*, **19**(2), 1991, 309–323.

15 ELIASSON, B., KOGELSCHATZ, U., Nonequilibrium volume plasma chemical processing, *IEEE Transactions on Plasma Science*, **19**(6), 1991, 1063–1077.

16 MENGES, G., MICHAELI, W., LUDWIG, R., SCHOLL, K., Koronabehandlung von Polypropylenfolien, *Kunststoffe*, **80**, 1990, 11.

17 HERBERT, P.A.F., BOURDIN, E., New generation atmospheric pressure plasma technology for industrial on-line processing, *Journal of Coated Fabrics*, **28**, 1999, 170–182.

18 TESCHKE, M., KORZEC, D., FINANTU-DINU, E.G., ENGEMANN, J., Plasma cleaning of 'endless' substrates by use of a tandem dielectric barrier discharge, *Surface and Coatings Technology*, **200**, 2005, 690–694.

19 KIM, J., CHAUDHURY, M.K., OWEN, M.J., ORBECK, T., The mechanisms of hydrophobic recovery of polydimethylsiloxane elastomers exposed to partial electrical discharges, *Journal of Colloid and Interface Science*, **244**, 2001, 200–207.

20 NOVÁK, I., FLORIÁN, E., Influence of ageing on adhesive properties of polypropylene modified by discharge plasma, *Polymer International*, **50**(1), 2001, 49–52.

21 DORAI, R., KUSHNER, M.J., A model for plasma modification of polypropylene using atmospheric pressure discharges, *Journal of Physics D: Applied Physics*, **36**, 2003, 666–685.

22 ARPAGAUS, C., ROSSI, A., VON ROHR, R., Short-time plasma surface modification of HDPE powder in a Plasma Downer Reactor – Process, wettability improvement and ageing effects, *Applied Surface Science*, **252**, 2005, 1581–1595.

23 KLAGES, C.P., EICHLER, M., PENACHE, C, THOMAS, M., *Plasmagestützte Oberflächentechnik ohne Vakuum*, Geesthachter Innovations und Technologie Centrum (GITZ), Geesthacht, Germany, 2005.

24 VINOGRADOV, P., SHAKHATRE, M., LUNK, A., Diagnostics of the afterglow of a Dielectric Barrier Discharge running in Ar/CxFy mixtures, *Verhandl. DPG (VI)*, **40**, 2005, p. P14.3.

25 DANI, I., Atmosphärendruck – PECVD zur Großflächenbeschichtung, Workshop 'Oberflächentechnologie mit Plasma- und Ionenstrahlprozessen', Mühlleithen, March 2004.

Understanding and improving textile recycling: a systems perspective

J. M. HAWLEY, Kansas State University, USA

Abstract: Presented from a unified micro–macro or systems perspective, the textile recycling process impacts many entities and contributes significantly to the social responsibility of contemporary consumer culture. Because textiles are nearly 100% recyclable, attention needs to be given to the recycling and recyclability of textiles so that less ends up in landfills. By raising consciousness concerning ecological issues, channels for disposal, and environmentally conscious business ethics, steps can be taken toward a more sustainable use and disposal of post-consumer textiles.

Key words: textile recycling, post-consumer textile waste, recycled clothing, over-consumption, second-hand clothing.

8.1   Introduction

It is widely acknowledged that environmental consciousness is on the rise. In the 1960s, when environmentalism was in its infancy, the focus was on solutions that would reduce manufacturing waste and emissions. Today, the focus has moved to a more conscious, holistic approach or *systems* approach that takes into account all constituents and products of the product life-cycle. Edwards (2005) refers to this profound paradigm shift as the sustainability revolution. This global shift from environmentalism to sustained and system-encompassing consciousness impacts a broad spectrum of interests and fundamental values. As Edwards points out (p. 2), it shapes 'everything from the places we live and work to … the endeavors we pursue as individuals and communities'.

As we consider the case of textile and apparel recycling and the processes used in their life-cycle – from fiber to consumption to disposal – it becomes apparent that the process impacts many entities and contributes significantly, in a broader sense, to the social responsibility of contemporary culture. By recycling, companies can realize larger profits because they avoid charges associated with dumping in landfills, while at the same time contributing to goodwill associated with environmentalism, employment for marginally employable workers, donations to charities and disaster relief, and the movement of used clothing to areas of the world where clothing is needed.

Because textiles are nearly 100% recyclable, nothing in the textile and apparel industry should be wasted. Japanese buyers can be seen at recycling companies searching through piles of old clothes for finds such as Harley Davidson jackets and T-shirts, Grateful Dead concert T-shirts, Nanette Lepore dresses, or vintage purses from the 1940s, 1950s and 1960s. Men's neckties and suits go to India as young Indians seek jobs in Western-owned off-shore companies. Used sturdy shoes are shipped to Uganda, cotton sleepwear to Bolivia, cotton pants and shirts to Haiti. Levi's are coveted all over the world, and worn-out promotional T-shirts are made into shoddy or wiping rags. In 2003, it was projected there would be a 3–5% increase in world fiber consumption, which equals 2 million tons (1.8 million metric tonnes) per year (Bharat Textile, 2004; Estur and Becerra, 2003). This presents a double-edged sword in that while it stimulates the economy (projected to add 10–20 new factories to meet the world market demand), it also gives rise to the increased problem of apparel and textile disposal.

This chapter begins with an overview of systems theory then presents a model that illustrates the textiles recycling processes, particularly as it pertains to post-consumer apparel. A micro–macro model using social systems theory is then presented. Finally, a synthesis of how systems theory provides a useful tool for projecting future trends for the textile and apparel recycling process is presented. It is important to note that this work is based primarily on the processes as they are in the United States. The research is based on nearly 10 years of qualitative data collection on, primarily, apparel and other fashion products consumed throughout the United States and the rest of the world.

8.2   Systems theory

Systems theory provides a useful theoretical framework for understanding the textile recycling process. Taking a holistic view (Olsen and Haslett, 2002), systemic thinking helps to explain the connectedness, interdependencies, feedback processes, and integration of the textile recycling system. General systems theory (GST) was first presented in the 1950s by Bertalanffy. His intent was to provide a superstructure that could be applied to various scientific fields. Bertalanffy's work stimulated many theorists to apply systems theories to their own field in one form or another. As a result, GST has been applied to economics, biology, organizations, and engineering, to name a few. It has been only recently that systems theory has been applied to complex social (human) systems. Mayrhofer (2004, p. 1) pointed out that humans were an 'essential element in the system's environment'.

GST, as it applies to social systems, provides a way of better understanding human and social units that are not only distinct, but also interrelated.

Social systems theory offers a unified framework for the analysis of social reality at a higher level. The theory allows for the understanding of individual behavior in the context of the environment and situational factors. For example, rather than simply acknowledging the importance of environmental factors, social systems theory makes it clear that many things – such as economics, legal/political constraints, technological advancement, cultural perspectives, competitive environment, and infrastructure – must be considered. In the case of individual behavior of textile recycling, environmental factors such as local solid waste policies, convenience of local charity shops, and local attitudes toward recycling can all affect individual recycling behavior.

In this chapter, social systems means systems constituted mainly by human beings, ranging from the micro unit such as individuals, families, and friends, to macro groups such as family-owned companies, large corporations, governments, and entire cultures. The interrelationship between human behaviors and decisions, environmental concerns, policies, technology, infrastructure, and competition are considered.

8.3   Understanding the textile and apparel recycling process

Western lifestyle is a significant contributor to landfill waste. Not only are products consumed at a high level, but Western goods are often over-packaged, contributing even more to the waste stream. As landfill capacity continues to be scarce, the costs of dumping will continue to rise. These rising costs are of concern for businesses as they seek ways to reduce their overhead costs.

8.3.1 The problem of over-consumption

The notion of fashion itself compounds the problem of over-consumption. The very definition of fashion fuels the momentum for change, which creates demand for ongoing replacement of products with something that is new and fresh. In addition, fashion has reached its tentacles beyond apparel to the home furnishings industry. Fashionable goods contribute to consumption at a higher level than need, but without the notion of fashion, the textile, apparel, and home furnishings industries would realize even more vulnerability in an environment that is already extremely competitive. In the not-to-distant past, the fashion industry presented four to five new lines per year. Today, however, the globalized, highly competitive fashion industry has reduced prices to a level so low that clothing has become almost a disposable commodity. Often referred to as 'fast-fashion', affordable apparel, aimed at the youth market, contributes to the insatiable consumer culture, leaving a 'pollutant footprint' (Claudio, 2007) along each step of the apparel pipeline and mounds of used clothing in its wake. As consumers continue to buy, waste will continue to be created, further complicating the problem of what to do with discarded packaging, apparel, and home textile products.

To further compound the problem, clothes in today's marketplace are different from those of several decades ago, not only in design but also in fiber content. After synthetic fibers came on to the market in the twentieth century, textile recycling became more complex for two distinct reasons: (a) fiber strength increased making it more difficult to shred or 'open' the fibers, and (b) fiber blends made it more difficult to purify the sorting process. Nonetheless, the recycling industry must cope with everything the fashion industry generates.

8.3.2   Textile recycling statistics

The textile and apparel recycling effort is concerned with recycling, recyclability, and source reduction of both pre-consumer and post-consumer waste. According to the Environmental Protection Agency, the per capita daily disposal rate of solid waste in the United States is approximately 4.3 (1.95 kg), up from 2.7 pounds (1.25 kg) in 1960 (US Environmental Protection Agency, 2003). Although textiles seldom earn a category of their own in solid waste management data, the Fiber Economics Bureau (http://www. fibersource.com/feb/feb1.htm) reports that the per capita consumption of fiber in the United States is 83.9 pounds (38 kg) with over 40 pounds (18 kg) per capita being discarded per year. A recent report from the Chinese Chamber of Commerce shows that China's fiber consumption has increased by 5 kg (11 pounds) in the past 5 years to an annual consumption rate of 13 kg (22.6 pounds). As consumers in China and other countries gain discretionary income, global fiber consumption will continue to rise, further increasing the complexity of the disposal problem (*People's Daily Online*, 2006). This report points out that China will continue to have the fastest growing fiber consumption market for the next 10 years (http://english. peopledaily.com.cn/200602/06/eng20060206_240556.html).

It is well established that recycling is economically beneficial, yet much of the discarded clothing and textile waste in the United States fails to reach the recycling pipeline. The US textile recycling industry annually diverts approximately 10 (4.5 kg) per capita or 2.5 billion pounds (1.1 million tonnes) of post-consumer waste from the waste stream. These amounts represent only about 30% of the total post-consumer annual textile waste

(B. Brill, personal communication, 2007). As an example, although there are several well-established uses for denim waste, the denim industry still deposits more than 70 million pounds (31 750 tonnes) of scrap denim in US landfills annually (McCurry, 1996, p. 84). Furthermore, analysis of municipal solid waste indicates that unrecovered textile waste contributes to approximately 4.5% of the US landfills (Hammer, 1993).

According to the US Environmental Protection Agency (2003), this equates to 4 million tons (3.6 million tonnes) of textiles going to the landfills each year. While this may not seem like a large amount, it is when one considers that nearly 100% of the post-consumer textile waste is recyclable. Cognizant of this, the textile industry's current efforts, enthused by the American Textile Manufacturer's E3 – Encouraging Environmental Excellence – program, focus on trying to increase recoverable textile waste that would otherwise end up in landfills.

8.3.3 The textile recycling industry

The textile recycling industry is one of the oldest and most established recycling industries in the world; yet few people understand the industry, its myriad players, or reclaimed textile products in general. Throughout the world, used textile and apparel products are salvaged as reclaimed textiles and put to new uses. This 'hidden' industry (Divita, 1996) consists of more than 500 businesses that are able to divert over 1.25 million tons (1.1 million tonnes) of post-consumer textile waste annually (Council for Textile Recycling, http://textilerecycle.org). Furthermore, the textile recycling industry is able to process 93% of the waste without the production of any new hazardous waste or harmful by-products. The Council for Textile Recycling has indicated that virtually all after-use textile products can be reclaimed for a variety of markets that are already established (Ed Stubin, personal communication, July 17, 2001). Even so, the textile recycling industry continues to search for new viable value-added products made from used textile fiber.

Textile recycling material can be classified as either pre-consumer or post-consumer waste; textile recycling removes this waste from the waste stream and recycles it back into the market (both industrial and end-consumer). Pre-consumer waste consists of by-product materials from the textile, fiber, and cotton industries that are re-manufactured for the automotive, aeronautic, home building, furniture, mattress, coarse yarn, home furnishings, paper, apparel, and other industries. Post-consumer waste is defined as any type of garment or household article made from manufactured textiles that the owner no longer needs and decides to discard. These articles are discarded either because they are worn out, damaged, out-grown, or have gone out of fashion. They are sometimes given to charities or passed on to friends and family, but additionally they are deposited in the trash and end up in the municipal landfills.

8.4   Textile recycling companies

The textile recycling industry has a myriad of players that include consumers, policy makers, solid-waste managers, not-for-profit agencies, and for-profit retail businesses (Hawley, 2000). Textile sorting companies, known as 'rag graders', acquire, sort, process, export, and market pre- and post-consumer textile products for various markets. The primary focus in this chapter is on post-consumer apparel and textiles. Most rag-sorting companies are small, family-owned businesses that have been in operation for several generations (Allebach, 1993; S. Shapiro & Sons, 1961). However, start-up entrepreneurs have begun new textile recycling businesses because they perceive it as a low-cost, easily accessible form of entrepreneurship. What many of the start-ups fail to realize, however, is that this business is highly dependent on global contacts that take years of relationship building in order to develop global accounts to sell their sorted goods. As one informant told me, 'I have spent as much as a year at a time away from my family while I developed and nurtured markets across Africa, Asia, and Latin America. Now that these business contacts have been established, I can pass the contacts on to my son who will be taking over the business soon.' An informant from a different company said, 'Establishing contacts in Africa is particularly difficult. But once those contacts are made, the bond between us is very strong and full of respect.' And an international broker from Europe stated: 'Buying and selling in Africa is an underground business. The used-textile brokers in Africa are substantially wealthier than many of the citizens who are the consumers of the used clothing. They must hide their wealth in order to maintain credibility among the citizens. One of our buyers has a beautiful burled-wood and gilded office that is located underground. When we go to Africa to do business we have to be secretly escorted underground to conduct business!'

Consumers often take apparel that is worn, out of fashion, or the wrong size to charity organizations such as Goodwill or the Salvation Army. Charity agencies then sort the clothes, choose items for the retail store, and the 'leftovers' or excess is sold as 'industrial rags' to rag graders for cents on the pound. The price per pound of used clothing is dependent on current market value, but often ranges from 3–6 cents per pound. At regularly scheduled times, trucks are dispatched to pick up the merchandise.

Textile recycling companies are often located in large metropolitan areas because transportation costs must be kept to a minimum. It has been found that transportation and sorting costs can be the decisive criteria for profitable business (Nousiainen and Talvenmaa-Kuusela, 1994). Clothing excess from the charity shops is then transported to the recycling facilities, emptied on to a sorting deck, and the sorting process begins. Depending on the current economic climate (primarily associated with materials availability, current value-added markets, and the current commodity price for used textiles), for-profit rag-sorting companies realize both success and hardship. As the volume of used clothing continues to rise based on over-consumption and fast-fashion, the prices for used textiles have fallen.

Although the primary goal for these small businesses is to earn profits, the business owners are also very committed to environmental principles and values and take pride in their contribution to waste reduction. As one informant offered:

> 'This is not a particularly lucrative, nor glamorous, business. The profit margin is so small that when the commodity prices increase, policy makers put up barriers, or the market becomes too saturated, it is very difficult to make a living. But, we in the textile recycling industry also take great pride in the role we play in improving the environment.'

These business owners continue to seek, develop, and nurture markets for reclaimed textiles to not only increase their company profits, but also to increase the amount of pre- and post-consumer textile goods diverted from the landfills. In recent years, companies are searching for value-added alternatives beyond second-hand clothing exports to developing nations. Many of the textile recycling companies in the United States are third or fourth generation. But as the competitive nature of the business has increased, and profit margins are threatened, the younger generations have opted for careers different from their parents. The result has been the recent closing of several textile recycling companies (personal communication with informant).

Many markets exist for used textiles and apparel. Sorting companies have had to evolve with the markets and remain sensitive to its requirements, whatever they may be. A recent discussion at an annual meeting of the members of SMART (Secondary Materials and Recycled Textiles) Association focused on the need for the textile sorting industry to consider ISO 9000 certification. *Daily News Record* reported that the International Organization for Standardization (ISO) agreed to 'craft norms for second-hand, defective or recalled [apparel] products ... to benefit both suppliers and consumers and help close a gap in global commerce' (B. Brill, personal communication, April 13, 2004). The report went on to say that 'second-hand clothing shouldn't mean inferior or shoddy.' Therefore, the intent is to develop ISO certification that would facilitate quality assurance in used textiles. Many US recyclers recognize the importance of this, especially in light of competition from their European counterparts who have already adopted ISO classification.

8.5   The sorting process

Crude sorts include the removal of heavy items such as coats and blankets, then the sorting, for example, of trousers from blouses from dresses. As the process advances, the sorts get more and more refined. For example, once all trousers are picked, they are further sorted based on women's or men's, fabric (e.g. woolens go to cooler climates, while cottons and linens go to hot climates), condition (e.g. tears, missing buttons, and discoloration), and brand/quality. Certain brands and styles (e.g. Levi's, Tommy Hilfiger, and Harley Davidson or Boy Scout uniforms and bowling shirts from the 1950s) are sorted because they are considered 'diamonds' based on the premium prices they bring in certain markets. (Interviews with several informants revealed that the special 'finds' in the sorting process are often referred to as 'diamonds'.) As the recycled goods are sorted, they are also graded to meet specific markets. It is not uncommon for a fully integrated rag sorter to have over 400 grades that are being sorted at any given time (P. Stubin, personal communication, March, 2007). It is often the quality of the grading process that establishes a competitive advantage of one rag sorter over another. One of the largest US sorting houses is in Texas where they sort a semitrailer load of post-consumer textiles per day. This adds up to over 10 million pounds (4500 tonnes) per year (personal communication with business owner, February 12, 2000).

Most rag sorters have a division of labor whereby the newest employees are trained to do the rough sorts, i.e. sorting into categories such as heavy outerwear and bedding, from the rest of the apparel items. As expertise increases, employees are promoted to more complex sorting and fine grading. For example, Marguerite, a head sorter and supervisor with several years of experience at one of the facilities in the United States can 'tell cashmere from wool at the touch of a hand.' One textile recycling facility employs a person with a Master of Fine Arts degree to forecast fashion trends in the vintage markets. Goods that are torn or stained are separated from the wearable goods and are used for a wide variety of markets as will now be explained.

8.6   The pyramid model

The pyramid model in Fig. 8.1 represents the sorting categories of textile recycling by volume. Sorting categories include sorts that are exported to developing countries, converted to new products from open recycling or redesign, cut into wiping and polishing cloths, dumped into landfills or incineration for energy, and 'mined' as 'diamonds'. (NB 'open recycling' refers to the process of mechanically or chemically 'opening' the fabric so as to return it to a fibrous form. Mechanically this involves cutting, shred-

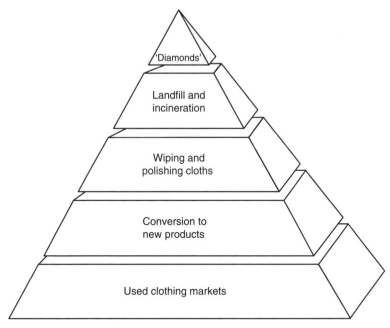

8.1 Pyramid model for textile recycling categories.

ding, carding, and processing the fabric. Chemically it involves enzymatic, thermal, glycolyse, or methanolyse methods. Once the post-consumer textiles are 'opened' they can be further processed into new products for renewed consumption.) For the most part, volume is inversely proportional to value. For example, exported second-hand clothing is the largest volume category and earns US$0.50–0.75 per pound whereas the more rare finds can bring several thousands of dollars per item, depending on their market and/or collectible value. Table 8.1 gives approximate volume to value estimations for used textiles.

8.6.1 Used clothing markets (48%)

The largest volume of goods is sorted for second-hand clothing markets, primarily for export to developing countries or for disaster relief. One informant reported that 'used apparel serves as the largest export from the United States based on volume' (personal communication with informant, May 10, 2004). The US exports to over 100 countries, and between 1989 and 2003, American exports of used clothing more than tripled to nearly 7 billion pounds (3.2 million tonnes) per year (International Trade Commission, 2004 (http://www.usitc.gov)). On many street corners throughout the developing world, racks of Western clothing are being sold (e.g. the author

Table 8.1 Volume to value estimation of used textile products

Category	Estimated total used textile goods by volume*	Estimated value*
Used clothing markets (export)	~48%	US$0.50–$0.75 per pound
Conversion to value added markets	29%	Value varies widely depending on product. Sold by weight
Wiping and polishing cloths	17%	US$0.80–$1.10 per pound
Landfill and incineration	<7%	Varies by location and/or rural/urban. Costs are by weight
'Diamonds'	1–2%	High value per item

*Values and volume vary over time depending on current markets.

has seen such racks in Taiwan, Thailand, India, Greece, and Mexico). The USA exports US$61.7 million in sales to Africa. One of its primary export sites is Uganda where a Ugandan woman can purchase a designer T-shirt for US$1.20 (Packer, 2002).

Western clothing is a highly valued commodity and perhaps serves as the only source of affordable clothing in many developing countries where levels of income are so low that food and clean water are the primary concerns. However, some have argued (Tranberg-Hansen, 2000) that the export of clothing to these nations has threatened the traditional dress for many indigenous cultures and at the same time may threaten the fledgling textile and apparel industries of those countries. While this is certainly an issue that should be taken into consideration, wearable, climate-appropriate, and affordable clothing is a valuable commodity for most of the population in less-privileged areas of the world.

One challenge for Western clothing exporters who export to low-income countries is the size differential between the larger Europeans and Americans and the low-income citizens in importing countries. In addition, the excessive surplus is usually in women's fashions, and the biggest demand is in men's clothing. In many parts of the world, cultural norms require women to wear traditional dress, whereas men are moving toward Western dress. For example, many women in India still wear the sari, but men are wearing Western-style dress.

Not all used clothing is exported to low-income countries. One informant shared that he has a new market in the United Arab Emirates (UAE), a country with one of the highest values of income per capita in the world. Used clothes in the UAE are not intended for the local population but, instead, for the immigrant labor from Bangladesh, Pakistan, and Indonesia who are in the UAE to provide servant labor for the rich; however the wages do not allow the workers enough discretionary income to purchase the designer labels that are offered in the local shops.

In recent years, rag sorters have realized that in order to stay viable, sort categories must be further refined to meet the demands of unique markets. They also work with textile engineers to engineer new products from used textiles. Available markets for used apparel flux in the marketplace. For example, wool has received renewed interest because European flammability legislation for upholstered furnishings and protective clothing has demanded higher wool content. Recycled woollens, therefore, command high prices.

Once sorted, the goods are compressed into large bales (usually 600–1000 pounds (270–450 kg)), wrapped, and warehoused until an order (often from a broker) is received. Several things are considered when sorting used clothing: the economic climate, relationships between the exporters and importers, and trade laws for used apparel.

8.6.2 Conversion to new products (29%)

Two categories of conversion to new products will be considered here. Shoddy (from knits) and mungo (from woven garments) are terms for the breakdown of fabric to fiber through cutting, shredding, carding, and other mechanical processes. The fiber is then re-engineered into value-added products. These value-added products include stuffing, automotive components, carpet underlays, building materials such as insulation and roofing felt, and blankets, to name a few. The majority of this category consists of unusable garments – garments that are stained, torn, or otherwise unusable. One informant, however, was sorting for 100% cotton sweaters because he was selling shredded cotton fiber to mix with sand for use in 'Punch-n-Kick' bags, made by one of the world's largest sporting manufacture companies. Used cotton sweaters, on the other hand, have several other value-added market options and therefore command a high market value.

A vast number of products are made from reprocessed fiber because much of this fiber is re-spun into new yarns or manufactured into woven, knitted, or nonwoven fabrications, including garment linings, household items, furniture upholstery, insulation materials, automobile sound absorption materials, automobile carpeting, and toys (Querci, personal communication, July 22, 2000). New yarn producers include, for example, those in

Prato, Italy who reduce cashmere sweaters to fiber, spin new yarns, and produce cashmere blankets for the luxury market. In 2007, Wal-Mart, the world's largest retailer, retired the blue smock that employees had worn as the company uniform. To keep the 1 million smocks from the landfill, the company partnered with Hallmark to process the old smocks into paper pulp that would provide 100 000 'thinking of you' greeting cards to active US troops, in Iraq and Afghanistan. Another portion of the vests was used to make lap blankets for veterans (S*Mart, 2007).

This process represents an economic and environmental saving of valuable fiber that would otherwise be lost to the landfill. Ironically, the most unusable and damaged of post-consumer textiles often have the highest level of specifications forced upon them by the end-use industries (e.g. building, auto, aeronautics, and defense). Another informant reports that used fibers are being utilized in the production of US currency, lining for caskets, and stuffing for pet beds.

The other category for conversion to new products is the actual re-design of used clothing. Current fashion trends are reflected by a team of young designers who use and customize second-hand clothes for a chain of specialty vintage clothing stores in the United Kingdom. Its offerings include 'cheap chic and occasional designer surprises' (Ojumo, 2002; Packer, 2002). As another example, a young designer in Dallas, Texas creates new from the old and sells wholesale to various trendy stores such as Urban Outfitters. This concept is common among boutiques with a youth-oriented target market. Sustainable apparel design using post-consumer recycled clothing has also been used increasingly in university curricula (Young et al., 2004).

8.6.3 Wiping and polishing cloths (17%)

Clothing that has seen the end of its useful life as such may be turned into wiping or polishing cloths for industrial use. T-shirts are a primary source for this category because the cotton fiber makes an absorbent rag and polishing cloth. Bags of rags can be purchased at retail stores such as in Wal-Mart's automotive department. One informant said that he sells wiper rags that he has reclaimed from the sorting process to a washing machine manufacturer for use-testing of the machines. Because of its excellent wicking and oleophilic properties, some synthetic fiber waste (particularly olefin) is cut into wipers to serve in industries where oily spills need to be cleaned up or wiped. An informant sells oleophilic wipers to the oil refining industry. Another informant reported that oil spills are being cleaned up with large 'snakes' made with a combination of used oleophilic and hydrophilic fibers.

8.6.4 Landfill and incineration for energy (<7%)

This category has two components. For some reclaimed fiber no viable value-added market has been established, so the used goods must be sent to the landfill. Rag sorters work hard to avoid this for both environmental and economic reasons, because there is a charge per pound for goods that are taken to the landfill. In the United States, testing has just begun for the process of incinerating reclaimed fiber for energy production. Although emissions tests of incinerated used fibers are above satisfactory, and the British thermal unit (BTU)[1] is very respectable, the process of feeding the boiler systems in many North American power plants is not feasible (D. Weide, personal communication, March 20, 2004). The incineration of used textiles as an alternative fuel source is more commonly done in Europe than in the United States.

8.6.5 'Diamonds' (1–2%)

In May 2001, an anonymous seller placed a pair of century-old Levi's on the Ebay auction platform. Believed to be the oldest in existence, the jeans (technically denim waist coveralls) were found buried in the mud of a mining town in Nevada, in fair to good condition. The anonymous seller opened the bid on May 17, 2001 for $25 000. One week later, after a frenetic final few hours of bidding, Levi Strauss & Co. won the bid and paid $43 532 for the 120-year-old dungarees (Lynn Downey, Levi's historian, personal communication, July 23, 2001). This is believed to be the highest price ever paid for denim jeans. Although the jeans were classified by the Levi's historian to be in 'fair to good condition', they provide a paragon for 'digging for diamonds'.

The 'diamond' category in the model accounts for approximately 1% of the total volume of goods that enter the textile recycling stream, yet this category also accounts for the largest profit center for most textile recycling companies. One informant told me that, 'When you find them [the diamonds], they are still diamonds in the rough, but once they are cleaned, pressed, and packaged, they are worth a lot in the marketplace.' Categories of 'diamonds' in the United States include couture clothing and accessories, Americana items such as Harley Davidson and Levi's, uniforms such as those worn by Boy Scouts, certain branded items, trendy vintage clothes, luxury fibers (e.g. cashmere and camel hair), couture clothing, and antique items. Many of the customers for diamonds are well-known designers or wealthy individuals. Ralph Lauren and Donna Karan both have vintage

[1] In North America, the BTU is used to describe the heat value or energy content of fuels. When used as a unit of power, BTU per hour is understood.

collections. Other 'diamond' customers include vintage shop owners who sell their 'diamonds' in retail boutiques or, increasingly, on the Internet. In recent years, dresses from American dressmakers that no longer exist have become collectible; styles from Nelly Don and Jerrell of Dallas are examples.

Many 'diamonds' have global markets, as evidenced by the fact that used goods move from country to country. For example, Americana items are highly prized in other parts of the world. When collecting data at one of the US rag dealers, five Japanese buyers were rummaging through piles of 'diamonds' to select what they wanted to buy. The owner of the business said that on many days of the month Japanese buyers are in-house making selections. Japan is the largest importer of used American 'diamonds' and has proved to be very interested in Americana items such as authentic Harley Davidson clothing, Ralph Lauren Polo clothing, and other American designers and items that represent American culture. After the September 11, 2001 terrorist tragedy, the second-hand signature red/white/blue-branded logo of Tommy Hilfiger goods realized increased interest in the global market. But perhaps the one item that has had consistent global interest is Levi's jeans, particularly certain older styles. One rag-sorter found a pair of collectible Levi's and sold them on the Paris auction block for US$18 000. Another rag-sorter sold a collectible find for US$11 000 to the Levi's corporation. One informant claimed that he has found enough collectible blue jeans to 'pay for my three kids' college education'. However, it requires a special eye and a sense of trend forecasting to be able to find 'diamonds' in the huge mine of used textiles that rag-sorters must sort.

Many owners of vintage shops are members of the National Association of Resale and Thrift shops (NART). Founded in 1984, this Chicago-based association has over 1000 members and serves thrift, resale, and consignment shops and promotes public education about the vintage shop industry. TRAID (Textiles Recycling for Aid and Development) is a charity organization that finances itself through the sale of quality second-hand clothing. As evidenced here, even though the 'diamond' category consists of only 1–2% of the volume of reclaimed goods, the profits for these 'diamonds' can make a big difference to used clothing businesses.

8.7   Textile recycling constituents

As has been shown, the textile recycling industry occurs along a pipeline with various activities and numerous constituents that function within a socio-cultural system that affects attitudes and behaviors. Each of these players functions within a cultural system that bounds his/her attitudes and behaviors. Without the inter-relatedness of the constituents, the system fails to operate to full potential – or perhaps even ceases to exist. As Elliott

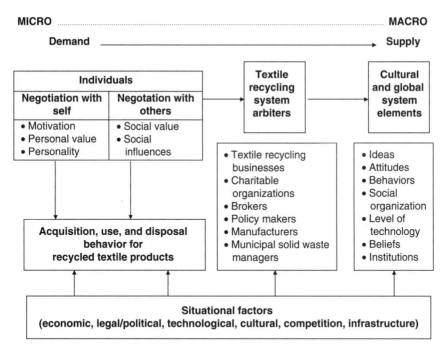

MICRO .... MACRO

Demand ————————————————————→ Supply

8.2 The textile recycling social systems model.

(1995, p. 222) pointed out, in order to make progress in the textile recycling process, 'all parties must be on common ground of understanding about how to reach an attainable goal'. Hamilton (1997, p. 167) asserted, 'if most research is grounded in paradigms that focus on only one or a few levels of analysis, then the resulting body of knowledge ... is incomplete to that extent and integration among levels is virtually impossible'. Extending Hamilton's (1997) micro/macro heuristic tool, data for this project suggest that the textile recycling system could be illustrated as in Fig. 8.2. The model provides one way of understanding the disposal of textile products. The continuum represents three positions ranging from the micro (the consumer) to the macro (the cultural system). A more detailed description of the framework follows and useful examples from the data are included.

8.7.1   The micro level: individuals

Here the focus is on individual ideology that determines recycling behavior. Scholarly research in textile recycling and its relationship to consumption of apparel remains limited and exploratory at best. Shim's (1995) exploratory work looked at the relationship between consumer environmental attitudes and their clothing disposal patterns. Kim *et al.* (1997) examined whether or not environmental concerns influenced consumers' responses to fashion advertisements.

Kim and Damhorst (1998) focused on the knowledge consumers have with regard to textiles and the environment and its relationship to consumer behavior. Although all of these works have contributed importantly to the literature, all focused on consumers – or what DeWalt and Pelto (1985) would identify as the micro component of analysis. Consumers, as individuals, have idiosyncratic determinants that affect their attitudes and behavior towards textile recycling. The findings from the study revealed that many consumers have positive attitudes toward environmentalism, yet when it comes to discarding their clothing, economic benefit often takes precedence over environmental attitudes. One consumer who had been recently divorced shared that it was difficult to part with her things because of economic uncertainty. She said that she feared that she would 'never be able to have nice things again'.

Other informants shared that they often had difficulty in parting with their things. For example, one informant said that after she cleans out her closets she 'carries the old things around in the trunk of [her] car for several weeks before [she] can emotionally handle the drop-off at Goodwill'. After another informant donated her deceased father's clothing to a charity, she sat in the car and cried as she watched the employees put the clothes in a dumpster because the charity was so overwhelmed with surplus stock. After a few minutes, she crawled into the dumpster, got her father's things, put them back in her car and drove away. This parallels the work of McCracken (1988) and Csikszentmihalyi (1981) which explains divestment of possessions and how consumers deal with the disposal of their things. McCracken (p. 87) explained that individuals make 'an attempt to erase the meaning that has been invested in the good by association'.

Although many consumers have developed home recycling practices, few if any consider the recyclability of used apparel except in terms of donation to charity organizations. Even though this is a viable disposal choice for many usable garments, there remains a limited response to the recycling of textile products (US Environmental Protection Agency, 1997 (http://www.epa.gov)). In a study done in the spring of 2008, data revealed that consumers' disposal choices are based on convenience even more than economic or environmental concerns (Heidebrecht and Hawley, 2008).

Individuals negotiate not only with themselves, but also with others in their near environment or the social world in which they function. Americans live in a throwaway, high-consumption society where an individual's worth is often measured by the clothes that they wear. This fast-fashion world is juxtaposed against public pressure to reduce consumption and behave in an environmentally friendly manner. How is it, then, that the American consumer negotiates the wearing of the latest fashions when their closets are already over-flowing with perfectly wearable merchandise? One way to justify their behavior is to donate their wearable, but slightly-out-of-fashion, clothing to charitable organizations. In this way, their appetites for fashion are satisfied and guilt is erased with benevolent acts through donations.

8.7.2 The macro level: textile recycling system arbiters

The second position in the micro–macro framework addresses the textile recycling system arbiters. These constituents include the various for-profit and not-for-profit businesses that drive the textile recycling processes. Approximately 200 companies at the primary processing level are currently recycling post-consumer textile waste. An additional 150 secondary processors, such as used-clothing exporters, wiper manufacturers, and fiber and fabric manufacturers are also part of the multifaceted industry (Council for Textile Recycling, 1997). Although these companies have historically received their inventory from charity surplus, they have recently begun to expand their base of suppliers by helping municipalities develop curbside and drop-off textile collection programs. This is in part due to supply and demand issues caused from new retail outlets being opened by charitable organizations.

Additionally, some contention has arisen from the increased competition, as both for-profit and not-for-profit entities compete for the same markets. Other arbiters include policy makers at the local, state, and federal level who are involved with setting policy and passing legislation that either supports or inhibits textile recycling. Many trade laws, for example, prohibit trade to certain countries. For example, recent negotiations between the US Department of Commerce and the Tanzanian Bureau of Standards via the US Embassy in Tanzania, are concerned with the following: (a) requirement of fumigation certificates; (b) ban on used undergarments, socks, stockings, and nightwear; (c) requirement that bales should not exceed 50 kg; (d) requirement for a health certificate to prove the country of origin is free from diseases; (e) certification of used garments; and (f) sampling of consignments (http://www.smartasn.org/news.html, retrieved March 31, 2008). Protectionists cite a list of concerns including infestation of harmful insects, chemicals, and microorganisms. The fact remains, however, that many people in developing nations, even those working in the fledgling textile and apparel industries, can not afford the clothing that is produced by those nations, particularly clothing that is manufactured for Western markets. Instead, they are thankful to be able to buy used clothing imported from developed nations. Even though trade policies prohibit the export/import of certain items, the industry still finds ways to continue trade in the global market while maintaining trade policies. As one informant revealed:

'India has developed a substantial industry of manufacturing wool blankets from used wool clothing. Trade laws between the United States and India do not allow the export of wool clothing from the United States. To meet this market demand, used wool clothing in the United States must be sent through a shredding machine that slashes the garment beyond wearable condition, yet keeping it in enough wholeness so that it can be more easily baled and shipped to India. Thus, the clothing is no longer 'clothing', but is, instead, 'used fiber'. Indian manufacturers process the fiber to a more fibrous state, into new yarns, and then into the manufacture of blankets.'

8.7.3 The macro level: cultural system arbiters

This level is concerned with the material world, social relationships, and shared cognitive repertoire, all of which embody clues that are critical to how the recycling industry plays out in our cultural system. Culture becomes such an integral part of human existence that it is the human environment, often making it difficult for attitudes and beliefs to change. Much of the concern here is the ideas, attitudes, behaviors, social organization, level of technology, belief systems, and institutional commitments of a society and includes ideas and attitudes put forth by government entities and environmental enthusiasts that contribute significantly to the attitudes and behaviors of textile recycling. At a time of record market demand for recycled textile products, a perplexing problem exists in that there is insufficient supply of raw textile product (K. Stewart, personal communication, April 27, 2009). This can be attributed to the cultural ethos that impacts consumer behavior, municipal solid waste management programs, or charitable organizations. For example, consumers have been encouraged by their social, economic, or political environment to recycle glass, aluminium, and plastics; however, textiles are seldom a category considered by municipal recycling programs. Furthermore, in Shim's (1995) exploratory study, results showed that even though consumers might have strong environmental attitudes and waste recycling behavior, this would not be an indicator for textile and apparel recycling.

In many states, waste reduction policies and programs are being implemented that facilitate voluntary pollution reduction. Although national statistics reveal that textiles constitute a significant portion (literature sources range from 4–8%) of US landfills, most states and municipalities do not consider textiles as a category worthy of solid waste management. Often it is the case that public awareness and marketing is the key for changing the culture so that textiles can become a marketable recycled refuse.

Interviews with consumers revealed that many consumers did not know all of their options for clothing disposal and very few knew that a textile recycling industry existed beyond the not-for-profit charitable organizations. As one consumer commented, 'I just set six trash bags full of old clothes out on the curb for the trash man to pick up. It was so out of fashion and so worn that I didn't think that even Goodwill would have a use for it.' This and other statements revealed that many consumers do not know how post-consumer apparel waste gets utilized and perhaps marketing efforts need to be implemented to educate the disposing consumer.

Around the world, cultures vary in their interest, values, behaviors and technological complexity regarding textile recycling. For the United States, environmental concern ranks high, yet Americans are among the highest consumers of apparel in the world. Undoubtedly, Americans consume more in quantity rather than quality, resulting in a plethora of used items. Often it is the case that we consume something new 'for the occasion'. In contrast, Europeans tend to consume higher quality goods in lower quantities, consuming something new 'for the season'.

The ebb and flow of situational factors such as global economics, international trade laws, technological and engineering advancements, cultural evolution, competitive environments, and infrastructure (including the availability of waste disposal options) are also important factors of the system model. These situational factors are constantly in flux and can impact at both the micro and macro levels of the system.

8.8   Discussion and future trends

The discussion of textile recycling as a system can not be concluded without attention being paid to the global nature of the system. Here there is a twofold condition.

1 Increased textile waste is being created throughout the world because of increased disposable income in developing nations. Thus, concerns for disposal must be considered from all parts of the world. This has implications for cross-cultural research.
2 Much of the market for used clothing is located in developing countries where annual wages are sometimes less than the cost of one outfit at retail price in the USA.

The developing country markets provide a venue where highly industrialized nations can transform their excessive consumption into a useful export. For many of these people, used clothing surplus provides a much-needed service. Unfortunately, global trade laws often hamper the free flow of used clothing.

As landfill space becomes scarce and costs continue to rise, so will the ethos for environmentalism. Those in the business sector of the micro–macro framework continue to make progress in creating markets for used textiles. At the same time, consumers must be provided with timely information of these markets so that they can make educated choices as to where and how they will dispose of their used textiles. In addition, laws and political environment must be adapted to make it easy for textiles to be recycled. It is for this reason that the macro level of phenomena most influences the textile recycling process, which, in turn, presents disposal choices to the individual consumer.

Certainly, it is a double-edged challenge. Consumers must be made aware that nearly 100% of their used clothing is recyclable and that numerous and various markets exist for used textile and fiber products. At the same time, an attitude shift toward purchase of garments made from recycled fibers and preferences for quality rather than quantity must be embraced in the United States as it has been for decades in Europe. By raising consciousness concerning environmental issues, channels for disposal, and environmentally conscious business ethics, steps can be made toward a more sustainable environment. Citizen concerns lobbied with municipalities will also increase the number of municipalities that offer textile recycling as one of the categories of their waste management process. To recycle successfully, consumers must embrace the system, not just make an occasional charitable donation. Meanwhile, arbiters must continue to develop new value-markets and market the after-use possibilities so that the system functions at full capacity and with commitment from all.

8.9   References

ALLEBACH, W. (1993), Making a pitch for textile recycling. *Neighbor*, **3**, July 3.

BHARAT TEXTILE (2004), World's biggest fiber consumer: China. Retrieved November 25, 2004, from http://bharattextile.com.

CLAUDIO, L. (2007), Waste couture: environmental impact of the clothing industry. *Environmental Health Perspectives*, **115**(9), A449–A454.

COUNCIL FOR TEXTILE RECYCLING (1997), Don't overlook textiles! unpublished manuscript.

CSIKSZENTMIHALYI, M. (1981). *The Meaning of Things: Domestic Symbols and the Self*. Cambridge University Press: Cambridge, UK.

DEWALT, B.R. and PELTO, P.J. (1985), *Micro and Macro Levels of Analysis in Anthropology: Issues in Theory and Research*. Westview Press: Boulder, CO.

DIVITA, L. (1996), Missouri manufacturers' interest in textile recycling. Unpublished Master's Thesis, University of Missouri, Columbia, MS.

EDWARDS, A. (2005), *The Sustainability Revolution*. New Society Publishers: Gabriola Island, British Columbia, Canada.

ELLIOTT, E.J. (1995), Textiles' role in the environment. *Textile World*, **145**, 221–222.

ESTUR, G. and BECERRA, C.A. (2003), Retrieved from http://www.icac.org/icac/cotton_info/speeches/estur/2003/fiber_cons_pattern.pdf.

HAMILTON, J.A. (1997), The macro–micro interface in the construction of individual fashion forms and meanings. *Clothing and Textiles Research Journal*, **15**(3), 164–171.

HAMMER, M. (1993), *Home Environment*. Institute of Food and Agricultural Sciences, University of Florida: Gainesville, FL.

HAWLEY, J.M. (2000), Textile recycling as a system: a micro/macro analysis. *Journal of Family and Consumer Sciences*, **93**(5), 35–40.

HEIDEBRECHT, S. and HAWLEY, J.M. (2008), What goes in, may not come out: A study of people's clothing disposal behaviour. Unpublished manuscript.

KIM, H. and DAMHORST, M.L. (1998), Environmental concern and apparel consumption. *Clothing and Textiles Research Journal*, **16**(3), 126–133.

KIM, Y., FORNEY, J. and ARNOLD, E. (1997), Environmental messages in fashion advertisements: Impact on consumer responses. *Clothing and Textiles Research Journal*, **15**(3), 147–154.

MAYRHOFER, W. (2004), Social systems theory as theoretical framework for human resource management – benediction or curse? *Management Revue. Mering*, **15**(2), 178–192.

MCCRACKEN, G. (1988), *Culture and Consumption: New Approaches to the Symbolism of Consumer Goods and Activities*. Indiana University Press: Bloomington, IN.

MCCURRY, J.W. (1996), Blue jean remnants keep homes warm. *Textile World*, **146**, 84–85.

NOUSIAINEN, P. and TALVENMAA-KUUSELA, P. (1994), Solid textile waste recycling. Paper presented at *Globalization – Technological, Economic, and Environmental Imperatives*, 75th World Conference of the Textile Institute, Atlanta, GA, September 27.

OJUMO, A. (2002), Charity shops are beating the high street at its own game. *The Observer*, **57**, November 24.

OLSEN, J.E. and HASLETT, T. (2002), Strategic management in action. *Systemic Practice and Action Research*, **15**(6), 449–464.

PACKER, G. (2002), How Susie Bayer's T-shirt ended up on Yusuf Mama's back. *New York Times*, **54**, March 31.

People's Daily Online (2006), China's per capita fiber consumption increases 5 kg, in five years (February 6). Retrieved March 31, 2008, from http://english.peopledaily.com.cn/200602/06/eng20060206_240556.html.

S. SHAPIRO & SONS INC. (1961), *Reclaimed Resources: A Handbook of Textile Fabrics and Fibres including Lists of Most Important Grades*. S. Shapiro & Sons: Baltimore, MD.

SHIM, S. (1995), Environmentalism and consumer's clothing disposal patterns: An exploratory study. *Clothing and Textiles Research Journal*, **13**(1), 38–48.

S*MART (2007, August). Vested venture: Associate vests recycled into cards, blankets. Retrieved April 27, 2009 from walmartstores.com/download/2628.pdf.

TRANBERG-HANSEN, K. (2000), *Salaula: The World of Secondhand Clothing and Zambia*. University of Chicago Press: Chicago, IL.

US ENVIRONMENTAL PROTECTION AGENCY (2003), Recycled textiles. Retrieved September 28, 2003, from http://www.epa.gov.

YOUNG, C., JIROUSEK, C. and ASHDOWN, S. (2004), Undesigned: a study in sustainable design of apparel using post-consumer recycled clothing. *Clothing and Textiles Research Journal*, **22**(1/2), 61–68.

Part II
Applications and case studies

9
Consumer perceptions of recycled textile fibers

M. RUCKER, University of California, USA

Abstract: This chapter provides an overview of consumer responses to recycled textile fibers set within the broader contexts of the environmental movement and the industry's development of a new generation of sustainable products.

Key words: recycled textile fibers, green consumerism, consumer perceptions.

9.1   Introduction

Surveys have suggested that the number of consumers with concerns about the environment has been getting larger and there is also more willingness to act on those concerns. At the same time, consumers are often unclear about what those actions should be. In the meantime, the textile industry has been responding to reports that the green market is growing with a plethora of new green products and an increase in green marketing efforts. In fact, Interstoff Asia Essential is a new trade fair intended for 'textile and garment manufacturers with a green conscience'. The task now is to make sure the industry activities are actually adding value for the consumer. When that is not the case, one can expect to see failed companies and unhappy consumers. This chapter is designed to provide an overview of some of the green consumption research with a goal of improving the production/consumption match.

9.2   Consumer characteristics related to attitudes toward sustainable products

A variety of consumer characteristics have been studied in efforts to classify and characterize consumers with positive attitudes toward sustainable products and practices. These characteristics have included both demographic variables and socio-psychological variables.

The demographic variables that have received the most attention include age, gender, education and marital status. Although there have been some findings to the contrary, the preponderance of the evidence suggests that green consumers are relatively young (Klineberg, McKeever and

Rothenbach, 1998; Roberts, 1993) (although Gilg, Barr and Ford (2005), reported that their environmentalists had the highest average age and their non-environmentalists had the lowest average age), female (Diamantopoulos, Schlegelmilch, Sinkovics and Bohlen, 2003; Shrum, McCarty and Lowrey, 1995; Stern, Dietz and Kalof, 1993), well educated (Klineberg *et al.*, 1998; Olli, Grendstad and Wollebaek, 2001) and married (Diamantopoulos *et al.*, 2003; Laroche, Bergeron and Barbaro-Forleo, 2001).

Some of the socio-psychological variables that seem to be correlated with green consumerism include being an innovator and/or opinion leader (Bhate and Lawler, 1997; Shrum *et al.*, 1995), holding altruistic values (Guagnano, 2001; Karp, 1996) and a belief that consumers' actions can affect the environment, i.e. perceived consumer effectiveness (Gilg *et al.*, 2005; Roberts, 1996).

9.3   External factors influencing consumers' attitudes toward sustainable products

While some scholars have been trying to characterize the green consumer, others have been analyzing external factors for their role in creating cycles in consumer concern about the environment. Media attention has been given credit for both increases and decreases in the size of the green market. Modern-day environmentalism has been in evidence in the United States since at least the early 1960s when *Silent Spring* was published (Carson, 1962). This book drew public attention to the environmental problems caused by the pesticide DDT (dichloro-diphenyl-trichloroethane). It was widely read, especially after becoming a Book-of-the-Month Club® selection (books chosen by the club's editorial board, which included respected literary figures, generally increased in prestige as a result of this selection), and made consumers more aware of the acute and persistent damage that could be caused by pesticides then on the market. More recently, the book by McDonough and Braungart (2002) entitled *Cradle to Cradle* has contributed to consumers' appreciation of resource depletion and environmental degradation caused by working against nature rather than with nature. Even more recently, the 2007 film called *An Inconvenient Truth* is credited by at least one source (Natural Marketing Institute, 2008) with causing a 'tipping point' in the sustainability movement. Following the release of that movie, other media such as NBC and the Discovery Channel have added material on sustainability to their programming.

It has also been proposed that well-publicized natural disasters have turned sustainability into a major force in consumer decision making in recent years. Some of the events mentioned in this regard include Hurricane Katrina, several large oil spills and evidence of melting ice at the poles (*Marketing News*, 2008).

On the other hand, the media have also been held accountable for a consumer backlash against green product marketing or at least some stagnation in enthusiasm for green products. Some analysts have faulted the advertisers, saying the backlash occurred in response to advertising claims that were inaccurate, misleading or outright falsehoods (Peattie, 1995; Wong, Turner, and Stoneman, 1996). Thøgersen (2006) argued that newscasters also had a negative effect by what they chose as newsworthy; as the green movement gained momentum, negative stories were judged to be more interesting than positive ones.

9.4   Measures of consumer attitudes toward environmental issues

Evidence from a variety of sources suggests that, even if there have been a few reversals, overall consumer concerns about the environment have been growing over the last two decades (Cortese, 2003; Donaton and Fitzgerald, 1992; Laroche *et al.*, 2001; McCoy, 2007; Osterhus, 1997; *Progressive Grocer*, 2008; Schwartz and Miller, 1991). For example, Osterhus (1997) noted that a Gallup poll found that 75% of Americans labeled themselves as environmentalists. More recently, a survey of United States adults cited in a *Progressive Grocer* article (2008) found that the number of adults surveyed who said they 'regularly' buy green products jumped from 12% to 36% in 16 months, whereas the percentage who reported 'never' buying green products dropped from 20% to 10%. Laroche *et al.* (2001) summarized data suggesting that, depending on the sample and the way in which the question was asked, willingness to pay more for green products was also rising in the early 1990s. In a study done in 1989, the majority of the sample were willing to pay a 5–10% premium for green products. That value increased to 15–20% in a study done in 1991 and to 40% in a study done in 1993.

9.5   Textile and apparel industry response to green consumerism

In order to satisfy a growing green consumer market segment and their own concerns about corporate social responsibility, along with the recognition that green can also be profitable, textile and apparel companies have been producing and using an expanding array of materials promoted as being green. These products range from virgin materials such as organic cotton and wool, corn, soy, bamboo and even seaweed and banana, to recycled materials including cotton, polyester and nylon (Mowbray, 2008; Swantko, 2006). Finishes are also being reformulated to be greener, e.g. oil- and water-repellent finishes that contain a reduced amount of fluorocarbon (*Ecotextile News*, 2008a) and textile screen printing inks that are polyvinyl chloride (PVC) free and phthalate free (*Ecotextile News*, 2008c).

Unfortunately, some companies in the textile and apparel industries also appear to have made some questionable claims for their products, engaging in 'greenwashing' along with companies in other fields. Furthermore, systems for assessing how eco-friendly a textile or other product is have become more complex. As Davis (2008) recently noted, terms used by the US Federal Trade Commission in their 1998 '*Green Guides*' were fairly simple and straightforward, e.g. reusable, non-toxic, biodegradable and compostable. Today, the concepts and terminology are much more complex, e.g. carbon offsets, renewable energy credits, cap and trade, and carbon footprint.

9.6   Confusion in the marketplace

As a result of the changes noted above, brands and retailers have difficulty evaluating products that claim to be eco-friendly, never mind the consumer. Mowbray (2008) reported that attendees at the most recent sustainable textiles seminars at Texworld USA had questions ranging from what is an organic product to what type of certification would be the best choice for a given product. Patterson (2008b) went further to stress the importance of considering the whole life cycle of a textile product but cautioned that life cycle analyses for textile products 'generally take on the form of multiple-stakeholder, long-winded, super-scientific studies that can yield results that are unfathomable but to all in the super-scientific fraternity'. He went on to point out that analyses in incomparable units are one source of confusion, e.g. relatively high use of water for one type of fiber versus relatively high use of pesticides for another. Wilson (2008) also noted that the competing claims of fiber producers can be confusing because companies will highlight those parameters that show their products to best advantage.

Setting the issue of sustainability aside, there is relatively little information available on other attributes of the newer eco-textiles. For example, Patterson (2008a) contends that virgin and recycled polyester are 'largely comparable, with the virgin fibre being slightly more robust'. Of course, attributes of the different types of fiber are modified as the technology continues to evolve. For example, a recent article in *Ecotextile News* (2008d) includes mention of the new range of recycled polyester yarns and fabrics from Teijin that are appropriate for high-quality suits. The new material is made from staple spun yarns for a cotton- or wool-like hand.

Once products with a particular set of attributes have been developed, adequate and reliable information may not be available for the consumer. For example, Teijin representatives have expressed concern that the United

States has no specific quality standard for recycled polyester clothing. There are general ISO (International Organization for Standardization) standards but no type of certification program.

The consumer may not even know whether the material in a garment is recycled or virgin. Some brands and retailers identify recycled material in their garments and others do not. JCPenney is one company that has initiated a 'Simply Green' logo for their private-label clothing and home accessories. The three categories carrying the logo are 'Organic', 'Renewable' and 'Recycled'. Products labeled as 'Recycled' must contain at least 25% recycled materials of some type (*Ecotextile News*, 2008b).

9.7   Meeting the needs of the green consumer

In the past, the usual approach to marketing sustainable products was to first segment the market by level of environmental concern and then target those at the highest levels. The assumption seemed to be that environmentally conscious consumers would purchase any type of product that ranked higher on sustainability than other offerings in that category. However, analysis of a variety of green products that were not successful suggests that this approach is a formula for failure.

Ottman (2003) summarized what she has learned from observations of green product failures this way: 'environmental attributes alone will not sell on the mass market. The marketing of a product, green or not, needs first and foremost to address the needs of the consumer'. Peattie (2001) concurred that one must understand the needs of consumers and what compromises they are willing to make for an increase in sustainability. Another factor he highlighted as important was the degree of confidence consumers had that a particular choice would actually have much of an effect on the environment.

9.8   Consumer perceptions of textile products

Perception refers to the way in which people filter, organize and interpret sensory stimuli. When information about a product is limited and marketing claims are viewed with skepticism, consumers generate meanings, including understandings of product attributes, based on needs, expectations and experiences with other similar products.

9.8.1 Consumer perceptions of traditional textile materials

There have been a number of studies of consumers' perceptions of traditional textile materials (Davis, 1987; Forsythe and Thomas, 1989; Hatch and Roberts, 1985; Schutz and Phillips, 1976; Workman, 1990). Both Davis

(1987) and Hatch and Roberts (1985) were interested in the effects of fiber content identification on the perceived quality of garments. Davis (1987) found that identifying a garment as 100% cotton led to higher ratings than labeling the garment as a polyester/cotton blend. Hatch and Roberts (1985) compared wool and acrylic labels at high and low prices. They found a price by fiber interaction; wool got a significantly higher rating but only at the higher price point.

Forsythe and Thomas (1989) compared perceptions of cotton, a polyester/cotton blend, silk and polyester across seven attributes. They concluded that perceptions were generally consistent with fiber performance. For example, polyester and polyester/cotton were rated relatively high in terms of ease of care relative to silk but relatively low on the attribute 'luxurious'. Workman (1990) compared responses to hangtags with cotton, polyester, cotton/polyester and no fiber information on perceptions of 27 attributes for jeans. Both polyester and cotton/polyester hangtags led to more negative attributions compared with the cotton and no information hangtags.

The most ambitious study in terms of fabrics and attributes was conducted by Schutz and Phillips (1976). They asked a sample of adult women to rate 46 different fabrics on 48 attributes. A main goal was to develop clusters of fabrics based on similar ratings on the attributes. Four clusters were found, accounting for 87% of the variance. Based on the factor loadings, these clusters were labeled cotton, wool, silk and synthetic. Consumers appeared to be using more familiar fabrics to organize their perceptions of the less familiar fabrics.

9.8.2   Consumer perceptions of recycled textile fibers

To date, there has been limited attention devoted to how consumers are responding to eco-textiles in general and the newer recycled fibers in particular. This gap needs to be addressed because, as Grasso (1995) pointed out, there has to be an integrated effort among manufacturers, retailers and consumers for recycling to succeed. Meyer (2001) has suggested that green clothing may be at a disadvantage in the marketplace as green products in general suffer from several negative stereotypes including higher prices, limited number of choices and aesthetic disadvantages. Lorek and Lucas (2003) specifically commented on the 'non-fashion' image of eco-textiles as a problem for this branch of the industry; they cited a study by GfK, a major market research company, showing that eco-textiles were viewed by the majority of consumers as having high prices, a lack of credibility, a lack of fashion and a limited supply. Similar findings emerged from a more recent study conducted by a marketing agency called EcoAlign; participants in that survey saw many forms of green technology, including recycled materials, as expensive and unattractive (*Environmental Leader*, 2008).

On the other hand, Meyer (2001) presented case studies supporting the contention that when an effort is made to understand consumer needs and perceptions in relation to market offerings, green clothing collections can be successful. A case study of Patagonia showed how this sportswear and outdoor clothing company has used market research information to position fleece products made of recycled plastic bottles. After finding that only 20% of customers say they care about the environmental impact of what they purchase, Patagonia positioned the synthetic fleece primarily in terms of durability, functionality and performance. The environmental benefit of the product completes the product positioning rather than being the major emphasis. To address the finding that customers know little about how clothing production can impact the environment, Patagonia added this type of information to its catalog and web site. Regarding price, the fleece costs more than the customers' reported willingness to pay by about 10%.

A recent survey of active American adults reinforces the proposition that recycled synthetic fibers can play an important part in the outdoor and sports apparel industry (Walzer, 2008). The survey, sponsored by *Textile Intelligence Magazine* and conducted by Leisure Trends, was put in the field during December 2007 and resulted in 270 completed surveys. In response to a question about what criteria distinguish outdoor and sports apparel that could be classified as eco-friendly, the majority of respondents mentioned materials, price and 'recycled'. Furthermore, over 40% reported that they 'always' looked for labels that identified fiber content and another group of about the same size said they 'sometimes' looked for fiber content labels. Moreover, over 20% of the respondents listed recycled materials in products as a characteristic that is 'most important' in evaluation of a company's green behavior.

Work reported by Hines and Swinker (1996) and Swinker and Hines (1997) focused on consumers' willingness to purchase recycled polyester textile products. These two papers should be considered together since the second is essentially an extension of the first; the second added data on carpet samples to the previously reported data on sweatshirts. The authors found that when both price and fiber varied, the majority of the respondents selected the lower priced product, regardless of whether the polyester was virgin or recycled. When price was the same for both types of polyester, the majority selected the recycled polyester sweatshirt. These data can be taken to reinforce the importance of price in the recycled fiber market. However, it should be noted that only price and type of fiber were varied in the research, perhaps exaggerating the importance of price. As demonstrated in the Patagonia case described above, thoughtful positioning of a recycled fiber product can overcome a certain amount of price resistance.

If recycled textile products are to be sold on attributes other than price, it is important to know what attributes consumers perceive to be significantly different between virgin materials and recycled materials. A study by Rucker and Haise (2007) looked at the perceptions of three types of recycled materials across 13 attributes. Three distinctively different types of materials were chosen – recycled denim, recycled polyester and recycled paper – to determine whether there are general stereotypes about recycled materials across a variety of fiber types. Nine other clothing materials were included in the study to avoid undue emphasis on comparing a recycled material with its virgin counterpart.

A summary of the analysis of variance results is shown in Table 9.1 in order of significance of the recycle effect. In all cases where there was a significant recycle effect, the difference favored the virgin fibers. However, it should be noted that the significant interactions suggest the stereotype has its limits. In general, recycled paper was viewed more positively than its virgin counterpart while recycled polyester and recycled denim were viewed more negatively than their virgin counterparts. One possible explanation is that consumers are more familiar with recycled paper products so are less likely to evaluate those products based on negative stereotypes of secondhand merchandise.

Another point to consider in examining the data in Table 9.1 is the *types* of characteristics where there is a significant recycle effect versus those where there is none. A comparison of the four highly significant effects with the bottom five non-significant effects suggests that recycled materials are

Table 9.1 Main effects and interactions in analysis of variance analysis (ANOVA) of fiber content and recycling information on 13 uses/characteristics

Apparel use/characteristic	Recycle F	Fiber F	Interaction F
High quality	62.13****	122.21****	32.87****
For dressy clothes	49.80****	24.23****	10.39****
Smooth	39.16****	18.09****	11.92****
Luxurious	38.99****	51.21****	14.62****
Unpleasant smell	9.85**	3.89*	10.13****
Easy to care for	9.55**	104.11****	10.14****
For yoga class	6.02*	66.06****	8.33***
Warm	5.20*	98.21****	11.81****
Can be bleached	3.95	3.82*	1.24
For playing tennis	3.22	43.46****	7.65***
Durable	3.04	178.24****	18.10****
Itchy	2.71	15.69****	2.44
Economical	0.41	3.12*	0.32

* $P < 0.05$.
** $P < 0.01$.
*** $P < 0.001$.
**** $P < 0.0001$.

seen as competitive on functional or utilitarian attributes such as durable and economical but lacking when it comes to more hedonic attributes such as smooth and luxurious.

To determine how contact with recycled materials may alter evaluations on attributes, follow-up research is in progress in which respondents have been provided with samples of the materials in addition to names. Preliminary analysis indicates that providing fabric swatches can modify evaluations for some attributes such as smooth. This finding is consistent with results of other studies on the effects of haptic (touch) information and should be given careful consideration in choosing retail channels and packaging strategies for new developments in recycled fibers.

9.9   Sources of further information and advice

* *Ecotextile News.*
* Fletcher, K. (2008). *Sustainable Fashion and Textiles: Design Journeys.* London: Earthscan.
* Hethorn, J. and Ulasewicz, C. (2008) *Sustainable Fashion: Why Now?* New York: Fairchild.

9.10   References

BHATE, S. and LAWLER, K. (1997). Environmentally friendly products: Factors that influence their adoption. *Technovation*, **17**(8), 457–465.

CARSON, R. (1962). *Silent Spring.* New York: Houghton Mifflin.

CORTESE, A. (2003). They care about the world (and they shop, too). *The New York Times*, July 20, Money and Business, 4–9.

DAVIS, L.L. (1987). Consumer use of label information in ratings of clothing quality and clothing fashionability. *Clothing and Textiles Research Journal*, **6**(1), 8–14.

DAVIS, V. (2008). 'Recyclable' is so last-century. *Environmental Leader*, April 23. Retrieved April 27, 2008 from http://www.environmentalleader.com/2008/04/23/recyclable-is-so-last-century/.

DIAMANTOPOULOS, A., SCHLEGELMILCH, B., SINKOVICS, R. and BOHLEN, G. (2003). Can socio-demographics still play a role in profiling green consumers? A review of the evidence and an empirical investigation. *Journal of Business Research*, **56**, 465–480. doi: 10.1016/SO148–2963(01)00241–7.

DONATON, S. and FITZGERALD, K. (1992). Polls show ecological concern is strong. *Advertising Age*, **63**, 49.

ECOTEXTILE NEWS (2008a). Green water repellant finish. *Ecotextile News*, **11**, February, 15.

ECOTEXTILE NEWS (2008b). JC Penney launches green label. *Ecotextile News*, **13**, April, 10.

ECOTEXTILE NEWS (2008c). PVC-free printing inks. *Ecotextile News*, **13**, April, 12.

ECOTEXTILE NEWS (2008d). Suitable for recycling. *Ecotextile News*, **11**, February, 19.

ENVIRONMENTAL LEADER (2008). Survey: Consumers have negative perceptions of green tech. *Environmental Leader*, January 15, Retrieved April 27, 2008 from http://www.environmentalleader.com/2008/01/15/survey-consumers-have-negative-perceptions-of -green-tech/.

FORSYTHE, S.M. and THOMAS, J.B. (1989). Natural, synthetic, and blended fiber contents: An investigation of consumer preferences and perceptions. *Clothing and Textiles Research Journal*, **7**(3), 60–64.

GILG, A., BARR, S. and FORD, N. (2005). Green consumption or sustainable lifestyles? Identifying the sustainable consumer. *Futures*, **37**, 481–504. doi: 10.1016/j.futures.2004.10.016.

GRASSO, M.M. (1995). Recycled textile fibers: The challenge of the twenty-first century. *Textile Chemist and Colorist*, **27**, 16–20.

GUAGNANO, G.A. (2001). Altruism and market-like behavior: An analysis of willingness to pay for recycled paper products. *Population and Environment*, **22**(4), 425–438.

HATCH, K.L. and ROBERTS, J.A. (1985). Use of intrinsic and extrinsic cues to assess textile product quality. *Journal of Consumer Studies and Home Economics*, **9**, 341–357.

HINES, J.D. and SWINKER, M.E. (1996). Consumers' willingness to purchase apparel produced from recycled fibers. *Journal of Family and Consumer Sciences*, **88**(4), 41–44.

KARP, D. (1996). Values and their effect on pro-environmental behavior. *Environment and Behavior*, **28**, 111–133. doi: 10.1177/0013916596281006.

KLINEBERG, S., MCKEEVER, M. and ROTHENBACH, B. (1998). Demographic predictors of environmental concern: It does make a difference how it's measured. *Social Science Quarterly*, **79**(4), 734–753.

LAROCHE, M., BERGERON, J. and BARBARO-FORLEO, G. (2001). Targeting consumers who are willing to pay more for environmentally friendly products. *Journal of Consumer Marketing*, **18**(6), 503–520.

LOREK, S. and LUCAS, R. (2003). Towards sustainable market strategies: a case study on eco-textiles and green power (Wuppertal Paper No. 130). Wuppertal, Germany: Wuppertal Institute for Climate, Environment and Energy.

MARKETING NEWS (2008). Going green. *Marketing News*, **42**, February 1, 14–15, 17–18.

MCCOY, M. (2007). Going green. *Chemical & Engineering News*, **85**(5), 13–19.

MCDONOUGH, W. and BRAUNGART, M. (2002). *Cradle to Cradle*. New York: North Point Press.

MEYER, A. (2001). What's in it for the customers? Successfully marketing green clothes. *Business Strategy and the Environment*, **10**, 317–330. doi: 10.1002/bse.302.

MOWBRAY, J. (2008). Learning curve. *Ecotextile News*, **11**, February, 22–24.

NATURAL MARKETING INSTITUTE (2008). Understanding the LOHAS Market Report™ Series: Introduction to the sixth edition series. Retrieved April 27, 2008 from http://www.nmisolutions.com/r_lohas.html.

OLLI, E., GRENDSTAD, D. and WOLLEBAEK, D. (2001). Correlates of environmental behaviors: bringing back social context. *Environment and Behavior*, **33**, 181–208. doi: 10.1177/0013916501332002.

OSTERHUS, T. (1997). Pro-social consumer strategies: When and how do they work? *Journal of Marketing*, **61**(4), 16–29.

OTTMAN, J.A. (2003). Lessons from the green graveyard. *In Business*, **25**(3), 31.

PATTERSON, P. (2008a). Moral fibre. *Ecotextile News*, **11**, February, 16–18.

PATTERSON, P. (2008b). Natural cycles. *Ecotextile News*, **13**, April, 22–23.

PEATTIE, K. (1995). *Environmental Marketing Management*. London: Pitman Publishing.

PEATTIE, K. (2001). Golden goose or wild goose? The hunt for the green consumer. *Business Strategy and the Environment*, **10**, 187–199. doi: 10.1002/bse.292.

PROGRESSIVE GROCER (2008). Americans turn 3 shades greener in 16 months: Survey. *Progressive Grocer*, March 19.

ROBERTS, J. (1993). Sex differences in socially responsible consumers' behavior. *Psychological Reports*, **73**, 139–148.

ROBERTS, J. (1996). Green consumers in the 1990's: profile and implications for advertising. *Journal of Business Research*, **36**, 217–231.

RUCKER, M. and HAISE, C. (2007). Consumer perceptions of recycled textile fibers. In First International Symposium on *Fiber Recycling*, June 20 and 21, Kyoto Institute of Technology, Kyoto, Japan.

SCHUTZ, H.G. and PHILLIPS, B.A. (1976). Consumer perceptions of textiles. *Home Economics Research Journal*, **5**(1), 2–14.

SCHWARTZ, J. and MILLER, T. (1991). The earth's best friends. *American Demographics*, **13**, 26–35.

SHRUM, L.J., MCCARTY, J.A. and LOWREY, T. (1995). Buyer characteristics of the green consumer and their implications for advertising strategy. *Journal of Advertising*, **24**(2), 71–82.

STERN, P., DIETZ, T. and KALOF, L. (1993). Value orientations, gender, and environmental concern. *Environment and Behavior*, **25**(3), 322–348. doi: 10.1177/0013916593255002.

SWANTKO, K. (2006). Buyers trend to green fabrics. *Knitting International*, September, 26–29.

SWINKER, M.E. and HINES, J.D. (1997). Consumers' selection of textile products made from recycled fibers. *Journal of Consumer Studies and Home Economics*, **21**(3), 307–313.

THØGERSEN, J. (2006). Media attention and the market for 'green' consumer products. *Business Strategy and the Environment*, **15**, 145–156. doi: 10.1002/bse.521.

WALZER, E. (2008). The green scene. *Textile Intelligence Magazine*, March/April.

WILSON, A. (2008). Home comforts. *Ecotextile News*, **11**, February, 26–27.

WONG, V., TURNER, W. and STONEMAN, P. (1996). Marketing strategies and market prospects for environmentally-friendly consumer products. *British Journal of Management*, **7**, 263–281.

WORKMAN, J.E. (1990). Effects of fiber content labeling on perception of apparel characteristics. *Clothing and Textiles Research Journal*, **8**(3), 19–24.

10

Eco-labeling for textiles and apparel

S. B. MOORE, Hohenstein Institute America, Inc., USA, and
M. WENTZ, Oeko-Tex Certification Body, USA

Abstract: The concept of an eco-label for products of commerce
continues to evolve. The textile industry in particular is now facing
an explosion of various textile labeling concepts. This chapter
examines the types, roles and reasons for eco-labels, how the process
of eco-labeling has evolved, and forecasts future developments.

Key words: eco-labels, sustainability, life cycle analysis, textiles, apparel,
chemicals, consumers.

10.1 Introduction

Eco-labeling is emerging worldwide as a differentiating factor in retail markets for textile and apparel purchases. Beginning as a response to consumer demand, eco-labels are emerging as a primary tool in marketing to more well informed and 'green' consumers. Surveys conducted in Germany, during the years of 2002 and 2006, indicated a substantial increase in the number of consumers concerned with harmful residues on textiles and the ecological impacts of textile production (Meding, 2008). A fifth of all interviewees in this survey found eco-labels influenced their purchasing decisions.

While there are many positive attributes associated with eco-labels, there are also negative consequences when such labels are used as trade barriers (Nimon and Behgin, 1999). Often 'greenwashing' results from companies making unsubstantiated environmental claims concerning products they produce or sell. Therefore, the various accepted types and characteristics of eco-labels and some of the emerging themes in regulation and standardization of eco-labels are worth investigation.

10.2 Key principles: eco-labeling and sustainability

Since the late 1980s, communication of sustainability has been framed around the Brundtland Commission's assessment that sustainable development involves 'meeting the needs of the present without compromising the ability of future generations to meet their own needs.' (Meadows and Club of Rome, 1974). The essence of sustainability for eco-labeling within the textile industry is the incorporation of life cycle assessment of goods and services into a lens envisioning 'value'. This requires defining what 'ecology' means in the context of textile products and production.

10.2.1 Concept of ecology in textiles

The general definition of 'ecology' in the context of textiles is complex and diverse, and 'ecology' is often used in a loose sense. This has created confusion in the usage of the term. Many references to 'ecological' or 'natural' textiles refer to unbleached textiles, or non-dyed textiles, or textiles with natural dyestuffs. However, surveys indicate that consumers do not want undesirable or potentially harmful substances in textiles and clothing. Actions by various governments in the EU and USA for example, have limited the use of hazardous dyes and lead in textile products (1997; 2001).

Beyond product ecology there is also a growing demand from the consumer to better understand the life cycle of textiles (Marin and Tobler, 2003) and how textile production relates to the physical environment and society. The term 'textile ecology' is easier to comprehend if explained in three parts: production ecology, human ecology and disposal ecology.

Production ecology

This refers to the process of production and manufacture of fibers, textiles and garments. *Sustainable* textiles should be environmentally friendly and should satisfy the rational conditions to respect social and environmental quality by pollution prevention or through installing pollution control technologies. Third-party certification bodies and governments have issued Restricted Substances Lists (RSLs) that link production ecology to human ecology as discussed below. Such lists provide stimuli to promote the use of safer chemical inputs and provide 'targets' for verification of cleaner production of textile products (see www.apparelandfootwear.org).

Human ecology or ecology of use

Human ecology focuses on the effects of textiles on the users and on their near environment or surroundings. According to our present methodolgy, concentrations of substances that could induce dangerous effects on humans during normal use must be understood, modeled and managed. Consumers are concerned with this aspect of textile human ecology. Risks have been addressed through the development of RSLs by governments, retail organizations, producers and non-governmental organizations (NGOs) which limit the use of certain chemicals in production. Substances on such lists

Table 10.1 Typical RSL chemical family comparison. Complete RSLs provide chemical names, CAS (Chemical Abstract Service) numbers, allowable limits and detection levels

Product testing	EU Flower label	Oeko-Tex Standard 100
pH		X
Formaldehyde	X	X
Heavy metals/organic tin compounds TBT	X	X
Pesticides	X	X
Chlorinated phenols		X
Phthalates		X
Banned arylamines	X	X
Carcinogenic dyes	X	X
Allergenic dyes	X	X
Chlorinated benzenes and toluolenes (carrier)	X	X
Biologically active products	X	X
Flame retardants	X	X
Color fastness	X	X
VOCS: volatile organic components	X	X
Emission of volatiles		X

must be analyzable on the final textile products used by people, and RSLs must be reviewed regularly as 'living documents'. Table 10.1 is a typical comparison of RSL chemical classes between two established standards.

Product analyses to detect and quantify RSL substances should be performed by accredited independent laboratories. Consensus-based test methods must be used to verify the absence or concentrations of harmful chemicals. The diverse and complex nature of global textile production requires analytical verification of the absence or concentration of restricted substances by accredited international laboratories. The modular concept of the Oeko-Tex Standard 100 certification at every stage of production (Fig. 10.1) has the advantage that intermediate textile components can be certified for eco-labeling. It prevents costly supply rejections at every step of the textile chain and supplements conventional quality assurance testing. The development of updated RSLs and the corresponding development of international third-party laboratory networks to verify RSL compliance is becoming an important tool for human ecology product assurance.

Disposal ecology

This concept is based on what happens at the end of the 'first use' of textile products. Disposal ecology addresses recycling, reuse, energy, disposal and/ or decomposition of textile products without release of harmful substances

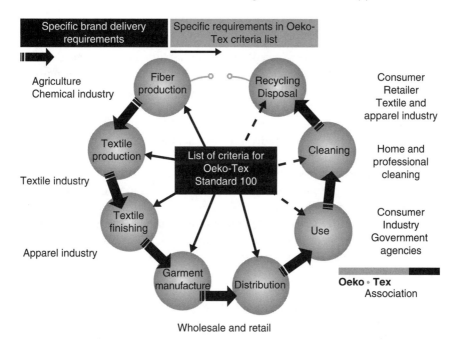

10.1 Modular approach to eco-labeling for components within the textile chain.

or thermal elimination without endangering air purity. Life cycle concepts incorporating production, human ecology and disposal ecology are well described and explained in the book *Cradle to Cradle* (McDonough and Braungart, 2002).

Ecology for textiles, and by inference 'eco-labels' for textiles, may address production, human and disposal ecologies. Because the textile industry is truly global in scope, products are made and sold throughout the world. Therefore, compliance with various countries' individual requirements can be a challenge. Some trade regulations have produced unified information label requirements that describe country of origin and fiber content. Eco-labels are now attempting to inform consumers additionally of the 'textile ecology' of the products they are buying.

Modern production technologies, analytical laboratories and rapid information dissemination can produce eco-certifications and labeling schemes that are transparent, accurate and cost-effective. Until recently, textile labels that addressed composition, care and origin were considered adequate. Human ecology, production ecology and life cycle information are now demanded by major international retailers. The eco-labels of the future will provide a myriad of information that encompasses social and environmental aspects of a product.

10.2.2 Sustainability, the new consumer and eco-labels

The rapid spread of the concepts of sustainability into enhanced value propositions for products and services has created a new consumer market segment: 'Lifestyles of Health and Sustainability – LOHAS' (http://www.lohas.com/). The growth in LOHAS shoppers has enabled retail merchants to expand the scope and value of 'organic' and 'natural' products. LOHAS now represents a global, rapidly growing segment of consumers that is increasing in numbers and importance as the middle class expands globally and sustainability increases in importance (Friedman, 2008). LOHAS gains importance as a consumer demographic owing to the emerging understanding that enterprises with intentional strategies linked to social and environmental responsibility (life cycles) create long-term value in their products and services. This is a consumer movement that is not strictly based on low-cost economics, but embraces the concept of a 'triple bottom line (TBL)' sustainable economic model,[1] (Senge, 2008; Willard, 2002).

Since the year 2000, financial analysts have been tracking the trends and the relationships between economic success and social and environmental responsibility (Hart, 2005). The seemingly positive synergies achieved through linking economic, social and environmental performance (TBL) are being translated into marketing campaigns by leading firms to inform stakeholders of 'green' innovations and compliances (Manring and Moore, 2009).

The Dow Jones Sustainability Index (DJSI) is one tool that tracks the relationship between verified sustainable practices and economic performance for large multinational companies (Willard, 2005). Firms that qualify as members on the DJSI advertise their inclusion in that index to promote the sustainable value of their companies. Firms such as IKEA, Marks and Spencer, C&A, Pottery Barn, Wal-Mart, Target and others, actively advertise the links between sustainable practices and increased shareholder value creation through sustainable production, eco-efficiency and increased innovation within the 'green' market space (Manring and Moore, 2009). There is a growing literature that links innovation and future success for enterprises that are on the path seeking sustainability (Hart, 2005; Hawken *et al.*, 1999; Hawken, 2007; Hollender and Fenichell, 2004; Manring and Moore, 2004; McDonough and Braungart, 2002; Schleg and Laur, 1998). These links are inspiring new concepts for eco-labeling. However, there are challenges with these new concepts as the claims made on eco-labels are

---

[1]The phrase was coined by John Elkington in 1994. It was later expanded and articulated in his 1998 book *Cannibals with Forks: the Triple Bottom Line of 21st Century Business* (Elkington, 1998). Sustainability, itself, was first defined by the Brundtland Commission of the United Nations in 1987.

not always adequately verified. Therefore, the Federal Trade Commission (FTC) in the USA is revising the conditions and verification of eco-labeling and marketing green textiles (2008–2009).

Despite these challenges, eco-labels and certifications are used to link product ecology, production ecology, and corporate social responsibility directly to consumer products or services. Ultimately, this leads to changes in consumer behavior and decision making (OECD, 2000). Eco-labels can have the following goals or effects:

- Promotion of consumer choice through empowerment of the consumer to differentiate between products that are harmful to the environment and those deemed to be more compatible with environmental objectives (Robert *et al.*, 2002).
- Eco-labels can enable 'market-based' beneficial changes rather than reliance on slow and inefficient regulatory or policy change mechanics (Dowell *et al.*, 2000).
- Eco-labels can help to develop markets via signaling greater environmental protection.
- Eco-labels produce continuous improvement as consumer expectations of enterprises increase towards reducing environmental and social impacts over time (Hawken, 2007).
- Eco-labels play an educational role as well as being a seal of ecological approval. An eco-label makes a product stand out on the shelves of stores. Obtaining an eco-label will induce producers to re-engineer products and if successful, this will eliminate incumbent practices (Christensen, 1997; Christensen and Overdorf, 2000; Christensen and Raynor, 2003; Christensen *et al.*, 2004).
- Eco-labels provide a window to substantiate claims outside of regulatory processes, which increases transparency within the international market (Dowell *et al.*, 2000).
- Eco-labels can be an indicator of innovation and entrepreneurship since 'green' innovation produces higher future value (Christensen *et al.*, 2004).

10.3   Standards and eco-labeling defined for textiles

The textile industry, always a leader in value creation of all sorts, has begun to use the tool of 'eco-labeling' to declare its adoption of a variety of sustainable practices leading to *eco-differentiated textile products*. To quote from the Global Ecolabelling Network (GEN): an 'ecolabel is a label which identifies overall environmental preference of a product within a product category based on life cycle consideration'. This label is awarded by an impartial third party to products that meet established environmental leadership criteria. Such labels are voluntary declarations and are a clear attempt to set a new, differentiated 'standard' for labeled textile products. (Global Ecolabelling Network (GEN), www.globalecolagelling.net)

A brief discussion on how value is created within a standard-setting process may clarify why eco-labels have emerged as one way to differentiate goods and services. Standard setting, in this case using product labeling in defining what is 'ecological', has the potential to create market power and enhance the market value of a technology or product *by reducing the number of close substitutes*. This may be the most important factor driving firms to adopt eco-standards.

The adoption of a standard can occur entirely within the organization or market (*de facto* – without outside monitoring), or through third-party standard-setting bodies (*de jure*). The consumer and market accept whatever standard can produce increased market share. The often-cited example of VHS emerging as the *de facto* (market-based) 'standard' in consumer video tape format rather than the technically superior BetaMax (Sony Corporation) format, demonstrates the power of standards (http://www.mediacollege.com/video/format/compare/betamax-vhs.html).   Whether standards are set *de facto* or *de jure*, the emergence of a standard can create market power. For the case of eco-labeling of textiles, which as a practice is less than 20 years old, the credibility of eco-labels remains driven by policy setting, whether this is done by third-party standard institutes governments, or the marketplace. The role of a credible eco-labeling framework is to produce intentional change in consumers via eco-labels and to avoid unintentional or deceptive consequences on consumers and markets. This makes the standards-setting process important and begs the question as to what role an eco-label plays in defining value for consumers or producers?

When the value of differentiating technology between products is not high within a product category – because the innovation is incremental or because there are many economically comparable inventions – formal standard setting (eco-labeling) has the potential to enhance value by informing the consumer about the differences in the products. Another possible way to explain this is that when patents and trademarks cannot be used to differentiate and protect a product or process, standards can be used to create differentiation. This is the emerging role of 'eco-textile' labeling: defining the new product category of eco-textiles.

Eco-textiles are a new (rediscovered?) product category or technology segment that represents in many cases, *minimal technical product advances over pre-existing alternatives*. Certainly, this assumption is not always true, and there are expansive, radical eco-textile innovations. However, as always within a product category, there are more incremental improvements than radical innovations (Ziman, 2000). Eco-textiles may be cleaner and greener and may incorporate life cycle analysis into their value propositions but, in many cases, eco-textiles are not a radical reinvention or technology advance from 'non-eco' textiles. However, they may be cleaner, greener or produced under that are conditions environmentally more desirable, and this can be an important differentiation to consumers, if these differences can be verified and clearly and fairly communicated.

Patents and trademarks are tools that protect significant technical or design advances within textile products and technologies. However, advances, as mentioned previously, can also be differentiated and communicated via eco-labeling standards. Answering the following question provides an indication of the role of eco-labeling for any particular product innovation: If an ecologically superior product and a traditional counterpart are not eco-labeled, can the consumer tell the difference? Answering 'no' to this question presents the opportunity for value creation within an evolutionary product development framework. 'Eco-labeling' can be used to inform the consumer of improvements, albeit not readily apparent, in textile ecology.

The user's requirements for most textile products within a product category are high. This is especially true for apparel. For example, whether underwear is eco-labeled or traditional, there are certain standards of fit and quality required. If one has ever worn underwear that does not fit, the need for such standards become immediately obvious!

Eco-labeling that incorporates human ecology information for garments meeting nominal compatibility standards produces product differentiation. In an example case, one sweater has an eco-label and a sweater that is similar in composition, style and fit does not. The only real differentiation is that one product is certified and labeled to be safer or greener than the other. The consumer then decides whether this information is valuable to him/her or not. This immediately produces the obvious question: What is the validity of such eco-labels and how are they verified?

Commonly, to differentiate between eco-textiles and traditional textiles a *de jure* standard is desirable, i.e. using a third party to certify and verify that the product is as labeled. The unverified, market-based self-labeling, the *de facto* approach for textile eco-labeling, invites fraud due to lack of third-party verification. Indeed, such flaws within voluntary, unaudited and unverified *de facto* processes may be the cause of recent international banking problems.

With all the various definitions of 'green' in the marketplace, it is easy to self-proclaim such attributes, especially if the product in question is produced on the other side of the world to its ultimate consumer. The feedback loops within global networks remain a challenge (Watts, 2003).

Unsubstantiated claims have been the subject of a recent workshop on 'green' labeling by the US FTC (2008). The potential for less than accurate labels, produced under *de facto* conditions, are being addressed in the

USA by the FTC's Green Labeling Guides (www.ftc.gov). These guides provide guidance on valid label claims. Other government-sponsored eco-labels, such as Blue Angel, ECO-Flower, etc., set policy beyond *de facto*, reducing the risk of fraud and setting legal remedies and penalties. This practice moves eco-labeling beyond voluntary actions and into the legal domain.

Voluntary eco-labeling communicates the differences between eco- and non-eco textiles. Without such labels to educate the consumer, the differences between similar products would not be apparent at retail. Its purpose is to create a higher value definition beyond simple product quality and simple economics. The formerly plentiful, less expensive, ready substitutes (traditional textiles) can be disqualified by the eco-label. Accurate eco-labeling enables the consumer to make an informed choice between an eco-product and its counterpart produced with less environmental focus. If the values touted on the eco-label differentiate the product and induce a purchasing decision, then the customer votes with that purchase for the value touted on the label.

When third parties set standards for voluntary eco-labeled textiles, it is a *de jure* labeling concept, which produces a new value proposition for textile products. *De jure* processes are verified via certification and communicated via labeling. Textile brands and manufacturing firms are willing to pay for certifications and the right to use an eco-label to remain compatible and in compliance with a consensus standard. The third-party accreditation organization is paid to verify and to certify compliance with its standard.

The process of producing and verifying an eco-labeling standard is a consensus-building process that has the goal of increasing the perceived value of eco-textiles. Textiles that are labeled 'Eco' without the oversight of verifiable third-party standards, i.e. if there were no valued third-party certification and labeling standards, would lose value in the market with any breach of performance. Therefore, the quality and credibility of the standards, i.e. the process of verification and certification, via independent consensus-based standards are very important. This has drawn standards organizations such as the International Organization for Standardization (ISO) into the process of defining eco-label standards and labels to achieve alignment and credibility with existing quality systems standards. (www.iso.org/iso/iso_14000_essentials).

The ISO goals for eco-labels are to provide verifiable and accurate information on the environmental aspects of products and services, to encourage demand for such goods, and to stimulate the potential for market-driven continuous environmental improvement. The ISO has defined three types of environmental performance labeling: Types I, II and III.

- Type I is voluntary, based on multiple criteria, third-party programs that award a license, which authorizes the use of environmental labels on products indicating environmental preferability within a category based on life cycle considerations (*de jure*). Type I programs can also be categorized as 'multi-criteria practitioner programs'.
- The Type II labels are informative self-declarations of environmental claims (*de facto*), these are self-declarations based on common terms, definitions and symbols.
- Type III labels are voluntary and provide quantifiable environmental data under pre-set categories, which are produced by a qualified third party and verified by that or another, qualified third party (*de jure*). Such programs provide quantified product information–report cards of performance in multiple areas of qualification such as social responsibility, ecological performance, toxic residues, etc.

The ISO labeling standards are principle-based and requirements include:

- accurate labeling – verifiable, relevant and non-deceptive;
- relevant information concerning attributes must be available and their derivation transparent to purchasers;
- labels must be based on scientific methods that are reproducible and based on agreed standards of practice;
- transparency for information and methods should be insured for all stakeholders and interested parties;
- labeling should include the life cycle of the product or service;
- administration of the eco-labels should not be burdensome;
- labels should not create unfair trade restrictions;
- labels should not inhibit innovations that improve ecological performance;
- label criteria should be developed by consensus.

In many ways, all voluntary eco-labels are an experiment in 'action science', where information and ideas are clearly biased to induce sustainable changes and advertised on the textile product (Argyris *et al.*, 1985). Eco-labeling educates the consumer, differentiates the product and the targeted market, provides a sustainable connotation for the producer or seller, and develops a higher or different perception for the product in the eyes of the entire supply chain. However, eco-labels can be used as market-based trade barriers, and some research indicates that while a global, transparent eco-labeling system benefits markets, regional eco-labeling can limit market access and reduce global competition (Hyvarinen, 2008; Marin and Tobler, 2003).

As part of an action approach to sustainable change, *de jure* eco-labels can provide a great deal of credibility and build value into the concept of eco-textiles. Lack of credible third-party verification and lack of science-based verification can lower the perceived quality of eco-labeling. Therefore, ISO values the following methods of verification for credibility:

- conformity declarations – the producer self-declares conformance;
- evidence of conformity through providing a review of supporting documentation;
- evidence of conformity with manufacturing declaration (auditing);
- evidence of conformity through product testing.

When these criteria are met and the auditing and verification processes used to examine the labeled products are transparent to all stakeholders, eco-labels have the potential to inform and differentiate products in a very valuable way.

10.4   Examination and anatomy of eco-labels

The US FTC uses the following language in their marketing guide under the heading: 'Eco-seals, seals-of-approval and certifications' (http://www.ftc.gov/bcp/edu/pubs/business/energy/bus42.pdf):

> Environmental seals-of-approval, eco-seals and certifications from third party organizations imply that a product is environmentally superior to other products. Because such broad claims are difficult to substantiate, seals-of-approval should be accompanied by information that explains the basis for the award. If the seal of approval implies that a third party has certified the product, the certifying party must be truly independent from the advertiser and must have professional expertise in the area that is being certified.
>
> The FTC analyzes third-party certification claims to ensure that they are substantiated and not deceptive. Third-party certification does not insulate an advertiser from Commission scrutiny or eliminate an advertiser's obligation to ensure for itself that the claims communicated by the certification are substantiated.

Figure 10.2 is an example of a certification label issued for the production of textile materials. The label is a registered trademark of the Oeko-Tex Association (www.oeko-tex.com). It clearly identifies the standard and the laboratory that performed the certification testing, and has a unique number that associates this particular certification to a particular product or producing company. This labeling format allows the consumer to recognize the logo of the standard immediately, in this case the Oeko-Tex 'Confidence in Textiles'. The label provides all the information necessary to get more

10.2 Generic Oeko-Tex Standard 100 label with design that shows certificate test number and institute performing the testing.

10.3 The new Global Organic Textile Standard (GOTS) logo, the label must also include identifiers for certificate numbers and product.

data on the standard and the company involved. The Internet leads the consumer to the Oeko-Tex website (www.oeko-tex.com) where the details of the standard, the testing methods and the certification process are available and transparent.

As discussed earlier, in a *de jure* ISO Type I or III standard, transparency within the certification process is very important. The design and information provided on eco-labels must enable the consumer to readily obtain information on the agencies that manage and certify the standard. This is required under the US FTC Green Guides as cited earlier.

The newly announced labeling standards for the Global Organic Textile Standard (GOTS) also follow the previous conventions. Figure 10.3 shows the GOTS logo that represents the standard; underneath this logo, information will be given that explains details of the certification (organic, percentage organic, etc.) and discloses the certifying body (http://www.global-standard.org/).

When examining an eco-label or considering the design of an eco-labeling scheme, the mark must meet the minimum legal requirements for the countries in which the products are produced or sold. In addition, in concert with ISO standards, the name of the certifying agency, a traceability code or number, and a clear identification of what is being certified, are necessary. Conversely, labels that do not meet such criteria should be considered suspect. The following advice from the US FTC (http://www.ftc.gov/bcp/edu/pubs/consumer/general/gen02.shtm) provides a good conclusion to this discussion of eco-labels, advising that consumers look for transparency, verification and traceability.

> When you evaluate environmental claims in advertising and on product labels, look for specific information. Determine whether the claims apply to the product, the packaging, or both. For example: if a label says 'recycled' check how much of the product or package is recycled. The fact is that unless the product or package contains 100 percent recycled materials, the label must tell you how much is recycled.

The ultimate role of any 'standard' in the consumer's decision-making process is both increasing the value of the 'brand' for the producer, retailer and increasing the security and value for the consumer. This idea was discussed during a recent FTC workshop in Washington, DC on 'green labeling' (FTC, 2008). In this workshop, a poll of an expert panel of textile industry representatives found that if the eco-label did not realistically represent the product, and this resulted in a consumer complaint, it was not the product producer or the eco-label that the consumer held most directly responsible, it was the retailer. Thus, one of the most important stakeholders in the eco-labeling standards development process should be the retailers, who must thoroughly examine the credibility of any eco-labeling schemes.

10.5   Future trends

Eco-labeling is a trend that continues to grow as a method of incorporating ecological and social information into a concise educational message to consumers. There are several emerging concepts of information transfer and labeling from other industries that may influence textile eco-labeling.

10.5.1 The use of the Internet

REACH (Registration, Evaluation, Authorization and Restriction of Chemical substances) is a new registration and communications program for chemicals and consumer products that contain chemicals within the EU. A brief summary of REACH can be found on the Internet at (http://ec.

europa.eu/environment/chemicals/reach/pdf/2007_02_reach_in_brief.pdf). One important aspect of REACH is that detailed information on chemicals found in consumer products must be available to the consumer on the Internet. This requirement has profound implications for the future of labels for consumer products and particularly for eco-labeling schemes. Oeko-Tex and other eco-labels for textiles already use certificate numbers on consumer labels to expand the information available to consumers. Therefore, it is not much of an extension to envision that eco-labeling conventions will expand to include off-label information published on the Internet. The implications of expanded information availability to consumer can be found in *'The Wisdom of Crowds'* by James Surowiecki, who discusses the phenomena of information in grass-roots *de facto* evolution of standards (Surowiecki, 2004).

Eco-labels as co-branding

As certain eco-labeling standards become recognized by LOHAS or other *'crowds'* of consumers, symbiotic relationships may develop between major retail brands of textile products and the organizations that have developed the eco-labeling schemes. Such relationships could move beyond endorsements and actually begin to be a significant part of the value identity for a particular product line. The best example of this could be the label 'organic' which has a particular definition and quality connotation. For example, when applied to a commodity product such as milk, the 'organic milk' label now allows milk with this label to command higher prices in the market than milk without the organic label. As life cycle transparency becomes an accepted standard label criteria, such information will increasingly be used for product differentiation.

10.6   Summary and commentary

In the previous sections, eco-labeling, its history and nomenclature have been reviewed. Eco-labels, whether *de facto* or *de jure*, appear to be here to stay as the primary means of defining 'green' textiles versus those that are not 'green' in the market place. When the term 'sustainability' is no longer a market differentiation and 'green' becomes the status quo, such labels can still inform consumers and define products. Perhaps in the future, we will label products as 'black', to differentiate those that are not sustainable and need to seek sustainability. We have a long sustainable path to follow before, if ever, we arrive at a time when a non-sustainable product is the exception other than the rule.

Eco-labels are becoming an important information source for consumers, who are increasingly using the labels to gain access to further product information on the Internet. This embedding of information in labels is a trend that will continue to escalate, supported by governmental policy and consumer demands. Eco-labels that are the result of a third-party certification, a *de jure* standard, appear to represent the most reliable and verifiable type of labeling scheme. However transparency in the standards development process, transparency in auditing and verification of performance and conformity, are extremely important.

There are emerging eco-labeling standards claiming 'proprietary processes' and 'black-box' techniques for evaluating production, human and disposal ecologies. This is a very risky approach as standards must be transparent to be viable. If there are valuable proprietary processes developed, the patent and trademark system should be used to protect and disclose such inventions. Otherwise, such 'black box approaches present a great potential for misuse of eco-labeling to create trade barriers, differentiation and 'green washing' that cannot be verified. 'New and improved' claims without transparency are only marketing tools, not verifiable improvements.

Non-transparent testing methods and questionable certification processes have the potential to damage the credibility of eco-labeling. They diminish rather than increase the value of 'eco' products. Eco-labeling must promote sustainability and responsible decisions by the retailers and consumers. The best techniques and practices must be used to produce eco-certification and -labels to allow for continuous improvement. Eco-labels that are not transparent in all aspects are suspect and should be avoided. Eco-labels that do not clearly allow for complete traceability within the defined frameworks of the declared certification standards are not adequate.

Eco-labels are very important to the development of a sustainable textile industry and a credible textile industry. A lack of transparency and good science in producing eco-labels for textiles will lead to government requirements and policy standards that are narrow and confining, and may restrain innovation in a dynamic textile and apparel industry. Industry and consumers should scrutinize certifications and labels, seeking good science, good practices and complete transparency. Anything less is just 'green washing', not eco-labeling.

10.7   Sources of further information and advice

Below are some helpful Internet links.

- The Oeko-Tex Association – http://www.Oeko-Tex.com.
- The Blue Sign Organization – http://www.Bluesign.com.
- Global Organic Textile Standard – http://www.global-standard.org/.
- A listing of all eco-labels and information on eco-labeling standards – http://ecolabelling.org/.

- Information on EU eco-labels – http://ec.europa.eu/environment/ecolabel/index_en.htm.
- USA Federal Trade Commission's Green Guides on eco-labels – http://www.ftc.gov/bcp/grnrule/guides980427.htm.

10.8 References

(1997) German Consumer Goods Ordinance. Germany, www.etad.com/documents/Downloads/publications/etad_information_19th_amendment.pdf.

(2001) Federal Register: September 28, 2001. *CFR*.

ARGYRIS, C., PUTNAM, R. and SMITH, D.M. (1985) *Action Science*, San Francisco, California, Jossey-Bass.

CHRISTENSEN, C. and OVERDORF, M. (2000) Meeting the challenge of disruptive change. *Harvard Business Review*, **78**, 66–76.

CHRISTENSEN, C.M. (1997) *The Innovator's dilemma: When New Technologies Cause Great firms to Fail*. Boston, Massachusetts Harvard Business School Press.

CHRISTENSEN, C.M., ANTHONY, S.D. and ROTH, E.A. (2004) *Seeing What's Next?: Using the Theories of Innovation to Predict Industry Change*, Boston, Massachusetts, Harvard Business School Press.

CHRISTENSEN, C.M. and RAYNOR, M.E. (2003) *The innovator's Solution: Creating and Sustaining Successful Growth*, Boston, Massachusetts, Harvard Business School Press.

DOWELL, G., HART, S. and YEUNG, B. (2000) 'Do corporate global environmental standards create or destroy market value?' *Management Science*, **46**, 1059–1074.

ELKINGTON, J. (1998) *Cannibals with Forks: The Triple Bottom Line of 21st Century Business*, Gabriola Island, BC, Canada, New Society Publishers.

FRIEDMAN, T.L. (2008) *Hot, Flat, and Crowded: Why Need a Green Revolution–and How it Can Renew America*, Waterville, Maine, Thorndike Press.

FTC (US FEDERAL TRADE COMMISSION) (2008) http://www.drinkerbiddle.com/files/Publication/8431a0a8-69ad-4dbb-a6f7-01d8e4b3a048/Presentation/PublicationAttachment/544e7f34-58fb-496c-ac05-0654eb5abf59/FTC2009.pdf.

HART, S. (2005) *Capitalism at the Crossroads: The Unlimited Business Opportunities in Solving the World's Most Difficult Problems*, Indianapolis, Indiana, Wharton.

HAWKEN, P. (2007) *Blessed Unrest: How the Largest Movement in the World Came into Being, and Why No One Saw It Coming*, New York, Viking.

HAWKEN, P., HUNTER LOVINS, L. and LOVINS, A. (1999) *Natural Capitalism: Creating the Next Industrial Revolution*, New York, Little Brown and Company.

HOLLENDER, J. and FENICHELL, S. (2004) *What Matters Most: How a Small Group of Pioneers is Teaching Social Responsibility to Big Business, and Why Big Business is Listening*, New York, Basic Books.

HYVARINEN, A. (2008) Eco-labeling and environmentally friendly products and production methods affecting the international trade in textiles and clothing, http://www.intracen.org/textilesandclothing/eco_labelling.htm.

MANRING, S. and MOORE, S. (2004) Creating and managing a virtual Inter-organizational learning network for greener production: a conceptual model and case study. *Journal of Cleaner Production*, **14**, 8.

MANRING, S. and MOORE, S.B. (2009) Sustainable strategy development in small and medium sized enterprises for sustainability and increased value creation. *Journal of Cleaner Production*, **17**, 276–282.

MARIN, A.W. and TOBLER, M. (2003) The purpose of LCA in environmental labels and concepts of products. *Journal of Life Cycle Analysis*, **8**, 115–116.

MCDONOUGH, W. and BRAUNGART, M. (2002) *Cradle to Cradle: Remaking the Way We Make Things*, New York, North Point Press.

MEADOWS, D.H. and CLUB OF ROME (1974) *The Limits to Growth; A Report for the Club of Rome's Project on the Predicament of Mankind*, New York, Universe Books.

MEDING, J. (2008) Specialist retailers' questionaire: textile quality marks, Stuttgart, Germany, BBE Retail Experts.

NIMON, W. and BEHGIN, J.C. (1999) Eco-Labels and international trade in textiles. Working Paper 99-WP 211, Center for Agricultural and Rural Development, Iowa State University, Lowa.

OECD (2000) *Creative society of the 21st century*, in Preceedings of Conferece on *21st Century Social Dynamics: Towards the Creative Society*, Berlin, Germany, OECD.

ROBERT, K.-H., SCHMIDT-BLEEK, B., DE LARDEREL, J.A., BASILE, G., JANSEN, L.J., KUEHR, R., THOMAS, P.P., SUZUKI, M., HAWKEN, P. and WACKERNAGEL, M. (2002) Strategic sustainable development – selection, design and synergies of applied tools. *Cleaner Production*, **10**, 197–214.

SCHLEY, S. and LAUR, J.S. (1998) *Creating Sustainable Organizations*, Waltham, Massachusetts, Pegasus Communications, Inc.

SENGE, P.M. (2008) *The Necessary Revolution: How Individuals and Organizations Are Working Together to Create A Sustainable World*, New York, Doubleday.

SUROWIECKI, J. (2004) *The Wisdom of Crowds: Why the Many Are Smarter than the Few and How Collective Wisdom Shapes Business, Economies, Societies, and Nations*, New York, Doubleday.

WATTS, D.J. (2003) *Six Degrees: The Science of A Connected Age*, New York, Norton.

WILLARD, B. (2002) *The Sustainability Advantage*, Gabriola Island, British Columbia, New Society Publishers.

WILLARD, B. (2005) *The Next Sustainability Wave: Building Boardroom Buy-in*, Gabriola Island, British Columbia, New Society Publishers.

ZIMAN, J.M. (2000) *Technological Innovation as an Evolutionary Process*, Cambridge, UK, Cambridge University Press.

11

Organic cotton: production practices and post-harvest considerations

P. J. WAKELYN, National Cotton Council of America (retired), USA, and M. R. CHAUDHRY, International Cotton Advisory Committee, USA

Abstract: Cotton is the most important natural textile fiber produced in the world and at least 5000 ha are grown in about 60 countries. Cotton is produced using a wide range of farming practices and cotton production is highly technical and very difficult. All methods of producing cotton have impacts that are not necessarily environmentally friendly but are necessary to produce the crop. The goal of cotton production should be to produce the best quality product for textile mills with the lowest environmental impact, in an economically viable manner. There continues to be worldwide interest in organic cotton because of environmental and economic reasons. In 2007/2008 production of organic cotton increased to over 0.5% of world cotton production, mainly owing to increased production in India, Turkey, Syria, China and some African countries. However, growing cotton using organic production practices is not suitable for all countries and all farmers. Organic production is not necessarily any more or any less environmentally friendly than current conventional cotton production. For the textile consumer, there is no difference between conventionally grown cotton and organically grown cotton with regard to pesticide residues. Growing organic cotton is more demanding and more expensive than growing cotton conventionally. Organic production can be a real challenge if pest pressures are high but with commitment and experience, it could provide price premiums for growers willing to meet the challenges. Conventional and organic cotton production can co-exist. Profitability will drive decisions on farms and throughout the supply chain as to what cottons are produced and used.

Key words: organic cotton, cotton production practices, post-harvest processing, environmental issues, textiles.

11.1 Introduction

Cotton is the most important natural textile fiber and cellulosic fiber produced in the world (Wakelyn *et al.*, 2007). Cotton is grown in about 80 subtropical and tropical countries (59 grow at least 5000 ha) in the world, primarily between 37°N latitude and 32°S latitude but can be grown as far north as 45 °N latitude in China, using a great diversity of farming practices. Cotton production is highly technical and difficult because of pest pressures and environment, e.g. drought, temperature and soil nutritional conditions. If a life cycle (from raw materials to end of life) assessment (LCA) of cotton production practices is considered, all methods of producing cotton cause environmental impacts that are not necessarily 'environmentally friendly' but are necessary to produce the crop. The goal of cotton production is to produce the best quality product for textile mills with the lowest environmental impact in an economically feasible manner.

Cotton grown without the use of any synthetically compounded chemicals – i.e. pesticides, plant growth regulators, defoliants, etc. – and fertilizers, and using an agreed upon organic production plan is considered 'organic' cotton (International Cotton Advisory Committee (ICAC), 1996; Chaudhry, 1998, 2003; Myers and Stolton, 1999a; Le Guillou and Scharpé, 2000; Guerena and Sullivan, 2003; Kuepper and Gegner, 2004; Pick and Givens, 2004; NOP, 2005). However, chemicals considered 'natural' and some acceptable synthetic chemicals can be used in the production of organic cotton (USDA, 2008b, 7 CFR Part 205.600–606; Synthetic substances allowed for use in organic crop production) as well as natural fertilizers. Different certification organizations have similar lists for allowed chemicals (e.g. see *Official Journal of the European Union*, 2006). *Bacillus thuringiensis* (Bt), a naturally occurring soil bacterium, can be used as a natural insecticide in organic agriculture (Zarb *et al.*, 2005). Bt produces a variety of proteins that are toxic to specific insects but harmless to fish and mammals, including humans. Scientists can transfer these Bt genes into plants to produce insect-resistant biotech cottons and other plants. However, biotech/transgenic cottons containing Bt genes and other genes added using recombinant DNA technology are not allowed to be used for the production of organic cotton under the general reasoning that the materials are synthetic and not natural.

There are several principles that characterize certified organic farming: biodiversity, integration, sustainability, natural plant nutrition and natural pest management (Kuepper and Gegner, 2004). The US National Organic Standards Board adopted the following definition of 'organic' agriculture (NOP, 2002): 'Organic agriculture is an ecological production management system that promotes and enhances biodiversity, biological cycles and soil biological activity. It is based on minimal use of off-farm inputs and on management practices that restore, maintain and enhance ecological harmony.'

The production of cotton using organic farming practices seeks to maintain soil fertility, to use materials and techniques that enhance the ecological balance of natural systems, and to integrate the parts of the farming system into an ecological whole. Under the US Department of Agriculture

(USDA) – National Organic Program (NOP), Organic Production and Handling Requirements (USDA, 2008a; 7 CFR Part 205.201–206) include land requirements, soil fertility and crop nutrient management practices, seed and planting stock practices, crop rotation and crop pest management.

'Organic' is a labeling/marketing term. All fibers sold as organic in the USA must meet the USDA NOP (2002) rule for crop production: *For cotton to be sold as 'organic cotton', it must be certified by an independent organization that verifies that* it meets or exceeds defined organic agricultural production standards (see Section 11.9). To produce 'organic cotton textiles', certified organic cotton should be manufactured according to organic fiber processing standards/guidelines (Wilson, 2008a; see Section 11.10). Regulations are critical to standardize criteria for organic cotton production. Guidelines for post-harvest handling/processing that can be certified by an independent third party to a defined set of criteria will facilitate domestic and international trade.

A three-year transitional period from conventional to organic cotton production (i.e. elimination of synthetic chemicals and fertilizers) is required for certification. Cotton produced during this three-year period is described by many organizations as 'transitional', 'pending certification' in California and elsewhere or 'organic B' in Australia. In order to increase the amount of organic cotton that they trade in and process, to satisfy the growing demand they see for organics, a giant spinner in Japan (i.e. Itochu Corporation) is teaming up with an agricultural cooperative in India, and a major retailer in the USA (i.e. Wal-Mart) is cooperating with growers in the USA to purchase 'transitional' cotton at a 10–20% premium. There are some other organic cotton alternatives (Wilson, 2008b) that are produced in the USA in very small amounts, these alternatives claim reduced use of chemical pesticides and fertilizers.

Label claims such as 'green', 'environmentally friendly', 'clean' or 'natural' are used by some manufacturers (Myers and Stolton, 1999a) but are not approved for use as claims by the environmental labeling requirements in the International Organization for Standardization (ISO) standards ISO 14020 (ISO, 2000) and ISO 14021 (ISO, 1999), the US Federal Trade Commission 'Green Guides' (2008a) and the UK Department for Environment, Food and Rural Affairs (Defra) Green Claims Code (2009). The claims are general claims with no government or official definition. When such claims are used to sell products they are considered too vague to be meaningful to the consumer and consumer perception and substantiation issues may arise. In contrast, the agricultural commodity 'organic cotton' has a specific definition, i.e. cotton produced without the use of synthetic chemicals and following the organic production plan for modern organic farming techniques outlined by USDA NOP, that is covered by mandatory government regulations.

Demeritt (2006) suggests that the word 'organic' for most consumers is primarily a marker – a word that symbolizes a lifestyle that they wish to be part of. Certification, the organic regulation itself or the 'science' behind organic products is not what many consumers care about when buying organic products. At the 2008 IFOAM (International Federation of Organic Agriculture Movements) Organic Fibers and Textiles Conference in Italy, Wolfgang Sachs, Senior Fellow at the Wuppertal Institute for Climate, Environment and Energy, noted that for the organic textiles market there was no obvious benefit to the consumer, unlike the consumers perception for organically produced food (Anonymous, 2008).

Based on consumer survey research carried out by Synovate, consumers in ten markets globally – Brazil, China, Colombia, Germany, India, Italy, Japan, Thailand, Turkey and the UK – are less concerned about use of genetic modification (GM) technology or input use for fiber than for food (Terhaar, 2008) and a significantly higher percentage of consumers responded that they would pay more for clothing made of natural fibers (64%) than for clothing made with organic cotton (37%). Furthermore, 52% of consumers said they had no awareness of organic cotton at all, and between 30 and 40% of consumers in such developed markets as Germany, the UK and Japan said they would not expend any extra effort to find 'environmentally friendly' clothing. Regarding perceptions of fibers and their relative environmental safety, consumers in the surveyed countries, including the USA, put cotton as the most environmentally safe fiber with other natural fibers viewed positively as well. Synthetic fibers, including polyester, were consistently ranked much lower on the environmentally friendly scale and tended toward a perception of being environmentally harmful.

11.2 World organic cotton production

Certified organic cotton was introduced around the 1989/1990 crop year and as of 2008 close to 30 countries have reported production of organic cotton (Table 11.1). With the help of international and national institutions, mostly from Europe and the USA, serious efforts have been made to produce organic cotton in many African countries including Burkina Faso, Benin, Mali, Tanzania and Uganda; insecticides and fertilizers were either not used or minimally used in conventional production in these countries (Ratter, 2004; Shell Foundation, 2006). There have been small projects in Mali (35 metric tons (MT) in 2003/2004, 238 MT in 2006/2007), Kyrgyzstan and some other developing countries (Anonymous, 2004; Traoré, 2005). Some countries have already stopped organic cotton production for economic reasons while others are expanding their production based on the global trends of a growing demand for organic cotton products.

The world production of organic cotton since 2000/2001 continues to increase but it is a very small part of world cotton production. By even the most optimistic estimates and projections it is unlikely that organic cotton production will exceed 2–3% of world cotton production. In 2001/2002 about 11 countries in the world produced about 10 148 MT (46 579 480 lb US bale equivalents) of organic cotton, according to the Organic Exchange, with Turkey, the USA and India accounting for about 75% of production. In 2004/2005, the world production of organic cotton, according to the Organic Exchange (2005) and Garrott and Paratte (2008), was about 25 394 MT (116 391 US 480 lb bale equivalents), about 0.1% of world cotton production, which was about 25 million MT (120 million US 480 lb bales). In 2006/2007 it was about 57 730 MT (265 515 bales; 0.22% of world cotton production); and in 2007/2008 it was estimated to be about 145 865 MT (668 581 bales; 0.61% of world cotton production) (Organic Exchange, 2008b) (see Table 11.1). But in 2008/2009 it dipped to 117 021 MT (537 126 bales) (Anonymous, 2009). The big increases since 2001/2002 have been a result of increased production in India, Turkey, Syria and China who currently account for over 90% of world organic cotton production. Syria, which produced no organic cotton before 2006/2007, listed its organic cotton production at 8185 MT for 2006/2007 and about 26 000 MT for 2007/2008 (M. Nayef, Director of Cotton Research Administration, Syria, personal communication, 2008). About 20 other countries around the world now produce organic cotton. In some countries the cotton is pre-contracted, while in other countries it is not pre-contracted but sold through agents. It is, however, sometimes difficult to determine whether the cotton is really certified (Anon., 2008). The major retail consumers of organic cotton in the world in 2008 are Sam's Club/Wal-Mart (Arkansas, USA), Nike (Oregon, USA), WoolWorths (South Africa), Coop Switzerland, Otto (Germany) and C&A.

11.3   Why organic cotton?

The emphasis today is on environmentally friendly production practices and 'sustainable' production through the entire textile chain. Consider the following: is organic cotton more 'sustainable' than conventional cotton; an environmentally preferable product, of added benefit to the environment, farmers, and consumers; or is it essentially a marketing tool or is it both? Proponents of organic cotton and those who market organic cotton products incorrectly promote the perception that conventional cotton is not an environmentally responsibly produced crop (Myers, 1999; Yafa, 2005; Organic Exchange, 2008a; Patagonia, 2008; Hae Now, 2008). Some of the reasons used to support their contentions are that conventional cotton production greatly overuses and misuses pesticides/crop protection products that have an adverse effect on the environment and agricultural

Table 11.1 World organic cotton production

Country	1990/ 91	1992/ 93	1993/ 94	1994/ 95	1995/ 96	1996/ 97	1997/ 98	1998/ 99	1999/ 00	2000/ 01
Argentina	—	—	75	75	—	—	—	—	—	
Australia		500	500	750	400	300	300	—	—	—
Benin		—	—	—	—	1	5	20	20	30
Brazil		—	1	5	1	1	1	5	10	20
Burkina Faso										
China (PRC)										—
Egypt	14	50	153	600	650	625	500	350	200	200
Greece				300	150	125	100	75	50	50
India		200	250	400	925	850	1 000	825	1 150	1 000
Israel		—	—	—	50	50	20	140	180	530
Kenya		—	—	—	—	—	5	5	5	—
Kyrgyzstan										
Mali										
Mozambique					100	75	50	—	—	—
Nicaragua				20	20	20	20	—	—	—
Pakistan										
Paraguay		—	100	75	50	50	50	—	—	—
Peru		200	675	900	900	600	650	650	500	550
Senegal					1	1	10	50	146	200
Syria										
Tanzania		—	—	—	10	100	100	100	200	250
Turkey		125	200	600	725	850	1 000	1 200	2 000	1 750
Uganda		—	—	25	75	300	450	250	246	275
USA	330	1 000	1 950	2 400	3 350	1 500	1 300	1 878	2 955	1 625
Zambia										
Zimbabwe							1	5	5	—
Total (MT)	344	2 075	3 826	6 150	7 482	5 507	5 562	5 575	7 545	6 480
(bales)		9 510	17 535	28 188	34 292	25 240	25 493	25 552	34 581	29 700
% of world production										

Source: Organic Exchange, 2008a, 2008b; Garrott and Paratte at Paul Reinhart AG (2006, 2007, 2008), Turkey, Israel, Mali 2004/05–2006/07 and Zimbabwe 2005/06–2006/07; M. Nayef M, personal communication, 2008; Syria, 2006/07; C. M. Djaboutou, Centre de Recherches Agricoles Coton et Fibres l'Institut National des Recherches Agricoles du Bénin (INRAB), Benin, private communication, 2008; Benin, Burkina Faso, Mali, Senegal, Tanzania, Uganda, Zambia, 2005/06–2006/07.

2001/ 02	2002/ 03	2003/ 04	2004/ 05	2005/ 06	2006/ 07	2007/ 08
—	—	—	—	—	—	—
—	—	—	—	0	4	—
38	46	25	67	67	207	223
—	—	—	—	17	20	82
		—	45	200	143	436
106	596	1601	1870	2532	4079	7354
200	122	122	240	240	250	761
—	—	—	—	0	59	72
696	855	2231	6320	12483	18790	73702
<425	390	380	436	600	370	313
—	—	—	2	6	3	3
		—	65	60	150	194
	19	35	296	153	258	335
—	—	—	—	0	0	—
—	—	—	—	0	0	65
	256	400	600	1000	271	206
—	9	60	70	184	238	105
300	300	404	813	1603	2017	1339
—	6	6	27	33	32	75
				—	8185	28000
400	380	600	1213	649	1662	2852
5504	12865	11625	10460	14460	23152	24440
250	500	740	900	1000	1798	2545
2227	1571	1041	1968	2512	1918	2716
		—	2	23	6	59
2	3	—	0	0	40	—
10148	19270	17645	25394	38840	64125.5	145865
46579	88449	80991	116391	178298	294336	668581
0.047%	0.10%	0.084%	0.096%	0.14%	0.24%	0.61%

workers and that conventionally grown cotton fiber/fabrics/apparel has chemical residues on the cotton that can cause cancer, skin irritation and other health-related problems to consumers (Environmental Justice Foundation, 2007). Factual documentation for many of the statements expressed by proponents of organic cotton is lacking and some global corporations base their marketing programs around undocumented, misleading, incorrect information.

Proponents of organic cotton also feel that organic cotton is a more sustainable approach (Myers and Stolton, 1999a) to cotton production and suggest that organic production is equivalent to sustainable. Organic cotton production is not equivalent to sustainable – either organic or conventional cotton production practices may be sustainable or unsustainable. The word 'sustainability' is used often in the environmental debate but is often misunderstood because there are no parameters and limits that can be defined objectively (Gleich, 2008). There are some reasonable definitions of sustainability but there is no universally accepted definition. The United Nations (1987) widely used and accepted international definition of 'sustainability'/'sustainable development' is that the use of resources for the needs of the present should not compromise the ability of future generations to meet their own needs. According to the US Environmental Protection Agency (US EPA, 2008e):

> Sustainability has many definitions but the basic principles and concepts remain constant: balancing a growing economy, protection for the environment, and social responsibility, so they together lead to an improved quality of life for ourselves and future generations.

'Sustainable agriculture' was addressed by the US Congress in the 1990 Farm Bill. Under that law, the term 'sustainable agriculture' means (Farm Bill, 1990):

> ... an integrated system of plant and animal production practices having a site-specific application that will, over the long term:

- satisfy human food and fiber needs;
- enhance environmental quality and the natural resource base upon which the agricultural economy depends;
- make the most efficient use of non-renewable resources and on-farm resources and integrate, where appropriate, natural biological cycles and controls;
- sustain the economic viability of farm operations;
- enhance the quality of life for farmers and society as a whole.

Sustainable agriculture has three long-term concurrent goals: (a) social/ quality of life goals (i.e. to satisfy personal, family and community needs for health, safety, food and happiness); (b) environmental quality goals (i.e.

to enhance finite soil, water, air and other resources); and (c) economic goals (i.e. to be profitable). The most sustainable choice is the one for which the net effects come closest to meeting these goals. Sustainable production must supply the world's demand for natural fiber and food; it must maintain environmental quality and the natural resource base upon which the agricultural economy depends; and it must sustain the economic viability of cotton farming operations. If a production system requires significantly more land and significantly more labor to produce the crop and production costs are significantly higher, it is questionable whether such a system is sustainable.

Various countries and labeling guidelines do not permit the use of the term 'sustainable'. The UK's Advertising Standards Authority (ASA, 2008) has indicated that since there is no universally accepted definition of the term 'sustainable', its use in an advertisement is likely to be ambiguous and unclear to consumers. The term 'sustainable' is also not approved for use by the US Federal Trade Commission, 'Guides for the use of environmental marketing claims' (2008a) and the environmental labeling requirements in ISO 14021/ISO 14020 (ISO, 1999, 2000), a recognized international environmental management system.

In summary, 'sustainability' should be considered as the achievement of a successful balance between three concerns: environmental protection/ health, economic profitability and social needs. Sustainability can neither be determined exactly nor defined objectively. It should also not be imposed as a strategy for all situations. If the term 'sustainable' is used as a marketing claim, it should be referenced to a set of criteria/metrics that can be verified by an independent third party. The United Nations and the US EPA definitions can be used as appropriate guides for 'sustainable' production but each industry and society has to weigh the priorities between different values, which can conflict with one another.

11.3.1 Some cotton production/farming practices for consideration

As discussed in Section 11.4, some aspects of conventional farming practices used in the past have been shown to be environmentally unsound. Irresponsible use of insecticides and other pesticides/crop protection products can lead to serious consequences in agriculture and, while some of the effects are long term, others are reversible. Excessive use of insecticides can significantly affect the natural biological control system, at the same time as it increases production. Over time as the population of insects increased, the number of sprays increased. Researchers have tried to compensate the natural biological control with artificial rearing of natural enemies, but significant success has been limited in most countries.

Biotech cotton varities, which officially (i.e. legally grown) accounted for over 50% of world cotton production in 2007/2008 (ICAC, 2008) (Table 11.2) but unofficially for over 60% based on production reports from China, India, Pakistan and Brazil where production is higher than officially reported, are not allowed to be used in organic cotton production. Officially, 48% of the cotton traded internationally in 2007/2008 was biotech cotton produced in nine countries. Biotech cottons have been shown to reduce the use of insecticides and herbicides, and minimize adverse effects on non-target species and beneficial insects (Fitt *et al.*, 2004; Wakelyn *et al.*, 2004). With the widespread cultivation of biotech cottons there has been significant return of beneficial insects to the fields, further reducing the number of insecticide applications necessary to control insects. The use of insect-resistant cotton plants with built-in resistance (e.g. 'Bt cotton' with insect toxin genes in the plant) reduces the use of harmful insecticides needed to control certain insect pests in the crop. Use of plants tolerant to a specific broad-spectrum herbicide ('HT cotton') allows this herbicide to be used to remove a broad range of weed species without destroying the genetically modified crops themselves. This type of weed management reduces the need for a greater number of spray treatments with a variety of herbicides that are each specific for a single or a few weed species.

The cost of crop protection products and the cost of conventional cotton production relative to the selling price of cotton, demands that insecticides, herbicides and other crop protection products are applied judiciously and environmentally responsibly. Pesticides must be used only when and where necessary to protect the crop, using integrated pest management (IPM) practices, weed management, and in some cases remote sensing/precision farming techniques (global positioning system (GPS) and satellite technology) and computer-aided crop and pest management (Cotton Australia, 2009). Reducing input costs drives many producer decisions today. It is encouraging to note that insecticide use on cotton worldwide is declining. Data show that the sale of insecticides (by value) used to control insects on cotton has not increased since 2000. This is partly a result of the extensive use of biotech cotton but it is also due to fewer and fewer sprays being used on cotton in non-biotech-cotton-producing countries because of the use of IPM and other modern practices of agriculture.

Crop protection products, fertilizer use and tillage practices are some examples of how organic and conventional production practices differ.

Crop protection products

Today, it is incorrect to say that pesticides/crop protection products are overused and misused in conventional cotton production in developed countries and in many developing countries. Because of better management

Table 11.2 Biotech cotton production in different countries

Country	1996/97	1997/98	1998/99	1999/00	2000/01	2001/02	2002/03	2003/04	2004/05	2005/06	2006/07	2007/08
Argentina	10		0.8	3.9	6.1	4.6	8	10	10*	20*	25	25
Australia		15	20	25	30	30	30	60	60*	90*	90	90
Brazil											<1	1?
China		1	2.4	14.2	25	32	51	58	65	70	71	69
Colombia								0.5	17.3	35.3	41.8	45.1
India							<1	1.1	6.1	14.1	40.7	61.4
Indonesia							<1	<1	<1	—		
Mexico	0.29	7.84	14.3	12.5	33.4	27.4	37.5	41.4	60.6	57.4	59	60
South Africa			12	28	24	74	84	95	95	90	90	90
Pakistan												?
USA	13	26	46	60	72	71	77	76	80	82	87.6	90.1
World area (%)	2.1	4.3	6.8	12.4	15.7	17.1	20.6	21.93	25.07	29.43	36.5	44
World production (%)							28.8	30.0	34	40	45	51
World export (%)							33.9	36.1	35	39	40	48

Source: ICAC (2008).

practices, the use of crop protection products is decreasing in all countries and those that are used are heavily regulated in developed countries and regulated to some degree in developing countries. In 2006, about 8.5% of all global crop protection chemical use, based on sales data in millions of US$, was on cotton (Cropnosis, 2006). Fruits and vegetables consumed about 29% and cereals crops, including rice and maize, about 35% (Tables 11.3 and 11.4). Some organizations incorrectly claim that about 1680 kg/ha (1500 lbs/ac) of insecticide and herbicide are used to grow cotton – a more accurate figure for US cotton production in 2007 was about 0.87 kg/ha (0.78 lbs/ac) of insecticide and 2.9 kg/ha (2.6 lbs/ac) of herbicide (USDA-NASS, 2008). Total pesticide usage in the USA cotton for 2007 (USDA-NASS, 2008) was about 5.7 kg/ha (5.1 lb/ac) – 5.6 g/kg (0.09 oz/lb) of cotton fiber produced. It is unfortunate that incorrect, misleading and undocumented data regarding the use of crop protection products are sometimes used as a basis for marketing organic cotton by some corporations and organizations.

Table 11.3 World pesticides sales in 2004 by crop

Crop	Total sales (US$M)	Percentage
Fruit and vegetables	9 298	28.7
Cereals	5 136	15.9
Soybeans	3 409	10.5
Rice	3 084	9.6
Maize	3 002	9.3
Cotton	2 745	8.5
Sugar beet	611	1.9
Oilseed rape/canola	482	1.5
Other crops	4 582	14.1
Total	32 349	100

Source: Cropnosis (2006).

Table 11.4 World pesticides sales in 2004 by product type

Product type	Total sales (US$M)	Cotton sales (US$M) (% of total sales)
Herbicides	14 849	777 (5.2%)
Insecticides	8 635	1 618 (18.7%)
Fungicides	7 296	70 (1.0%)
Others	1 569	280 (17.8%)
Total	32 349	2 745 (8.5%)

Source: Cropnosis (2006).

Human health and environmental hazards associated with insecticide application, disposal of insecticide containers and other aspects of insecticide storage and handling have been reduced through the development of new insecticide products that are less persistent and less toxic to humans. New crop protection products, on average, cost millions of US$ and take years to develop. In the USA each new crop protection product is subjected to about 120 separate tests and costs about US$180–220 million and 8–9 years to develop from discovery to first sales (CropLife International, 2005). 'Tolerances' (term used in the USA) or 'maximum residue limits' (MRLs; term used in many other countries), the limits on the amount of pesticide that can remain in or on a product, are usually set at as low as 1000 times lower than the 'no observable effect level' (NOEL) depending on the risk factors (e.g. cancer risk, infants and children's exposure, etc.) used (US EPA, 1999, 2008b, 2008c (40 CFR Part 158, Subpart F)). Since cotton is a fiber and food crop, any crop protection products that are used in the production of cotton must meet the same regulations as any food crop. In addition, countries like the USA have strict regulations for approval and use of crop protection products as discussed above (US EPA, 1999, 2008b, 2008c) as well as strict worker protection standards for application of crop protection products and field re-entry after application (US EPA, 2008a), and for storage and disposal of crop protection products and used containers (US EPA, 2008d). This oversight greatly reduces health risks to workers and the environment. No crop protection products registered for use on US cotton and cotton from many other countries are on the list of restricted products that have to be tested to comply with the EU Eco-label for Textiles (*Bremen Cotton Report*, 1993; EU, 2002). According to the US EPA, US pesticide safety is the highest in the world. Biotech (transgenic) cottons containing Bt genes can also greatly reduce the health risks to workers from agricultural chemicals (Fitt *et al.*, 2004; Wakelyn *et al.*, 2004) but are not allowed for organic cotton production even though Bt can be used directly for insect control for organic cotton production.

Organic cotton production does not use synthetically compounded chemicals but can use 'natural' chemicals (USDA, 2008b) like sulfur dust, Bt sprays and other biological control agents in pest management, and can use organic acid-based foliar sprays (e.g. citric acid) and nitrogen and zinc sulfate in harvest preparation (see Section 11.5.9). These natural chemicals used in harvest preparation are not as effective for leaf drop, which can lead to slower harvesting, reduced grade of the cotton and increased cost of ginning (Swezey and Goldman, 1999), when cotton is machine picked. Hand picking does not require defoliation whether it is organic or conventional production but there can still be green leaf trash in hand-picked cotton that can affect grade, if the plant is not defoliated before harvesting.

Fertilizer use and tillage practices

Plants require water and nitrogen (N) to grow. In 2007 in the USA, total fertilizer use on cotton was 168 kg/ha (150 lbs/ac) of N (USDA-NASS, 2008). Organic and conventional cotton production vary greatly in the kind and amount of fertilizer used to supply N for the plant. To replenish N in the soil, organic production relies primarily on manure application, about 6.7–10.8 MT/ha (3–5 tons/ac) of poultry manure (Alabama Cooperative Extension System, 2006) or about 20–30 MT/ha (9–15 tons/ac) of cattle/dairy manure (Agronomy Facts, 1997). In contrast, conventional cotton production typically relies on about 365–495 kg/ha (325–440 lbs/ac) of a commercial nitrogen fertilizer, e.g. urea or UAN (liquid form of urea and ammonium nitrate), to supply N for the plant.

With commercial fertilizers the nutrients are readily available for plant uptake and can be applied when the crop most needs the nutrient but the loss of nutrients through leaching, volatilization and erosion can occur. Poultry manure, the preferred manure for organic cotton production, is a renewable resource but the use of manure can present a problem in containing nutrient run-off into streams (e.g. phosphates to surface water can cause eutrophication (Schmidt and Rehm, 2002)) or leaching into groundwater (e.g. nitrates in groundwater can be a mammalian toxin), and biohazards can be associated with manure (e.g. the H5N1 avian flu virus and pathogenic bacteria are carried in poultry manure). In addition, in the USA, the total amount of manure produced from animal agriculture is estimated to be only able to meet the fertilizer needs of about 20% of US agriculture.

Manure is applied to fields that have been worked with initial tillage operations, by spreading and incorporating into the soil with two additional discings (Swezey and Goldman, 1999). Increased tillage can have adverse effects on the soil and can lead to more soil erosion as well as the use of more fossil fuels. In contrast, over 60% of conventional cotton production, in countries like the USA, use conservation tillage practices. Conservation tillage leaves more crop residue on the soil surface and also reduces greenhouse gas emissions – by burning less fossil fuel and by sequestering carbon in the soil. Conservation tillage leaves crop residue (plant materials from past harvests) on the soil surface, reduces run-off and soil erosion, conserves soil moisture, helps keep nutrients and pesticides on the field, and improves soil, water and air quality. Conservation tillage is difficult or impossible to implement in organic cotton production systems because of the heavy reliance on mechanical cultivation and/or use of extensive hand labor for weed control.

Cost of production

The elimination of synthetically compounded insecticides, herbicides and fertilizers in organic production systems should reduce the cost of production for growing and harvesting one hectare of cotton. However, elimination of such important inputs could have serious consequences on yield and result in greater hand-weeding costs thus increasing the cost per kilogram of lint (Swezey *et al.*, 2007). Shifting from intensive farming practices to organic cotton production is more expensive. Results from a 6-year study in the USA showed that organic cotton production costs were 37% higher than those of conventional cotton (Swezey, 2002; Swezey *et al.*, 2007) and the yields for organic cotton were about 65% of conventional cotton yields. Organic cotton usually has yields that are about 30–70% lower than conventional cotton, as can be seen from recent US organic cotton production data (Table 11.5). Production in Uganda in 2007/2008 where organic production yields were less than half those of convention cotton (Muwanga, 2008) and in 2008/2009 where conventional cotton yields reportedly were an average of 900 kg/ac, while organic cotton yields were an average of 200 kg/ac (Kasita, 2008), indicates that organic cotton growers must be provided with an organic production technological package and access to better organic input products if organic cotton is to be produced successfully. Lower yields mean organic cotton requires more land to produce the same quantity of cotton. Organic cotton can have lower lint quality (color grade and leaf rating) and grades, further reducing economic returns (Swezey *et al.*, 2007). Organic cotton requires significantly more labor to produce, primarily for weed control, i.e. many workers with hoes to kill weeds. In addition, organic cotton can require more energy than conventional cotton production because of increase tillage and increased trips to the fields to transport the large amounts of animal manure required.

Dogan (2008) of Akdeniz Exporters Association, Turkey reported that production of organic cotton is 50% more expensive and requires more land, water and labor. Matthews and Tunstall (2006) believe that while there may be a 'market for "organic" cotton production, it would be a retrograde step to ignore the technological advances that enable much higher yields to be obtained economically'.

11.3.2 Fiber/fabric/apparel

Another major marketing point used by organic proponents is that conventional cotton contains pesticide residues that are potentially harmful to consumers (Environmental Justice Foundation, 2007) and that certified organic means 'pesticide residue-free'. Such a claim is not supported by data and is a common misconception (Kuepper and Gegner, 2004). In fact,

Table 11.5 Yields for US organic cotton versus conventional cotton

Year	USA organic acres	USA organic bales	lbs/acre organic	ha organic	MT organic	kg/ha organic	Total USA acres (×1000)	Total USA bales (×1000)	USA lbs/acre	Organic conventional yield (%)	USA % organic	World % organic	World organic cotton production (MT)
2008/09	9279			3757			7727	13 040	810				
2007/08	8510	12 466	703	3445	2716	788	10 492	19 033	879	80%	0.065%	0.61%	145 865
2006/07	5968	8803	708	2416	1918	794	12 731	21 588	814	87%	0.04%	0.24%	57 731
2005/06	9537	11 530	580	3859	2512	651	13 803	23 890	831	70%	0.05%	0.14%	37 799
2004/05	9213	9033	471	3728	1968	528	13 057	23 251	855	55%	0.04%	0.10%	25 394
2003/04	9875	4778	232	3996	1041	261	12 004	18 255	730	32%	0.03%	0.08%	17 645
2002/03	10 551	7211	328	4270	1571	368	12 417	17 209	665	50%	0.04%	0.10%	19 270
2001/02	11 456	10 222	428	4636	2227	480	13 828	20 303	705	61%	0.05%	0.05%	10 148
2000/01	15 027	7559	273	6081	1625	306	13 052	17 188	632	43%	0.05%	0.06%	12 035

Source: Organic Exchange (2008b) and USDA-NASS (http://www.ers.usda.gov/briefing/organic/).

from a pesticide residue-free standpoint there is essentially no difference between conventionally grown cotton and organic cotton as documented by the following publications.

- In response to numerous inquires concerning agricultural chemical residues on cotton, the Bremen Cotton Exchange, since 1991, has tested for pesticides residues (herbicides, insecticides, fungicides) on US and other cottons. The results showed that all cottons, including US cottons, satisfy the eco-label standard (i.e. no contamination according to Oeko-Tex Standard 100 (2008a)) and easily pass the regulations for foodstuffs – 'Thus cotton under German law theoretically can be used as a foodstuff' (Bremen Cotton Report, 1993). These studies have continued through 2007 and 2008 with the same results (Bremen Cotton Exchange, 2008; Weckman, 2008). In 2008, China and Pakistan were added to the program.
- An example of the misinformation concerning conventional cotton and pesticide residues is a 2007 publication by the Environmental Justice Foundation (2007) which stated that 'hazardous pesticides applied during cotton production can also be detected in cotton clothing'. They reference as their source, a Polish paper (Rybicki *et al.*, 2004) but have misrepresented the paper, which actually stated that the results confirm that the heavy metals and pesticides did not exceed the permissible values allowed by Oeko-Tex Standard 100 (2008a) for textile products used in direct contact with the skin. The main concern of the paper was not with raw cotton, but with possible heavy metal contamination from dyeing and other textile finishing.
- US cotton and cottons grown in some other countries meet the requirements for the current EU Eco-label for textile products (EU, 2002) without testing. This is because none of the pesticides that have to be tested for are registered by the US EPA for use on US cotton; they are also cotton not registered for use on cotton in many other countries. Therefore, if these pesticides are used, they are being used in violation of regulations.
- 'The customer demand for organically grown cotton is not a residue-free issue' (Fox, 1994), since there is no difference between organic and conventionally grown cotton from a residue standpoint.
- In preparation for dyeing and finishing, cotton fabrics undergo scouring and bleaching treatments that would remove any pesticide residues, if they were present (Kuster, 1994).

11.3.3 Summary

Differences in cotton production practices should be considered when assessing the 'sustainability' of organic cotton and conventional cotton.

Organic cotton production is no more or no less 'environmentally friendly' or 'sustainable' than current conventional cotton production. However, it can require more land to produce the same amount of cotton, it requires more labor and more tillage, and it costs significantly more (37–50% more) to produce. Furthermore, from a pesticide residue standpoint, there is no difference between conventionally grown cotton and organically grown cotton.

11.4 Cotton production practices: historical background

Before mechanized agriculture and the use of synthetic crop protection products, which began in the 1930s, cotton production practices of planting and cultivation were performed with mule-drawn farm equipment, weeds were controlled by the hoe, the use of fertilizer was sporadic and the crop was harvested by hand (Lee, 1984). From 1926 through 1945 it required 175 man-hours to produce a 218 kg (480 lb) bale of US cotton; in 2004, 3 man-hours were required per bale. Tables 11.6 and 11.7 show US cotton acreage and production in the 1930s and US cotton production from 2000/2001 to 2007/2008, respectively. Yields were less than a quarter of what they are today and it took a large amount of land to produce much less product. In 1930 in the USA, 3 159 000 MT (14 517 000 480 lb bales) were produced on 17 534 296 ha (43 329 000 acres) while in 2007/2008, 4 147 000 MT (19 033 000

Table 11.6 US cotton acreage and production, 1930–1940

USA crop year	Acreage (×1000 acres)		Production (No. of 480 lb bales, ×1000) (metric tons)
	In cultivation July 1*	Harvested	
1930	43 329	42 444	14 517 (3159)
1931	39 110	38 704	17 809 (3877)
1932	36 494	35 891	13 545 (2949)
1933	40 248	29 383	13 591 (2959)
1934	27 860	26 866	10 037 (2185)
1935	28 063	27 509	11 081 (2412)
1936	30 627	29 755	12 916 (2812)
1937	34 090	33 623	19 735 (4296)
1938	25 018	24 248	12 441 (2708)
1939	24 683	23 805	12 309 (2680)
1940	24 871	23 861	13 090 (2849)

*Planted acre data not available.
Source: USDA-Economic Research Service (ERS), Statistics on cotton and related data.

Table 11.7 US cotton acreage and production, 2000/01–2007/08

USA crop year	Acreage (×1000 acres)		Production (No. of 480 lb bales; ×1000) (metric tons)
	Planted	Harvested	
2000/01	15517	13053	17188 (3742)
2001/02	15769	13828	20303 (4420)
2002/03	13958	12417	17209 (3746)
2003/04	13480	12003	18255 (3974)
2004/05	13659	13057	23251 (5062)
2005/06	13900	13700	23885 (5204)
2006/07		12731	21588 (4808)
2007/08		10492	19033 (4147)

Source: USDA-NASS.

480 lb bales) were produced on 4245882 ha (10492000 acres) – a reduction of about 13290000 ha (33 million acres) to produce about 30% less cotton. Beginning in the 1930s, when the first steps toward mechanization began and modern cotton production practices were started, yields began to rise because of the increased use of synthetic fertilizer and attempts at insect and weed control, which allowed the crop to be managed better.

Agricultural production has changed considerably in many ways, since the 1930s. Cotton production has shifted from not using synthetic compounded chemicals to the currently chemical-based production system and higher yielding varieties with improved quality have been introduced. Organic fertilizers (animal manure) and green manure (composed of plant matter) were the only sources of replenishment of soil nutrients. Insect pressures from the boll weevil and other insects and plant diseases were mainly controlled through agronomic operations, crop rotations and mixed cropping, in addition to natural biological control. Table 11.8 indicates the size of the yield losses caused by insects and plant diseases in the 1930s before synthetic crop protection chemicals were widely used for insect and weed control.

Cotton is a major cash crop in many countries in the world and cotton production and processing are an important source of income. This has led to increased cotton production. The demands for increased production since the 1930s have mainly been met by increasing yields through the intensive use of chemical inputs and irrigation, and the use of higher yielding varieties. Improvements in cotton production have benefited farmers but have involved some environmental and social costs. Organic fertilization and agronomic operations could not cope with the needs of higher cropping intensity that depleted soils. Lower soil fertility resulted in lower

Table 11.8 Percentage reduction from full yield per acre of cotton, by stated causes

USA crop year	Deficient moisture	Excessive moisture	Other climatic causes	Plant diseases	Boll weevil	Other insects
1930	27.7	2.8	6.3	1.7	5.0	1.9
1931	8.3	2.6	3.5	2.0	8.3	1.8
1932	8.0	3.9	6.1	3.2	15.2	3.1
1933	6.8	2.6	3.7	2.3	9.1	2.2
1934	20.7	1.9	7.3	1.9	7.3	1.6
1935	9.2	3.7	6.5	2.2	8.1	5.0
1936	16.2	1.9	8.4	2.2	4.9	3.0
1937	5.7	1.5	4.1	2.2	5.3	3.0
1938	6.8	3.3	4.0	1.9	9.9	4.2
1939	10.1	4.2	5.9	1.8	8.7	2.2
1940	5.5	6.5	6.5	2.0	6.5	1.9

Source: USDA-ERS, Statistics on cotton and related data.

yields. Higher cropping intensity provided continuous availability of host plants and favorable conditions for insects to multiply at faster rates. The insect population started building up and researchers turned from natural chemicals to the use of synthetically compounded chemicals to control insects, without, necessarily, adequately considering the long-term consequences. Herbicides and other crop protection products started to be used. Organic fertilizers were supplemented with inorganic synthetic fertilizers. Economic conditions in the developed countries allowed them to embrace the new chemical-based production system faster than developing countries. Governments of developed and developing countries provided subsidies to promote the use of fertilizers and pesticides. In hindsight, the earlier overemphasis on the use of fertilizers and pesticides occurred without fully understanding the consequences.

11.5   Organic cotton production practices

The goal of all cotton production is to produce the highest quality product, in an economically feasible manner with the lowest environmental impact. An ideal crop production system consists of many components that are critical to yield, fiber quality and crop profitability (National Cotton Council, 2007). If an LCA of cotton production is considered, it can be seen that all methods of cotton production have some practices that are not necessarily environmentally friendly but are necessary to produce the crop. Cotton is produced in the world under a great diversity of farming practices. Cotton production (Hake *et al.*,

1996) is considered highly technical and difficult because of both crop vulnerability to a variety of pests and sensitivities of quality and yield to environmental (e.g. drought and temperature) and soil nutritional conditions. Conventional cotton production in the USA and other more developed countries utilizes the highest technology practices whereas cotton producing in less-developed countries can have more adverse impacts on the environment. However, production practices in all countries continue to improve.

Current practices in conventional and organic cotton production are similar in some ways but differ in other ways depending upon the operation (Myers and Stolton, 1999a; Guerena and Sullivan, 2003). If all conventional practices were followed in organic production, organic production would be ineligible for certification as organic cotton. Organic cotton production does not allow the use of most synthetically compounded chemicals (fertilizers, insecticides, herbicides, growth regulators and defoliants) that are registered for use for conventional cotton production or biotech cotton varieties. In contrast, current conventional cotton production depends on appropriate use of crop protection products/ chemicals and chemical fertilizers. Since these products are critical inputs to conventional production, as are biotech cotton varieties in some countries, alternative methods are necessary for organic production. Organic production can be a particular challenge on some soils and if pest pressures are high. The transition or conversion period for a site to become certified organic from a conventional system may take longer than three years. Although the certification period requirement is only three years, in fact examples such as the one from Louisiana have shown that a sustainable and dynamic organic system may take many more years to produce to the required standards and, thus, may be difficult to justify economically when comparing the costs of organic versus conventional cotton (Myers and Stolton, 1999b).

Both organic and conventional cotton production employ crop rotation. In the USA, rotation of cotton with maize, peanuts, soybeans, wheat, alfalfa, rice, etc. is standard practice. For organic production all other crops grown on the organic fields would have to be grown using organic production practices. In addition, no non-natural crop protection products can be used on organic fields for three years prior to the start of organic production, i.e. the 'transition' period.

It should also be noted that many of the environmentally friendly production practices that are used in some organic cotton production are also used in producing conventional cotton – cover crops, trap crops, strip cropping, wind breaks, biological control of insects (including pheromone trapping and mating disruption), etc. (Wakelyn, 1994; Wakelyn et al., 2000; Cotton Australia, 2009). In conventional cotton production, computer-aided management modeling systems, such as COTMAN, an expert system for cotton plant management based on in-season plant monitoring (ICAC, 2005), are used by some to help manage and monitor crop development as well as precision farming/remote sensing (Smith, 1996). Organic cotton production uses some IPM systems, which have shown a high potential for success both ecologically and economically, but does not and can not consider all available IPM techniques used in conventional cotton production. About 60% of US cotton acreage uses IPM (USDA-NASS, 2001).

Organic farming will require a whole systems approach that not only considers soil fertility – including microbes, nutrient supply and residue management – but also pest control to continue to keep fields productive with minimum input selection and availability (Tu *et al.*, 2006). Conventional and organic cotton systems can exist side-by-side as demand requires and both may be needed globally depending on the farm size, hand labor availability, policy, research and development available to producers in different parts of the world (Williamson *et al.*, 2005; SECO, 2006). Furthermore, the ideas and programs for organic cotton are supported, particularly in third-world countries, by global companies that wish to show environmental efforts and input toward research that will have global implications (Shell Foundation, 2006). 'Growing organic cotton is demanding, but with commitment, experience, and determination, it can be done' (Guerena and Sullivan, 2003).

11.5.1 Planting seed

Unlike most other field crops, seed cotton can not be stored for a year and planted the next year. Seed cotton is a perishable commodity that must be separated into fiber and seed by ginning (Wakelyn *et al.*, 2005). Farmers therefore usually have to sell their seed cotton and thus lose the ownership of the seed. Obtaining the seed from the gin or through a middleman can cause the seed purity and germination to be questionable. Planting seed for organic production is not different to planting for conventional production, except that it can not be a biotech cotton variety. Mechanical or acid delinting (i.e. an allowed synthetic chemical) are options for smooth flow of Upland cotton (*Gossypium hirsutum*) seed during planting. Extra Long Staple (ELS) cotton (*G. barbadense*) seed is usually smooth and does not require delinting. In addition, the planting seed for certified organic production can not be treated with synthetic fungicides, insecticides or any other synthetic chemicals, as used in conventional cotton planting seed. Both organic and conventional planting seed contain calcium carbonate (to neutralize the seed), biological fungicides and a polymer coating. Seed treatments are necessary to protect the seed from fungi and bacteria that can affect germination and lower yields.

11.5.2 Soil/land preparation (tillage) and planting

Organic agriculture is often incorrectly characterized as being addicted to *maximum tillage* (Kuepper and Gegner, 2004). Soil and seedbed preparation can be about the same for organic production as for conventional but usually involves more tillage for preparing soil for planting. Organic production conditions require soil rich in organic matter and seedbed preparation that begins with working the ground with initial tillage operations. Higher organic matter could also be helpful in better germination, better plant stands and higher yields, particularly if other resources are not available for preparing the seedbed. Plant densities are sometimes lower in organic systems than in conventional systems without necessarily sacrificing yield. The resultant changes in the microclimate around the plants can reduce pest populations and improve plant growth parameters (Van Elzakker and Caldas, 1999).

11.5.3 Soil fertilization

Mineral nutrition of cotton requires proper management of the crop and usually fertilizer use. Rates of fertilizer use depend on soil condition and yield target. Meeting the nutrient needs of the cotton plant as closely as possible assures high yields if the crop is properly protected against losses due to pests. Recommended rates for N, an important nutrient for cotton production, vary by soil type and other parameters, e.g. from 84 kg/ha (75 lbs/ac) on silt loam soils up to 100–112 kg/ha (90–100 lbs/ac) on clay soils (Huber, 2008). Dryland and irrigated cotton take up between 15.7 and 22.4 kg of N/ha (14–20 lbs of N/ac) to produce each 45.4 kg (100 lbs) of lint (Snyder and Stewart, 2008). Since soil fertilization needs vary for different soils in the world, the N rate to lint output range is much wider. For example, in Australia, conventionally grown (irrigated) cotton works on a different N fertilizer to lint production ratio, typical N rates do not exceed 200 kg/ha (224 lbs/ac) for lint yields in excess of 1500 kg/ha (1340 lbs/ac).

Organic farming and conventional farming vary greatly in soil fertilization practices. Organic cotton production primarily relies on manure application for N. Conventional farming depends on commercial fertilizers, although manure may also be used in areas where cattle/dairy or poultry production makes the manure economically advantageous. On a global scale, the nitrous oxide (N_2O; a greenhouse gas), emissions from commercial synthetic and manure-based fertilizer sources appear to be about equal (Minami, 2000).

Organic production must start with a good soil fertility status. Soil fertility in an organic production system is a long-term undertaking that typically includes crop rotation, cover cropping, animal manures and use of naturally occurring rock powders (Guerena and Sullivan, 2003). Organic production of cotton increases the soil levels of phosphorous (P), potassium (K) and micronutrients but the N level has to be augmented. Livestock manure is the key N source for organic farming. The manure should be composted before it is applied to the field. Manure is typically low in N content ranging from 2 to 5% but this varies with a number of factors, including: animal type, bedding, feed ration, storage and handling, environmental conditions, field application method and age of the manure. Because of the low N levels, large amounts of manure must be applied – ranging from 6.7–10.8 MT/ha (3–5 tons/ac) of poultry litter (16.5–32.6 kg N/MT (37–73 lbs N/ton)) to about three times that amount, 20.2–33.6 MT/ha (9–15 tons/ac) for cattle/dairy manure (4.5–4.9 kg N/MT (10–11 lbs N/ton)) (Agronomy Facts 55, 1997; Alabama Cooperative Extension System, 2006) – as the basic fertilizer, along with green manure, crop rotation, nitrogen-fixing crops and incorporation of cotton stalks. The total N in poultry manure is slightly less effective compared with N in an ammonium nitrate fertilizer (Alabama Cooperative Extension System, 2006). Poultry manure contains more N than cattle manure and is preferred for organic cotton production; however, N loss can be significant due to volatilization of ammonia. In organic production there is no way to manage the N availability in the soil, which is highest at the time of planting and reduced as the season progresses, in accordance with plant needs. Besides the inability to control nutrient availability and the large amounts that must be applied, the other disadvantage of manure use is the accumulation of phosphorous and heavy metals in the soil. These nutrients may also cause surface water contamination through run-off, leaching into groundwater and biohazards.

Conventional cotton production uses about 89.7–168.1 kg/ha (80–150 lbs/ac) of N, 0–89.7 kg/ha (0–80 lbs/ac) of P and 0–89.7 kg/ha (0–80 lbs/ac) of K from commercial fertilizers. Urea (46% N) and UAN (~34% N) are typical commercial fertilizers used to supply N, so about 195–365 kg/ha (174–326 lbs/ac) of urea or 212–494 kg/ha (189–441 lbs/ac) of UAN would need to be applied to meet the N needs. The advantages of commercial fertilizers are that there are a variety of means of application, including through irrigation and precision nutrient delivery, meaning that the nutrients in commercial fertilizers are readily available for plant uptake and can be applied when the crop most needs the nutrient. For example, N uptake is limited for cotton before squaring and the majority of N is taken up after first bloom with as much as 2–3 lb/ac per day taken up during fruiting. Split applications of N – with 10–20% of the total being applied preplanting and at or before first bloom, and the remainder during boll development – should increase the chances of N needs being met during the crop's peak demand periods. Organic production does not have this option because manure is used. Cotton bolls have a high demand for K. Approximately 70% of the K uptake occurs after first bloom, peaking at about 2.2–3.4 kg/ha per day (2–3 lb/ac per day). The disadvantage of commercial fertilizers, aside from escalating costs, is the loss of nutrients through leaching, volatilization and erosion.

11.5.4 Water use/irrigation

Cotton is a drought- and heat-tolerant crop that does not requires excessive amounts of water to grow. Cotton's global water footprint is about 2.6% of the world's water use, lower than soybeans 4%, maize 9%, wheat 12% and rice 21% (Hoestra and Chapagain, 2007). Cotton reportedly takes about 4000 l of water/kg of lint to produce, most of which comes from naturally occurring rainfall (Zwart and Bastiaanssen, 2004). Almost half of the world cotton area has no assured irrigation water supply and depends on natural rainfall. Cotton requires about 610–660 ha mm (24–26 acre inches) of water during the growing season as rainfall or supplemental irrigation to produce about 750 kg lint/ha of cotton. The yields achieved, which depend not only on water and method of irrigation (e.g. sub-surface drip, furrow, flood irrigation) but on many other factors, determine the water use efficiency (WUE) (ICAC, 2003). According to Orgaz et al. (1992), cotton WUE is 2.7 kg lint/ha mm. WUE varies considerably between countries – about 227 kg lint/Ml water in Australia, about 139 kg/Ml in California, about 136 kg/Ml in Egypt and only about 50 kg/Ml in Pakistan (ICAC, 2003).

Cotton production uses about as much water as a Bermuda grass lawn. Proper irrigation management is essential for cotton, and is used to balance vegetative growth with boll development, as well as to manage disease and insect populations. Much of US cotton is produced in areas where annual average rainfall exceeds the crop water requirements ('green' water). When irrigation is used, it results in yield stability and maximizes land use efficiency. For US farmers who do irrigate, tools to determine when irrigation is needed and advanced methods of water application are utilized, greatly reducing the amount of irrigation water used in cotton production. About 35% of the US acreage receives supplemental irrigation (USDA-NASS, 2004) and about 55% of world cotton production comes from irrigated land (ICAC, 1998, 2003).

Variety selection and site will determine irrigation needs for cotton production. Consideration of more advanced irrigation techniques may help in planning for organic cotton. Timing and quantity of water affects cotton quality and quantity. Under drip irrigation, organic cotton could have less pest problems and possibly improved quality characteristics. In areas that are rain-fed, management of organic cotton immediately becomes more complex and difficult. In addition, the cost of intensive irrigation systems in more arid climates may take longer for an organic grower to recoup than a conventional grower owing to additional input costs and possibly lower yields in organic production. More fine-tuning of irrigation input would help sustain organic cotton production if the infrastructure could be built and maintained over several years with profitable sales of organic cotton. Even so, more coordinated management would be required under the organic system and more intense evaluation of all inputs would be necessary.

The delivery method, number of applications and the amount of applied surface water varies from location to location. Total applied water requirements depend on the soil type, residual soil moisture and water availability. For cotton, furrow irrigation is the most common application method. Methods should be used to improve irrigation efficiency by minimizing evaporative water loss and reducing labor costs. Without access to irrigation technology to stabilize and optimize cotton production, many more millions of hectares of land would be required to maintain current levels of world production. The water use requirements per hectare for conventional cotton production are usually less than for organic cotton. If yields are reduced with one system relative to the other, then WUE would also be reduced.

11.5.5 Weed control/management

Control of weeds presents one of the more difficult problems in organic cotton production (McWorter and Abernathy, 1992; Swezey and Goldman, 1999). Cotton production can be affected by a wide variety of weed pests, including velvetleaf (*Abutilon theophrasti*), pigweed (*Amaranthus* spp.), tropical spinderwort (*Commelina benghalensis*), Bermuda grass (*Cynodon dactylon*), yellow nutsedge (*Cyperus esculentus*), purple nutsedge (*Cyperus rotundus*), crabgrasss (*Digitaria sanguinalis*), morning glory (*Ipomoea* spp.), common purslane (*Portulaca oleracea*), foxtail (*Portulaca oleracea*), Johnson grass (*Sorghum halepense*), cocklebur (*Xanthium* spp.), etc. The elimination of synthetic herbicides in organic production requires a different weed management approach in organic farming versus conventional farming. Weed control is no different in organic production if weeds are removed manually or mechanically which is common in most developing countries where labor is still abundant and inexpensive.

Cultural practices are an important component of organic farming practices. Delayed planting in organic production systems has been utilized so that additional mechanical weed control can be obtained prior to planting. The use of cover crops or mulches may also be used to produce some allelopathic weed suppression on fields prior to clearing the location for planting (Daar, 1986). Since cotton germinates at a soil temperature of 61°F (16°C) at a depth of about 2 inches (Head and Williams, 1996), delaying planting until the soil temperature reaches 66°F (19°C) allows the crop to emerge rapidly and uniformly and more vigorously, giving the crop more of a competitive advantage against weeds. Delayed planting also allows additional growth on any winter cover crops used. However, this strategy can also be detrimental, allowing more of a risk of increased damage from certain insect pests such as boll weevil (where still prevalent), tobacco budworm and cotton bollworm.

Generally, heavy reliance on mechanical cultivation/tillage or use of extensive hand labor are the most commonly used methods to control weeds in organic systems. This means that conservation tillage practices commonly used in conventional cotton production – with their environmental benefits such as less soil erosion and reduced use of fossil fuels – are difficult or impossible to implement in organic cotton production systems. Studies have shown that a preplant tillage, a blind tillage (during the pre-emergance to early post-emergance period) with a seeding rate increase in the crop of 5–10% to compensate for any losses to blind tillage, and one or more inter-row cultivation timed when weeds are just germinating or as soon as possible after rain or irrigation are necessary to control weeds adequately in organic cotton production.

Crop rotation, used in both conventional and organic cotton production, reduces the weed problem to some extent. Other current methods of weed management for organically produced cotton, in addition to mechanical cultivation, include flame weeding (Sullivan, 2001; Diver, 2002) which has been used to some extent but is expensive in equipment, fuel, and labor (McWilliams et al., 2007), and other cultural practices, such as hand hoeing. However, these manual and cultural operations may not be adequate to produce clean fields (Guerena and Sullivan, 2003). Furthermore, weed control using mechanical cultivation, hand hoeing and other manual practices are the main source of ergonomic worker health problems in agriculture.

Established weeds in a cotton field may harbor insects, steal nutrients from cotton and compete for water and light thus reducing the cotton yield. Smaller grower operations can more readily handle the weed management issue but it can be a concern for large organic growers.

11.5.6 Insect control/management

Under organic growing conditions, the crop is going to be attacked by the same insects that affect conventional crops. Cotton is attacked by a wide variety of insects (King et al., 1996) including the American cotton bollworm (*Helicoverpa armiger*), pink bollworm (*Pectinophora gossypiella*), tobacco budworm (*Heliothis virescens*), spiny, Egyptian and spotted bollworms (*Earias* spp.), red or Sudan bollworm (*Diparopsis* spp.), cutworms

(*Agrotis* spp.), beet armyworm (*Spodoptera exigua*), armyworms (*Spodoptera* spp.), boll weevil (*Anthonomus grandis*), cotton aphid (*Aphis gossypii*), silver leaf whitefly (*Bemisia argentifolii*), sweetpotato whitefly (*Bemisia tabaci*), lygus (*Lygus* spp.), alfalfa looper (*Autographa californica*), cabbage looper (*Trichoplusia ni*), thrips (*Thrips* spp.), spider mites (*Oligonchus* spp. and *Tetranychus* spp.), etc. Synthetic chemical insecticides used in conventional cotton production are not allowed in organic production. This prohibition is one of the most significant changes/differences in production practices with organic farming. Elimination of synthetic insecticides can potentially be a big saving for organic producers. These savings could be used to improve other field operations and management of the crop. Organic cotton producers not only save on the cost of insecticides but also the spray equipment.

Pest management in organic farming is directed toward enhancing and utilizating the natural balance of useful pests. The primary insect and mite pest management tools of organic growers are: field strips of vegetation as beneficial insect habitats (i.e. trap cropping, strip cropping and border vegetation); regular and systematic monitoring of the population levels of pests, natural predators and parasites; and the release of biopesticides/biological control agents (e.g. bacteria, such as Bt, viruses and fungal insect pathogens) (Swezey and Goldman, 1999; Guerena and Sullivan, 2003). Organic cotton production also may use 'natural' chemicals like sulfur dust.

The first two years of transition away from the use of synthetic insecticides could be difficult but by the third year, natural balances usually build up, thus reducing the need for insecticides. The need for laboratory rearing of biological control agents and utilization of other cultural control measures is more necessary in organic farming than in conventional production. Conventional cotton production also uses IPM systems that apply a number of different controls on insect pests, including insect-resistant biotech cotton, systematic monitoring of the population levels of pests, beneficial natural predators, field strips of vegetation as beneficial insect habitats and pesticides.

Organic systems may maintain more beneficial insects within the cotton field than conventional production, where insecticides can be sprayed, including but not limited to spiders within the cotton crop canopy (Bundy *et al.*, 2005). Cultural, biological and other non-chemical means of insect control are environmentally friendly and desirable, yet, the fact remains they can not eliminate insecticide use in every site situation. Certainly, they have the potential to reduce the use of insecticides. Under the circumstances, there is going to be higher insect pressure on the plant in the absence of insecticides in most production locations.

Some examples of organic insecticides and management practices used in organic cotton production are described below (McWilliams *et al.*, 2007).

1   Neem-based pesticides from extracts of the Indian neem tree include the pesticide azadirachtin which is capable of controlling over 200 types of insect pests, as well as some species of mites and nematodes. The neem-based extracts are harmless to birds, mammals and beneficial insects.

2   The Nursery Crops Laboratory in Maryland, USA has developed a product called Biosoap which has potential as an insecticide against whitefly. Biosoap is made from the extracts of tobacco plants and is environmentally safe and not harmful to beneficial insects. Biosoap killed all whiteflies in greenhouse trials and killed 94% and 78% of the whiteflies in two field trials compared to unsprayed control plots.

3   Vegetable oils are used as safe alternatives to synthetic insecticides since there are few human hazards. They are simple to use, generally inexpensive and insects and mites do not develop resistance to oils. Oils primarily act by physical action, but also as behavior modifiers in the control of insects. Efficacy of an oil for the combined activities of behavior modification and lethal effect has to be ascertained before it is used as an insecticide.

4   Microbial insecticides such as bacteria, fungi, nematodes, protozoa and viruses can be used as insecticides to control selected target insect pests. Viruses, bacteria and nematodes have been used extensively in a number of countries. Registered bioinsecticides for commercial use are available, though not popular for cotton.

5   Sex pheromones are chemicals that are used to influence the behavior of organisms. They produce a scent similar to that produced by females to attract males for the purpose of mating. The pheromones are used in such a large quantity in the field that the males are confused and are unable to locate females for mating, resulting in reduced progeny. Pheromones have been successfully used in cotton on a number of insect pests, such as pink bollworm, in the USA and in many cotton-producing countries. They are highly species-specific and safe for beneficial insects.

6   Trap cropping has been used to reduce the cost of boll weevil control by 30% and can be further enhanced by the use of such pheromones as Grandlure™ within the trap crop, particularly on earlier- or later-planted cotton. A Mississippi study showed that sesame used as a trap crop was more attractive than cotton to cotton bollworm. Oklahoma studies have reported more predators in the sorghum planted nearby cotton than on the cotton, helping the cotton crop in this strip cropping experiment, while some nearby planted alfalfa kept lygus bugs at bay.

7   Managing border vegetation to fields has also helped limit insect problems. For example, wild geranium is a particularly good host to tobacco budworm.

A number of other products like papaya leaves and red pepper have also been used in a number of countries. There is a need to develop better, less expensive and more effective organic products.

Unfortunately, not all locations can easily convert to organic cotton. Insect pressure and effective weed control measures will be predominant factors in that decision. While boll weevil and pink bollworm programs have helped limit these primary insect pests, secondary pests such as lygus have gained greater hold over some fields. So perhaps only better managed, cleaner, weed-free fields, and certain soil types, as well as areas where less detrimental or more beneficial insect populations exist, should be considered initially for organic cotton production practices.

11.5.7 Disease control/management

Plant diseases occur if the pathogen is present, the host is susceptible and the environment for disease development is favorable (Guerena and Sullivan, 2003). The disease is prevented if any of these factors is eliminated. Cotton may require management for seedling disease, soil disease, boll rot and foliar diseases. The causes of cotton diseases include fungi, bacteria, viruses and nematodes. Seedling diseases are due to soil-borne fungi, primarily *Rhizoctonia solani, Phyium* spp., and *Thielaviopsis basicola*. The most common fungal soil diseases of economic significance are Fusarium wilt (*Fusarium oxysporum*), Verticillium wilt (*Verticillium dahliae*) and Texas root rot (*Phymatotrichum omnivorum*). Boll rots are a problem where bolls are starting to open or have been damaged by insects in areas of high rainfall and humidity. Foliar diseases include: bacterial blight (*Xanthomonas campestris*), common in areas with warm, wet weather during the growing season; Alternaria leaf spot (*Alternaria macrospore*); and cotton leaf crumple virus, which can be vectored by silverleaf whitefly (*Bemisia argentifolii*).

Most cotton diseases are controlled through the use of cotton varieties that are naturally resistant/tolerant to pathogens (host-plant resistance) or by creating conditions unfavorable for pathogens to thrive. Resistant varieties used in conventional production are the best option against most diseases. If the soil is well drained and proper crop rotations are followed, diseases may not be a major concern for organic production. Control measures under organic farming conditions, such as sanitation and planting when the soil is warm, are more important practices than they are conventional farming conditions, since synthetic crop protection products can not be used in organic production. Plant disease control substances allowed for use in organic crop production (USDA, 2008b, 7 CFR Part 205.601), if necessary, include calcium polysulfide, sulfur, *Bacillus pumilus*, various fixed copper-based materials, potassium bicarbonate and others.

Biocontrol methods (competitive exclusion) for reducing or eliminating aflatoxin contamination of cottonseed have been developed and approved by the US EPA (Cotty *et al.*, 2007).

11.5.8 Nematode control/management

Nematodes are soil-dwelling, worm-like animals causing damage that is usually considered a soil disease. The main nematodes affecting cotton are: reniform (*Rotylenchulus reniformis*), root-knot (*Meloidogyne incognita*) and Columbia lance (*Hoplolaimus columbus*). The purpose of nematode management is to keep nematode population densities at a low level during the early growing season to allow cotton plants to establish healthy root systems (National Cotton Council, 2008b). *The goal is* not the complete elimination of nematodes, because nematodes are a normal part of the soil microbial population and most are beneficial.

Use of tolerant varieties of cotton and cultural practices are the control measures used for organic farming. Many early resistant cotton varieties were developed primarily to control the root-knot nematode/Fusarium wilt complex. Currently, most commercial varieties have some tolerance to this complex, but very few varieties of cotton show significant tolerance to other major cotton nematodes. Effective cultural practices include tillage, water management, clean equipment and crop rotation. Conventional cotton production, in addition to using tolerant varieties of cotton and cultural practices, uses crop protection products when necessary; these products can not be used in organic cotton production.

Chemical controls include fumigants and non-fumigants. *Fumigants* are non-selective materials that vaporize when applied in the soil and, as gases, move up through air spaces in the soil, killing nematodes and other microorganisms. *Non-fumigants* are available in liquid or granular forms that are applied either in a band or in the seed furrow at planting. Non-fumigants protect plants early in the growing season allowing them to produce deep, healthy root systems.

11.5.9 Harvest preparation, boll maturation and harvesting

Hand picking is the harvesting method used in about 75% of world production. Hand picking of organic cotton is the same as for conventional cotton. Hand picking can result in cleaner cotton, if only the fiber is removed from the boll during hand harvesting, instead of removing the whole boll, and other contaminants are avoided; these contaminants may be from workers and the field (e.g. synthetic fibers from clothing and plastic), from natural 'organic matter' from inadequate defoliation and from 'inorganic matter' such as sand or dust. In addition, there can be seed-coat fragments and stickiness (insect and plant sugars). Efficient ginning of the cotton also makes a difference in trash/contamination in the lint. According to the Cotton Contamination Surveys 1999–2001–2003–2005–2007, by the International Textile Manufacturers Federation (ITMF, 2007), the most contaminated cottons originate in Mali, India, Turkey and Central Asia (Uzbekistan and Tajikistan); these are all hand-picked cottons. The cleanest cottons can be sourced from the USA, which is all machine harvested, and from Zimbabwe and selected West African countries (Senegal and Chad), which use hand harvesting. The bale wrapping can prevent contamination if the bale is fully covered but can also contribute contamination, e.g. cotton bagging that is actually 5–50% polyester fiber instead of 100% cotton can contribute polyester fibers to the yarns, which can be detected under ultraviolet light.

If machine picking (picker or stripper harvesting) is used to harvest the cotton, defoliation is usually necessary, particularly where timely frost does not occur. Lack of defoliation options is a significant obstacle to organic production (Guerena and Sullivan, 2003). Synthetic chemical defoliants are prohibited for use on organic cotton and no other effective practices are available for mechanized farmers to eliminate green leaves, except frost. 'Natural' chemicals like organic acid-based foliar sprays (e.g. citric acid) and zinc sulfate can be used in harvest preparation in organic production. Flamers can also be used to defoliate organic cotton. Research by Funk (2008) and Funk et al. (2006) demonstrates that cotton can be prepared for harvest without chemicals using thermal defoliation. The studies show no damage to lint value, reduced late-season pest populations and fuel costs that are similar to chemical costs. Ceasing irrigation can assist leaf drop and boll maturation in some growing areas. These techniques help boll maturation, plant desiccation and leaf drop but they usually do not achieve the same results as the synthetic materials used in addition to frost in conventional cotton production (Swezey and Goldman, 1999).

Green leaves must be removed (leaf drop) prior to mechanical harvesting because leaf materials slow the harvest, reduce the grade of the cotton and increase gin costs (Swezey and Goldman, 1999). Frost helps to drop leaves but it may not occur or, if it does, it may not be advisable to wait for the freezing temperatures when most bolls on the plant have already opened. Damage to cotton quality as well as increases in aflatoxin levels, in areas where it is prevalent, are related to harvest date. So it is always advisable to harvest as soon as possible after a sufficient number of bolls have opened. Development of naturally leaf-shedding varieties is another choice, but no such varieties have been commercialized to date. Proper water management could also help to enhance leaf shedding. However, in order for organic cotton to progress with large-scale growers, alternative defoliants need to be developed that are acceptable under organic conditions, since the current practices of using organic acid-based foliar sprays and nitrogen and zinc sulfate in harvest preparation are not always very effective (Swezey and Goldman, 1999).

11.6   Post-harvest handling/processing of organic cotton

Growers of organic fibers must meet obligatory standards but the subsequent processing of the fibers into textiles is, at this time, only covered by voluntary schemes (Wilson, 2008a) (see Section 11.10). In order to protect organic integrity, all stages of processing, storage and transport of organic cotton fiber products should be segregated and protected from co-mingling with conventional fiber products and not come into contact with prohibited materials or other contaminants. Seed cotton, which consists of cotton fiber (lint) attached to cottonseed plus plant foreign matter, obtained at harvest is a perishable raw agricultural commodity (Wakelyn *et al.*, 2005). The seed cotton is transported to the ginning plant in trailers or modules, or is stored in the field in modules. When the cotton is stored in the field in modules, the modules have to be properly covered to protect the cotton from wet weather to prevent loss of quality. Ginning, which is considered part of the harvest (Wakelyn *et al.*, 2005), is the process of removing and separating lint fibers from the seed and plant foreign matter (Anthony and Mayfield, 1994). Cotton essentially has no commercial value/use until the fiber is separated from the cottonseed and foreign matter at the gin.

In order to produce organic cotton textiles, certified organically produced cotton should be processed according to acceptable processing schemes based on a set of criteria that can be certified by an independent third party (Tripathi, 2005) (see Section 11.10). The three major post-harvest processes in the conversion of raw cotton fiber into a finished fabric are: (a) yarn manufacturing (spinning, or yarn making); (b) fabric manufacturing (weaving, knitting or non-woven); and (c) preparation, dyeing and finishing. The textile product should be spun, woven/knitted and processed with energy-efficient and toxic-free methods. All processing agents used must meet requirements for toxicity, degradability and environmental, health and safety (EHS) standards. All wet processing facilities should have water conservation and resource management in place and should conform with wastewater (e.g. the US EPA textile effluent guidelines (US EPA, 2008g)) and air emission (e.g. the US EPA National Emission Standards for Hazardous Air Pollutants (NESHAP): Printing, Coating and Dyeing of Fabrics and Other Textiles (US EPA, 2008f)) standards (Anderson, 2007). The principal hazardous air pollutants (HAP) emitted by printing, coating and dyeing of fabrics and other textiles include toluene, methyl ethyl ketone

(MEK), methanol, xylenes, methyl isobutyl ketone (MIBK), methylene chloride, trichloroethylene, *n*-hexane, glycol ethers (ethylene glycol) and formaldehyde (US EPA, 2008f), and the emissions of these need to be controlled. The goal for processing of all textiles is and should be to reduce textile processing impacts on the environment ('textile eco-metrics') by reducing energy use, water use, pollution to air and water, textile waste (i.e. re-use, recycling, disposal) and greenhouse gases (i.e. become climate-neutral). Patterson (2008) has developed a web-based tool for measuring the impact of textiles on the environment with thousands of different fiber/yarn/garment and wet processing combinations.

11.6.1 Ginning

Organic cotton has to be ginned separately from conventional cotton and therefore the gin has to be thoroughly cleaned prior to ginning organic cotton. Organic cotton must also be stored in a segregated section of the gin yard unless a gin is designated for only organic production. The movement and quarantine of gin trash/by-products is sometimes necessary to reduce transmission of soil/plant-borne pathogens from conventional and organically raised cotton. After cleaning the gin, the first bale of cotton ginned is considered 'conventional cotton' in case any conventional lint still remains in the ginning system. Ginning is no different for organic and conventional cotton unless there is high trash content. Ginning operations are normally considered to include conditioning (to adjust moisture content), seed-fiber separation, cleaning (to remove plant trash) and packaging. Upland cottons are ginned on saw gins, whereas roller gins are used for ELS cottons. Higher levels of trash in seed cotton necessitate additional cleaning (i.e. stick machines, incline cleaners) at the gin before ginning the seed cotton and additional cleaning (lint cleaners) after the lint fiber has been through the gin stand and before baling. Additional mechanical processing can have an effect on fiber quality as well as cause higher gin loss. In the case of mechanical harvesting, if the picker or stripper harvesters have picked green leaves along with cotton because harvest aid chemicals were not used where necessary, there could be a noticeable impact on quality.

11.6.2 Yarn manufacturing

The manufacturing textiles from organic cotton – including opening, blending, carding, drawing, combing, if necessary, roving, spinning into yarn and winding – is no different than for conventional cotton, particularly if no processing oils are used. Processing oils are not usually necessary because of the natural waxes on the surface of cotton. Sometimes biodegradable oils are used in ring spinning, but these oils are removed by fabric scouring and bleaching prior to dyeing.

Prior to yarn and fabric manufacturing all lines must be thoroughly cleaned to remove conventional cotton or lines have to be dedicated to organic cotton yarn and fabric manufacturing. The first step in textile mill processing is opening and blending (Wakelyn, 1997). The quality parameters of cotton (i.e. length, length uniformity, strength, micronaire (an indicator of fineness and maturity), color, leaf and extraneous matter) vary considerably from bale to bale and from growing region to growing region. In order to ensure consistency in processing efficiency and product quality, many cotton bales of similar quality are blended to produce a homogeneous mix. To do this, bales of cotton are arranged in a 'lay-down' so that sophisticated blending equipment can continuously remove some cotton from up to 100 bales of cotton at a time, thereby ensuring consistency of fiber properties along the length of the yarn. If the cotton is not blended properly there can be problems later with the dyed (e.g. barré) and finished fabric/textile. Since the supply of organic cotton is limited, it may be difficult to get sufficient bales of organic cotton, with similar properties from one growing area, to blend to avoid these quality problems.

11.6.3 Fabric manufacturing

Weaving, knitting or non-woven fabric manufacturing of organic cotton and conventional cotton should be similar as long as natural starch sizes are used. Weaving preparation (i.e. direct and indirect warping and sizing) is an important process in the production of woven fabric, especially for high-speed weaving, and will continue to be dependent on the future technology of woven fabric production (Diehl, 2004). It requires yarn to be coated with starch or some other 'size' prior to weaving for extra strength and abrasion resistance. Sizing is the only operation that may have some environmental concerns in weaving. The size has to be removed, prior to dyeing and finishing. If starch is used for sizing, it is normally removed with enzymes and scouring. The starch size is not recycled and has to be handled in a manner that does not cause water pollution or other environmental problems. Starch breaks down during the enzymatic treatment and is removed by washing. It consumes oxygen thus lowering oxygen (biological oxygen demand or BOD) in the wash-off water. Polyvinyl alcohol size, commonly used in conventional sizing, can be removed by scouring and can be recycled. It is, therefore, considered to be more environmentally friendly than 'natural' starch. There is research underway on size-less weaving, but there are no commercially available processes. Sizing is not required in the manufacture of knitted fabrics, but knitting oils are usually used. They are removed during normal scouring in preparation for dyeing and finishing.

Non-woven fabric manufacturing does not require the use of natural or synthetic sizes either but various additives when used would need to be considered natural if they were to be used in the processing of organic cotton.

11.6.4 Preparation, dyeing and finishing

Preparation

Typically, in preparation for dyeing or printing and finishing, raw (greige) fabric is singed, desized, scoured, bleached and mercerized (Cotton Incorporated, 1996; Wakelyn *et al.*, 2007). These treatments remove natural non-cellulosic constituents/impurities and increase the affinity of cellulose for dyes and finishes. These processes are, for the most part, similar for organic and conventional processing, except that some chemicals (see Appendix 11.1) are prohibited for use on textiles that comply with various optional/voluntary organic textile processing standards (Section 11.10). Caustic soda (sodium hydroxide) is commonly used in scouring to remove natural waxes and impurities from the fabric. Enzymatic scouring with alkaline pectinase has been reported to be an environmentally friendly alternative to conventional scouring that offers water and energy savings and improved whiteness (Ismal *et al.*, 2007).

To achieve a true white or predictable dyed color, bleaching is usually necessary. Bleaching is done with hydrogen peroxide but without optical whiteners at a temperature over 60°C (140°F), which enhances energy consumption. Peracetic acid (Steiner, 1995) and also a waterless bleaching system that uses oxygen gas (Mowbray, 2008) have been reported as alternative, environmentally safe methods for bleaching cotton. Ozone also can be used as another waterless process to bleach cotton. Bleaching removes residual impurities and changes the cotton fabric color to clear white rather than the off-white (i.e. various shades of yellow) it was before bleaching. If chlorine bleaching methods, which are very rarely used today, are used, it is sometimes incorrectly claimed (Green Cotton Thailand Co. Ltd, 2006 (www.greencotton.co.th)) that 'dioxins' are created. Cotton is not lignified cellulose and it has been shown that dioxins are not formed by chlorine bleaching of cotton (Wakelyn, 1994).

Mercerization, treatment of the fabric with a strong aqueous solution of sodium hydroxide, is performed to add luster, strength and dye absorption properties to the fabric. Fabric appearance and strength are greatly enhanced by mercerization.

Appendix 11.1 contains a list of some chemicals that are allowed and others that are prohibited for use in the preparation, dyeing, printing and finishing of textiles that follow various optional/voluntary organic textile processing standards (see Section 11.10; Organic Exchange, 2005, 2007) for wet finishing of organic cotton textiles.

Dyeing/coloration and finishing

Fabrics are dyed after preparation. Dyeing is also done on raw stock/fiber and yarn. For organic processing, all dyes should conform to the Ecological and Toxicological Association of Dyes and Organic Pigments Manufacturers (ETAD) Guidance Documents regarding residual heavy metals and aromatic amines found in finished products (ETAD, 1997; EU Directive, 2002). The dyeing of organic cotton should use dyes free of heavy metals, e.g. chromium, and formaldehyde or other potentially hazardous chemicals. A closed-water-circuit dyeing process is used to filter and neutralize the waste (Tripathi, 2005).

Conventional cotton textiles are dyed with an extensive number of dye classes, including reactive, azoic, direct, indigo, pigment, sulfur and vat dyes (Cotton Incorporated Co. Ltd, 1996). The first choice for dyeing organic fabric, where applicable, could be plant-based natural vegetable dyes. Natural dyes are often not safer or more ecofriendly than synthetic dyes and are less permanent and more difficult to apply. Cotton is not particularly suitable for natural dyes but some natural dyes will work on cotton, especially if mordanted with tannins or other metal-based chemicals. However, commercially available natural dyes are extremely limited and require much plant material which would require large quantities of land to grow, making them in most cases not cost-effective; natural dyes also usually require fixing agents (metal-based 'mordants' that are pollutants and require high temperatures), can have poor light and wash fastness (e.g. unfixed dyes can rub off or fade during washing) and can show variation in color tones (Buchanan, 1995; Glover, 1995). Natural dyes do not appear to be a viable alternative to synthetic dyes and may be impractical in today's world. The best choice could be low-impact dyes made from petrochemicals. Many dyes used in conventional dyeing of cotton fabric are synthetic and are prohibited for use on organic cotton textiles. However, some 'nontoxic' dyes that are used on conventional textiles, including some azo reactive dyes, are allowed to be used on organic cotton textiles. Low-salt reactive dyeing systems, low bath ratio dyeing machines and pad-batch dyeing systems are available. Azo dyes that by reductive cleavage upon degradation can release one or more of 22 aromatic amines (ETAD, 1997; EU Directive, 2002) are prohibited for use in organic processing. These dyes also are prohibited for use in the EU and are no longer used in the USA and most other countries in conventional textile processing. Pigment dyes, except for non-toxic, naturally occurring pigments – e.g. indigo and clays – are also prohibited in organic processing.

Only printing methods based on water or natural oils are allowed in organic processing. Printing using heavy metals as discharging agents is prohibited. Digital printing can be carried out to reduce chemical use.

After dyeing/printing/coloration, fabrics can be treated with chemicals to convey attributes such as flame resistance, water repellence, durable press/crease resistance, etc. Chemicals that can release or contain formaldehyde are prohibited. Stone washing and other environmentally harsh textile finishing processes that use auxiliary additives and large amounts of water are prohibited in organic processing. Eco-labels with established criteria are intended to be a market-oriented tool to guide consumers. However, eco-labels may not identify products that are 'better' for the environment than non-labeled products, because the process by which products are assessed can be flawed owing to difficulties inherent in LCA (Hardy, 1998).

It has been suggested that dyeing should be missed out to prevent pollution (Green Cotton Thailand Co. Ltd, 2006, www.greencotton.co.th). However, most dyeing and finishing processes can not be avoided if organic cotton textiles are to have the same aesthetics as conventional textiles, but they can be substituted with more environmentally friendly operations without the overuse of dyes or auxiliary chemicals (Mowbray, 2008). Also, as a more environmentally friendly practice, the use of laser light technology has been reported in place of stone washing of denim to produce the distressed look (Mowbray, 2008). Since some operations and treatments can not be avoided, even though they are not organic, it is important that energy-efficient processing with less use of toxic chemicals and low water usage be utilized. Large quantities of chemicals and water can be used in finishing of a fabric, but today most conventional and organic dyeing and finishing processes use low inputs for water, energy and chemicals. Foam finishing which uses low moisture at reduced rate is carried out to reduce water use. A system that uses an environmentally friendly plasma process to apply water-repellent, stain-repellent and moisture management properties to apparel and other textiles has been reported (Mowbray, 2008). This finishing system holds the potential to be more energy efficient and powerful than any system currently available because the film deposition (coating) takes place at atmospheric pressures.

The conventional dyeing and finishing processing systems at textile mills can pollute the environment if proper practices, including effluent guidelines and air emissions controls, are not followed. The resultant effluent should be treated in a wastewater treatment process at the dyeing and finishing plant prior to being emitted as an effluent. Some textile facilities are using biological wastewater treatment systems. The effect of chemicals on workers' health can also be a concern if proper labor/workplace occupational health and safety regulations are not followed.

11.6.5 Product assembly

For organic textiles, natural fibers such as cotton (organic if available) are preferred for use as sewing thread but synthetic sewing thread may be used. Labels made of natural fiber can be used.

11.6.6 Quality assurance

To meet consumer demands for good aesthetic and fastness/durability properties organic textiles should meet the same quality/fastness parameters for color uniformity, light and wash fastness, wet and dry crock fastness, and shrinkage (Tripathi, 2005) that consumers expect from conventional textiles.

11.7   Limitations to organic production

Growing cotton organically may not be suitable for all countries and all growing areas. Various issues hinder the adoption of organic cotton production. These include production problems (particularly insect and weed control) and marketing problems (particularly price variability and unstable, underdeveloped markets).

11.7.1 Problems with organic production practices

In 2002/2003, the Organic Trade Association (OTA) (Pick and Givens, 2004), in a project funded by Cotton Incorporated, attempted to identify limitations to organic cotton production in the USA. The Organic Fiber Council of the OTA surveyed all US organic cotton growers on production practices and problems associated with organic cotton. The ICAC collected similar information in 1994, in addition to information on the cost of production and the price premium for organic versus conventional production. These studies concluded that the main problems for organic cotton production are: weed control (as a result of prohibition of herbicides), defoliation (because of a lack of efficient natural products) and insect control (owing to lack of efficient natural/organic insecticides). Defoliation is a limitation in countries where most or all the cotton is machine harvested, because the plant needs to be treated for leaf drop (defoliated) before harvesting machines enter fields. Hand picking does not require defoliation but there can be green leaf trash in hand-picked cotton that can affect grade if there is not some defoliation. Some farmers also indicated that lack of seed treatment, which is not permitted in organic production, is a problem. Lack of availability of inputs needed for organic farming and additional paperwork/record keeping also hinder adoption of organic cotton production.

The boll weevil has caused heavy losses to cotton production in the USA since the 1920s (Dickerson *et al.*, 2001). The boll weevil eradication program, which has successfully eliminated the boll weevil from many parts of the US cotton belt, particularly in the southern states (El-Lissy and Grefensette, 2002), continues in many other parts of the USA (National Cotton Council, 2008a). Farmers are not able to grow organic cotton in these boll weevil eradication treatment areas. The same is true in the pink bollworm eradication areas. In the areas where eradication is completed, there should be fewer insect problems – this would be helpful to organic cotton production.

The USDA (2008a), at first did not allow acid-delinted seed, but had to amend its decision since no machines for planting un-delinted fuzzy seed are being manufactured. Acceptance of acid-delinted seed for organic planting did not increase organic cotton production, but was helpful to the planting of organic seed. In the OTA report (Pick and Givens, 2004), the NOP rule (USDA, 2008a) was identified 'as creating some difficulties for farmers, including some increased costs, the sourcing of agricultural inputs, increased paperwork, and inconsistencies in interpretation of the rule by certifiers.'

US and international organic standards do not allow the use of biotech cotton. In 2007, biotech cotton constituted over 50% of the cotton produced in the world (Table 11.2). The current biotech varieties convey insect resistance and herbicide tolerance (Fitt *et al.*, 2004; Wakelyn *et al.*, 2004). Insect resistance is conferred through the incorporation of genes from *Bacillus thuringiensis* (Bt) that produce Bt δ-endotoxins, naturally occurring insect poison for bollworms and budworms. The use of Bt biotech cotton reduces the use of insecticides and minimizes adverse effects on non-target species and beneficial insects. Herbicide tolerance (HT) enables reduced use of herbicides and use of safer, less-persistent materials to control a wide spectrum of weeds that reduce yield and lint quality of cotton. Biotech herbicide tolerance moves cotton weed management away from protective, presumptive treatments toward responsive, as-needed treatments. If biotech varieties were allowed for organic cotton production, it would be helpful to organic growers.

11.7.2 Other issues that hinder the adoption of organic cotton

Various other issues that are hindering the adoption of organic cotton include alternative inputs, crop rotation problems, ineligibility of transgenic cotton for organic certification, lack of organic cotton marketing information and organic certification issues. One of the most important aspects of organic cotton production and expansion is the requirement for

improvements in marketing, and market linkages between cotton producers and international organic cotton buyers, including access to market
information distribution channels. Improvements in all these areas are
needed to improve the promotion of organic cotton.

Lack of information on cost of production

Elimination of some conventional inputs is expected to lower the cost of
production per hectare, but may not mean lower cost of production per
kilogram of lint, if reduction in yield is greater than the reduction in input
costs. In addition, other organic practices, such as the control of weeds by
hand hoeing, increase labor costs. The cost of production varies greatly
among growers, production regions and countries, but unfortunately no
authentic data are available to compare cost of production of organic cotton
versus conventional production. In the absence of such information or
limited information, farmers can be reluctant to adopt organic agricultural
practices.

Price premium/unstable markets

Organic producers expect a price premium in exchange for the cost of
certification, risk of lower yield and possibly lower quality because of
spotted cotton or more trash in the cotton because of imperfect control
of bollworms or other insects and problems in harvesting. Spotted and
higher trash cotton – lower grades of cotton – could cause the farmer to
receive a discounted price rather than a premium.

There appears to be no fixed premium for organic production efforts.
The OTA survey (Pick and Givens, 2004) data indicated that in the USA,
'the average price per pound received by farmers showed a wide range,
from $0.69 to $1.40 for Upland [organic] cotton' compared with US$0.57–
0.76 in 2003/04 and US$0.48–0.57 in 2004/2005 for base grade conventional
cotton (A-index world price). The enormous variability in prices received
by farmers is an indication of how unstable the market is. Some projects in
Africa offer a premium for organic cotton of 15% over the market price
(N. Pattni, BioRe® Tanzania Limited, personal communication, 2008). The
premium is supposed to be enough to compensate for the loss in yield or
even to increase farmers' income over conventional production. No
authentic information is available for potential organic cotton growers on
how much premium an organic producer could expect if he/she shifts to
organic production and this has discouraged some growers from growing
organic cotton. Solid indications that price premiums will be received
would encourage organic production.

Development of national markets

Most organic cotton produced in Turkey and the USA is exported to other countries. Organic cotton production in Turkey was initiated by a multinational company, called the Good Food Foundation, in 1989, which was followed by a second project by a German company called Rapunzel. Currently, a number of companies are involved in organic cotton production in Turkey and almost all organic production is through contract farming. Companies contract growers to produce organic cotton for them and also arrange their own certification. In the USA, organic cotton certification is by the USDA. In spite of many years of organic production in both countries, local markets for the consumption of organic cotton have not developed, thus leaving organic cotton producers at the mercy of unknown international buyers scattered in many countries. Beyond production and local markets, the standards for eco-friendly textile products have been variable. There is a need to develop local markets for organic cotton closer to the production chain and harmonization of international labeling schemes for eco-friendly organic textile products.

Organic cotton should be a producer-driven initiative

Currently, organic cotton production is market driven. Farmers are producing organic cotton because organic cotton-consuming companies, environmental groups or groups with specific mandates contract them to produce organic cotton. Therefore, organic cotton production is mainly a type of contract farming. It is helpful that there is an assured market for much of the organic cotton produced in the world but it is not good that there is a limited market for organic cotton. Consequently, the organic cotton growers are for the most part not getting a sufficient price premium for their product. Countries and growers who have currently produced organic cotton on their own – without contracting a buyer – have faced problems in selling their organic cotton and getting a premium price. The market for organic cotton is limited and producers do not have enough options to negotiate price premiums. If organic production was a producer-driven initiative there would be more options for farmers to negotiate a better price for their cotton.

11.8   How to improve organic cotton production

If some of the factors that have limited the adoption of organic farming practices discussed in Section 11.7 and that restrict it to a niche market are removed, organic production could spread to more countries and production expanded in those countries. The chances of obtaining higher prices for organically produced cotton are much better if it is produced under contract for a committed business.

11.8.1 Suitable varieties

All of the 20-plus countries where farmers have tried to produce organic cotton have used conventionally grown varieties developed for that growing area. These varieties have been developed for high-input conditions, with synthetic fertilizer and insecticide usage. Varieties that perform well under optimum conventional conditions can not maintain their yield level when the conditions they have been developed for change so drastically. Varieties suitable for organic farming production conditions should be developed (Chaudhry, 1993, 2003).

11.8.2 Development and transfer of production technology

For conventional cotton production, breeders, agronomists, entomologists and pathologists jointly develop a technological package that includes the best use of inputs and production practices. The technology package has three aspects: (a) recommended production technology practices for achieving high yields; (b) the information is disseminated to cotton growers in many ways through federal and state extension workers and consultants, this communication/dissemination of the technology is equally as important as the development of the technology; and (c) the adoption of technology is at the discretion of individual growers.

The development, transfer and adoption of technology help to give the farmer some assurance that high yields can be obtained. For conventional cotton production this is the normal practice. Organic cotton production technology needs to be and should be developed and formally transferred by specialized extension workers to organic cotton growers.

11.8.3 Soil fertility

The soil has to be replenished to maintain the optimum supply of nutrients the plant requires, and if the soil is not replenished, growth and yields are affected. Nutrient needs change from minimum to maximum for N, P and K during the course of crop development. Matching N delivery to plants needs is more critical, because N moves in the soil and could leach out by the time the plant's needs reach their peak. This is the reason why N fertilizers are always split into doses for conventional production so that supply is close to the needs of the plant. Organic fertilizers (manure) and other permissible sources of nutrient for organic conditions fail to meet the plant's changing needs for N and, thus, limit organic cotton production.

Improved methods are required to keep soil fertility high and sufficient to meet the plant's needs over the growing season to obtain high yields.

11.8.4 Improved insect and weed (pest) control

Improved insect and weed control and defoliation technology could help increase organic cotton markets. The cotton plant is naturally vulnerable to a variety of insects. Unless efficient and effective control methods are available under organic conditions, yield is going to be affected negatively. Insects like bollworms/budworms and whitefly can lower cotton grade and depress prices received by farmers. Sucking insects at early stages of cotton plant growth affect physiological activities in leaves and result in the plant producing less fruit. Insecticides provide efficient and effective control but are not allowed. Biological control and non-conventional insect control insecticides like products from the Neem tree (*Azadirachta indica*) do not provide control equivalent to synthetic insecticides. Moreover, after synthetic insecticides are eliminated, the reduction in yield is greatest in the first year but yields may improve as natural biological defense builds. Many farmers can not sustain the loss in yields in the first few years. This is a disincentive for potential organic producers. Research must be conducted to keep the loss in yield to a minimum for the first two years, when loss in yield can be high and premiums are not available since the cotton is still in the transitional stage.

11.9 National obligatory standards for organic cotton and organic cotton certifiers

Certification is a prerequisite for a product that is to be sold as 'organic' cotton. Certification provides a guarantee that a specific set of standards has been followed in the production of the organic cotton. Cotton was first certified as organic in 1989/1990 in Turkey. Currently there are hundreds of private organic standards worldwide for all organic products. Organic standards have been codified in the technical regulations of more than 60 governments. There are at least 95 accredited certifying companies/organizations in the world dealing with cotton. The NOP (2005) lists 55 organizations located in the USA and 40 in all the other organic cotton producing, processing and consuming countries. EU regulations, International Federation of Organic Agriculture Movements (IFOAM) standards, and the US National Organic Standards (NOP) have helped to formulate organic farming legislation and standards in the world. Certifying companies develop their own standards but all are essentially comparable. Organic cotton producers have to commit to follow the standards set by the certifying organizations/companies, and this includes verification through field visits by independent third parties. The certifying agency must be accredited, and recognized by buyers, and the system must be independent and transparent. Self-certification, in the case of large growers, and involvement of local government agencies in certification, particularly in developing countries, could lower certification costs for organic producers. The fee for certification must be low enough that it does not add significantly to the cost of production, otherwise it can become a disincentive to grow organic cotton.

In many cases, cotton certification has been limited to fiber production. Yarn and fabric manufacturing, dyeing and finishing are carried out in the conventional way. Such efforts can facilitate organic production and at the same time recognize the difficulties faced in certifying the whole chain as organic. The OTA has adopted voluntary organic standards for fiber processing requirements for post-harvest handling, processing, recordkeeping and labeling (Murray and Coody, 2003; OTA, 2005b) as has the Organic Exchange. In 2007, fiber process certification standards were approved (see Section 11.9.8).

11.9.1 International Federation of Organic Agriculture Movements (IFOAM)

The IFOAM (2008) is an organic umbrella organization that was established in 1972 and unites member organizations in 108 countries. IFOAM implements specific projects to facilitate the adoption of organic agriculture, particularly in developing countries. The IFOAM adopted basic standards for organic farming and processing in 1998. IFOAM standards, revised over time, are not binding for any country/producer of organic agricultural products, but they do provide valuable guidelines for organic producers and processors. The IFOAM Organic Guarantee System enables organic certifiers to become 'IFOAM Accredited' and this allows their certified operators to label products with the IFOAM Seal, in conjunction with their own seal of certified production/processing.

11.9.2 European Union Regulation 2092/91, regulation of organic cotton (Europe)

The EU adopted Regulation 2092/91 in June 1991 for the organic production of agricultural products and labeling of organic plant products (EU, 1991; Dimitri and Oberholtzer, 2005, 2006). It came into effect in 1993. Since then it has been amended on several occasions. The regulation defines a minimum framework of requirements for organic agricultural products – including organic production methods, labeling, importation and marketing – for the whole of Europe, but each member state is responsible for interpreting and implementing the rules, as well as enforcement, monitoring and inspection. EU labeling of organic products is complex because some member states have public labels, while private certifiers in other member states have their own labels. According to this regulation, the minimum period to convert from conventional farming to organic production is two years (before planting) for annual crops and three years (before the first harvest) for perennial crops. According to the EU regulation, just as in the US regulations, biotech cotton varieties can not be used for certified organic production nor can ionizing radiation be used on products.

In March 2000, in an effort to improve the credibility of organic products, the EU introduced a voluntary logo for organic production bearing the words 'Organic Farming' (EU, 2000). It can be used throughout the EU by producers whose systems and products, on inspection, satisfy EU regulations. The logo assures that the product complies with the EU rules on organic production. EU regulation 2092/91 permits labeling a product with 'Organic Farming' if:

• at least 95% of the product's ingredients have been organically produced;
• the product complies with the rules of the official inspection scheme;
• the product has come directly from the producer or preparer in a sealed package;
• the product bears the name of the producer, the preparer or vendor and the name or code of the inspection body.

If 70–95% of the ingredients are from organic production conditions, the product may refer to organic production methods in the list of ingredients but not in the sales description. If less than 70% of the ingredients are from organic production conditions, the product cannot make any reference to organic production methods.

In December 2005, the European Commission made compulsory the use of either the EU logo or the words 'EU-organic' on products with at least 95% organic ingredients. Organic products from other countries can be imported into EU countries and freely moved within the EU countries if the organic production rules in the exporting country are equivalent to the EU regulations.

11.9.3 National Organic Program (NOP) (USA)

The 1990 US Farm Bill (1990) contained provisions requiring the USDA AMS to develop National Organic Standards. USDA promulgated the regulations/standards governing organic products in December 1997 and they became effective on October 21, 2002 (NOP, 2002; Dimitri and Oberholtzer, 2005, 2006). Organic certification in the USA is voluntary and self-imposed. According to the US standards, biotech products, irradiated foods and crops fertilized with municipal sewage sludge can not be certified as 'organic'. The USDA AMS is in charge of implementation of the standards. Any company involved in certification of organic products must have authority from the USDA to carry out certification activities. The NOP website (http://www.ams.usda.gov/nop/) has a comprehensive list of the USDA Accredited Certifying Agents (ACAs) organized alphabetically by state for domestic ACAs and by country for foreign ACAs. In the USA, if a cotton grower decides to produce organic cotton, he/she has to submit an 'Organic System Plan' to a USDA-accredited certifying company/department for approval. All natural materials are permitted to be used in organic production or processing unless prohibited on the national list.

For labeling, the USDA NOP standards allow:

- a product to be certified as '100 percent organic', if it is all organic or contains 95% organic ingredients;
- a product to be certified as 'made with organic ingredients', if it has at least 70% organic ingredients.

The NOP issued a statement on August 23, 2005 intended to clarify its position with respect to the issue of products that meet the NOP standards for organic products based on content, irrespective of the end use of the product (NOP, 2005). Agricultural commodities or products that meet the NOP standards for certification under the Organic Foods Production Act of 1990, 7 U.S.C. 6501-6522, can be certified under the NOP and be labeled as 'organic' or 'made with organic' pursuant to the NOP regulations, 7 CFR Part 205.300 *et seq*. Operations currently certified under the NOP that produce agricultural products that meet the NOP standards to be labeled as 'organic' and to carry the USDA organic seal, or which meet NOP standards to be labeled as 'made with organic', may continue to be so labeled as long as they continue to meet the NOP standards.

In the USA, there are the Textile Act and Rules, which are enforced by the US Federal Trade Commission and cover fibers, yarns and fabrics, and textile products made from them. If you advertise or sell clothing and household items containing cotton, the product labels must reflect the fabric content (US Federal Trade Commission, 2008b). A manufacturer can indicate that the product contains a certain percentage of organic cotton, if the organic cotton is certified under USDA NOP requirements (2008a). However, if the label of a product made from various kinds of cotton names a cotton type, it must also give the cotton's percentage by weight and must make clear that other types of cotton were also used to make the product. For example, a sheet that contains 5% certified Organic Cotton, 50% Pima Cotton and 45% Upland Cotton may be labeled '100% Cotton', '100% Cotton (5% Organic Cotton)', '5% Organic Cotton, 50% Pima

Cotton, 45% Upland Cotton', or '5% Organic Cotton, 95% Other Cotton'. If your product contains more than one kind of cotton, a content statement that claims the product is made of only one type of cotton is not acceptable. For example, when a sheet contains 50% Organic Cotton and 50% Upland Cotton, a fiber content label that reads '100% Organic Cotton' is unacceptable.

11.9.4 Japanese Agricultural Standard (JAS)

The production and processing of organic textiles sold in the Japanese market are regulated/certified by the Japanese Agricultural Standard (2001). These certified organic textiles can be identified with the JAS organic seal of the Japanese government.

11.9.5 Australian Organic Standard (AOS)

The Australian Organic Standard (AOS, 2006) outlines requirements for operators wishing to attain certification. This includes records required, inputs allowed and minimum practices required. Ultimately, certification to the AOS allows certified operators to use the Australian Certified Organic bud logo which is the customers' guarantee of organic integrity. It is the only organic standard that gives accountability of organic status for the organic industry in Australia. In addition to organic criteria, the standard includes certification for biodynamic production systems, as well as criteria for the registration of agricultural input products.

11.9.6 Organic Exchange

The Organic Exchange (2008a) is a non-profit organization in the USA committed to expanding organic agriculture, with a specific focus on increasing the production and use of organically grown fibers, such as cotton. It has sponsored work at the farm level in India and many countries in Africa and Latin America. The Organic Exchange is accredited by USDA NOP to certify organic cotton production. It also tracks world production of organic cotton. The Organic Exchange is funded through company sponsorships and revenues from its activities. It has an 'Organic Cotton' logo for use by its member companies to identify their products or family of products, from yarn to finished goods, that contain organic cotton.

11.9.7 Organic Trade Association (OTA)

The OTA (2008) is the membership-based business association for the organic industry in North America. The OTA develops guidelines but is not a certification organization (OTA, 2005a). Since there are already US and other organic fiber production standards in place concerning the on-farm production of raw fiber (cotton), the OTA does not have their own separate standards for organic cotton production. However, since there are no US standards for the processing of organic raw fiber from the time it leaves the farm to when a finished product is available for retail, the OTA helped developed organic fiber processing standards/guidelines (OTA, 2005b) that have evolved into the Global Organic Textile Standard (GOTS) (see Section 11.10).

The OTA's standards for claims that can be put on labels are the same as those in the USDA NOP standards. The four label categories are modeled after the standards for organic foods in the US organic regulations: '100% organic', 'organic', 'made with organic' and listing of the individual organic components on the ingredients panel. However, the USDA-proscribed system for labeling fiber products only allows one category for organic cotton textiles. The USDA rule says goods that utilize certified organic fibers in their manufacture may only be labeled as 'made with organic cotton'. The USDA has addressed the scope of the federal program. Certifiers are able to certify according to the OTA standards, using the OTA's labeling provisions. So long as the processed fiber product is certified clearly to the OTA's American Organic Standards, the OTA's language can be used to describe a product. The US Federal Trade Commission has indicated that companies may list the percentage of organic fiber content on a product's content label and include the word 'organic' to describe the fiber. Because there are no federal standards for finished fiber products (i.e. the retail product) in the USA and the OTA standards are voluntary, a company may sell organic fiber products certified under these or other standards in the USA as long as the certifier of the organic fiber is accredited by the USDA and the label claims are truthful.

11.9.8 Quality Assurance International (QAI)

Quality Assurance International (QAI, Inc.), an NSF International company, is an organic certification service. QAI currently offer organic certification under the USDA NOP and fiber certification under the OTA or other US organic fiber standards (http://www.qai-inc.com).

11.9.9 ECOCERT International

ECOCERT International, with operational offices in Germany, is an inspection and certification body accredited to verify the conformity of organic agricultural products against the organic regulations of Europe,

Japan and the USA. They currently perform such inspection and certification services in 70 countries outside the EU, on all continents. ECOCERT is accredited according to ISO Guide 65 (equivalent to European Norm EN 45011), as a member of the European Accreditation (EA) and the IAF (International Accreditation Forum) and thus is an internationally recognized accreditation body. In the USA, they have been accredited to the NOP standard by the USDA and in Japan to the JAS of organic agricultural products by the Ministry of Agriculture, Forestry and Fisheries (MAFF). ECOCERT certifies organic cotton production in many African and European countries. Their website does not mention certification or guidance for organic cotton textiles, although some organic cotton textiles carry the ECOCERT label (http://www.ecocert.com/index.php?id=about&1=en).

11.10 Optional/voluntary organic textile processing standards and eco-textile standards

Growers of organic fibers must meet the obligatory government-regulated standards discussed in Section 11.9, but at the current time the subsequent processing of the fibers/fabrics into textiles has no mandatory global or US processing standards (fiber to finished fabric) for organic or sustainable textiles. Only optional/voluntary processing schemes (i.e. process claims for a product based on a set of criteria) are available (Lackman and Lachman, 2006; Wilson, 2008a; Wilson and Mowbray, 2008), some of which are discussed below. There are many eco-labels for textiles and new ones are being developed. For a list of textile eco-labels see http://ecolabelling.org/type/textiles/ and *Eco-Textile Labelling* (Wilson and Mowbray, 2008).

11.10.1 Global Organic Textile Standard (GOTS)

The OTA (http://www.ota.com/index.html) developed voluntary organic standards for fiber processing (post-harvest handling, processing, record-keeping and labeling; 'Organic Trade Association's Fiber Processing Standards') in 2004 (Murray and Coody, 2003; OTA, 2005b). These standards evolved into GOTS, which was developed by The International Working Group on Global Organic Textile Standard made up of the OTA (USA), The Soil Association (UK), International Association of the Natural Textile Industry ('IVN', Germany) and the Japan Organic Cotton Association. GOTS is a voluntary third-party certification standard for 'process claims' for a product based on a set of criteria.

GOTS covers all post-harvest processing, from storage of organic fiber at the gin or warehouse, to yarn manufacturing, fabric manufacturing, wet finishing, quality assurance and labeling, and contains an extensive list of chemicals/materials permitted for, or prohibited from, use in organic fiber processing under the standards (see Appendix 11.1). It was approved in 2007. These standards (http://www.global-standard.org/) define the requirements to ensure the organic status of textiles, from harvesting of the raw materials, through environmentally and socially responsible manufacturing including labeling, in order to provide some assurance to the end consumer. The evaluation criteria are designed to minimize negative environmental effects and risks to human health. For example, materials allowed under the standards can not be known to cause cancer, genetic damage, birth defects or endocrine disruption. In addition, they must be biodegradable and meet strict requirements that limit toxicity. Examples of materials prohibited by the standards include chlorine bleach, formaldehyde, polyvinyl chloride (PVC) (often found in screen printing systems), alkylphenol ethoxylates (APEOs) (often found in detergent, soap or chemical mixtures as surfactant and in wetting agents for scouring) and some azo dyes.

The standard is valid for fiber products, yarns, fabrics and clothes; and covers the production, processing, manufacturing, packaging, labeling, exportation, importation and distribution of all natural fibers. GOTS can be considered a 'safe harbour' for a textile label process claim certified by a third party but can not be tested in a lab. With the publication of the revised Version 2.0 and the introduction of the logo and labeling system GOTS offers a high accountability for a reliable quality assurance concept.

The key criteria are:

- two label system: *Products sold, labeled or represented as 'organic'* – 95% or more of the fiber content of the products (excluding non-textile accessories) must be of certified organic origin; *Products sold, labeled or represented as 'made with x% organic materials'* – no less than 70% of the fiber content of the product (excluding accessories) must be of certified organic origin;
- all inputs have to meet basic requirements on toxicity and biodegradability;
- no toxic heavy metals, no formaldehyde, no genetically modified organisms (GMO);
- environmental policy required in manufacturing sites;
- social criteria (based on International Labour Organization (ILO) key criteria) are compulsory;
- dual system of quality assurance consisting of on-site inspection and residue testing.

Conditions for approval

There is a list of Approved Certification Bodies on the GOTS website. One of these is the Institute for Marketecology (IMO) in Switzerland (see below). GOTS requires certification by certification bodies, accredited according to ISO 65, approval by the International Working Group and a contract with it.

11.10.2 Institute for Marketecology (IMO)

The IMO (http://www.imo.ch/index.php?seite=imo_index_en) can certify processors of organic cotton for GOTS, which is a 'process or product' claim and is accredited by the Organic Exchange to certify two standards that deal with 'fiber only' claims (see Section 11.9.4): (a) the Organic Exchange 100 Standard which is applicable to companies that seek third-party certification of the 100% organic cotton fiber-content in their products; and (b) the Organic Exchange Blended Standard which ensures that organically grown cotton is being used to the claimed percentage in cotton blended goods. GOTS, which takes into account the socially responsible aspect of sourcing cotton goods as well as their environmental impact, is stricter than the Organic Exchange 'fiber only' claim standards. Not all producers can meet the more holistic requirements of GOTS.

11.10.3 The Institute for Market Transformation to Sustainability (MTS) SMART© Sustainable Textile Standard 2.0

The Institute for Market Transformation to Sustainability (MTS) announced its Consensus Unified Sustainable Textile Standard 2.0 on 3 January 2005. The purpose of the SMART© Sustainable Textile Standard 2.0 (Fabric & Apparel) (SMART©, 2008) for promoting sustainable textile achievement is to provide a market-based definition for 'Sustainable Textile', to establish performance requirements for public health and environmental issues, and address the triple bottom line – economic–environmental–social – throughout the supply chain. It is intended to help raw material suppliers, converters, manufacturers and end-users. The scope of the standard enables organizations throughout the textile supply chain to apply performance requirements to achieve sustainable attributes and certify compliance with levels of achievement through quantifiable metrics. The standard is based on LCA principles, and provides benchmarks for continuous improvement and innovation. Certification to this standard is intended to allow inclusive participation and encourage the progressive movement of the textile industry toward sustainability. This standard identifies six levels of sustainable attribute performance and four levels of achievement by which textile materials and products can be measured with respect to specific attributes that indicate progress toward sustainability.

11.10.4 bluesign® textile standard

The bluesign® textile standard (bluesign®, 2008) is an independent industry textile standard that covers the textile supply chain from an EHS perspective. It applies to all raw materials and components used to manufacture yarns, dyes and additives. Instead of testing finished products, the components and processes that meet the specified criteria are determined before production begins. All relevant aspects during manufacturing – from customer safety issues, to environmental aspects, to air and water emission – are taken into consideration. The full details of the bluesign® standard (details of which chemicals are restricted at which levels) are not public.

In this standard, textile components, production processes and technologies are divided into two categories, grey or blue, according to criteria of resource productivity, consumer safety, emission control, water protection and safety at work. The blue category designates raw material for use in a particular textile product that completely satisfies the bluesign standard. Where there is a generally applicable ban on raw materials and substances contained in them, these are already excluded. The grey category means that substances may be used, but that they are subject to restrictions. Substances that can not be replaced at the present time without significant impairment of functionality, quality and/or design are classified as grey provided that they can be tolerated under certain conditions without in any way compromising consumer safety.

SGS, a global testing, inspection and certification organization has acquired a 50% stake in bluesign® technologies. SGS operates a network of more than 1000 offices and laboratories around the world and should provide bluesign® with a much wider platform for bluesign® to grow its business in a number of regions. Currently, bluesign® counts brands such as Patagonia and Mountain Equipment Coop as partners. Also, bluesign® has signed a deal with DyStar, the world's largest supplier of textile dyes and auxiliaries, which allows their products to be published in the bluesign® bluefinder™ – a database of the ecologically and economically 'best available products' for the global textile industry. The bluesign® standard now has DyStar, Clariant and Huntsman signed up as supporters of the bluesign® standard which makes the bluefinder™ database more important.

11.10.5 Oeko-Tex Standard

The Oeko-Tex Standard 100 for textile products (Oeko-Tex, 2008a) is a process standard for textile and clothing products of all types that provides an objective assessment of harmful substances that are used in preparation, dyeing and finishing. It includes: (a) substances that are prohibited by law, such as carcinogenic dyestuffs; (b) substances that are regulated by law, such as formaldehyde, softeners, heavy metals or pentachlorophenol; (c) substances that according to current knowledge are harmful to health, but which are not yet regulated or prohibited by law – such as pesticides, allergy-inducing dyestuffs or tin-organic compounds; and (d) parameters such as colorfastness and a skin-friendly pH value, which are precautionary measures to safeguard consumers' heath. Until the introduction of the Oeko-Tex Standard 100 in the early 1990s there was neither a reliable product label for consumers to assess the human ecological quality of textiles nor a uniform safety standard for companies within the textile and clothing industry that enabled a practical assessment of potential harmful substances in textile products. The Austrian Textile Research Institute (ÖTI) and the German Hohenstein Research Institute jointly developed the Oeko-Tex Standard 100 on the basis of their existing test standards. To complement the product-related Oeko-Tex Standard 100, the Oeko-Tex Standard 1000 (Oeko-Tex, 2008b) is a testing, auditing and certification system for environmentally friendly production sites throughout the textile processing chain.

11.10.6 European (EU) Eco-label for textiles

The EU Eco-label for textiles (EU, 2002), revised in 2002, establishes the ecological criteria for awarding the European Community Eco-label to textile products. It is a voluntary scheme designed to encourage businesses to market products and services that are kinder to the environment.

11.10.7 Organic Exchange 100 and Blended standards

The Organic Exchange developed voluntary 'fiber only' claim guidelines (Organic Exchange, 2004) to help companies involved in the production of organic products containing certified organic cotton fiber understand how to track and document the purchase, handling and use of organic cotton in their products. The Organic Exchange 100 Standard (Organic Exchange, 2007) was developed and approved in 2004 (Organic Exchange, 2004) to allow companies to have their operations certified by a third party for products containing a 100% organic cotton fiber. The Organic Exchange Blended standard (Organic Exchange, 2005) is for products containing a percentage blend of organic cotton. Both of these standards set detailed criteria on how to prove the origin of organic cotton, how to separate and identify the organic cotton products and how the quantitative organic product flow has to be documented and monitored (i.e. have minimum production standards for yarns and blended yarns).

11.10.8 Better Cotton Initiative (BCI)

The Better Cotton Initiative (BCI, 2008) promotes improvements in the environmental and social impacts of cotton cultivation. The BCI plans to initiate global change in the mass market, with long-term benefits for the environment, farmers and other people dependent on cotton for their livelihood. Unlike organic cotton, the BCI allows the minimal use of pesticides and biotech cottons by farmers and will ensure they are used safely and responsibly. It is intended also to ensure that water use is optimized (both irrigated and rain-fed) and obtained legally without adversely affecting groundwater or water bodies. Minimum tillage of soil and the use of cover and rotation crops will also be addressed.

The BCI charter is being developed through a collaborative multi-stakeholder approach involving global buyers of cotton products to increase the demand for larger amounts of 'better cotton'. The BCI principles reflect the distinction between what is under the control of the farmer (Production Principles) and what is under the control of the BCI (Enabling Principles) (http://www.bettercotton.org/pics/BCI_Global_Principles_and_Criteria_v0.5_en.pdf). The BCI Principles and Criteria were published on July 7, 2008, for further consultation and detailed consideration with Regional Working Groups. In January 2009, Version 2.0 of BCI's Principles and Criteria and the wider Better Cotton System was published, for field testing in the 2009 growing season. After this period, a final review of the Better Cotton System will take place, to define a Final (2010) version. BCI will develop a Global Implementation Strategy, National Guidance Material, and Monitoring and Evaluation Programmes in collaboration with the Regional Working Groups, advisors, experts, partners, and other value chain actors in early 2009. These components will complete a Better Cotton System that BCI will start implementing in the 2009/2010 growing season.

11.10.9 Sustainable Clothing Action Plan

As part of their work on Sustainable Consumption and Production (SCP), the UK Department for Environment, Food and Rural Affairs (Defra) are developing a product roadmap, called the Sustainable Clothing Action Plan (2008), to reduce the environmental and social impacts across the life cycle of clothing products. The clothing roadmap focuses on garments and includes textiles used in the manufacture of clothing, but excludes accessories and commercial textiles.

11.11 Corporate social responsibility (CSR)/ethical production

Social responsibility is one of the three components/basic principles and concepts of sustainability (US EPA, 2008e). Companies manufacturing cotton textiles need to consider corporate social responsibility (CSR) as part of doing business. Definitions of CSR vary based on a number of factors including industry sector, organizational structure, location and relative business importance. Common elements of CSR are: (a) a reliance on meeting or exceeding the letter and spirit of legal, ethical, commercial and other business requirements; and (b) a focus on the impact of a company's operations, products and services on people, communities and the environment.

11.11.1 Some key corporate social responsibility issues

1 *Human rights and worker safety in the workplace* are a measure of the effect of business operations on direct employees and the employees of business partners, including suppliers and vendors in a company's supply chain. The United Nations Declaration on Human Rights and SA 8000 (see Section 11.11.2) describe the social, economic, civil and political rights of individuals and can be used as a guide for the examination of the impacts of a wide range of corporate behavior, including their operations, product development and supply chain.

   For cotton manufacturers, human rights issues are primarily focused on the labor conditions of workers in factories and farms, including but not limited to safe working conditions and issues such as child labor, forced labor, discrimination and freedom of association. OHSAS 18001 (see Section 11.11.2) and ILO conventions contain international workplace acceptable norms.

2 *Marketplace integrity* is the quality of a company's interface with its customers through its product or services. Marketplace integrity is defined by the extent to which a company's policies and procedures has a positive impact on: (a) product integrity in manufacturing quality and safety; (b) truthful disclosure, labeling and packaging of products; and (c) the ethical marketing, advertising, pricing, distribution and selling practices.

3 *Environment impacts* are increasingly being measured by companies that want to manage the effects of their operations, products and supply chains on the environment. In recent years, corporate environmental management has come to be seen as a business imperative that can create efficiencies and result in cost savings. Environmental

concerns include materials use, energy use, water use, biodiversity, transportation, emissions (including CO_2 emissions), effluents and waste from suppliers, products and services.

11.11.2 Some standards/initiatives

1   *AA 1000.* AA1000 was designed to help companies integrate stakeholder engagement into business decision-making processes and improve organizational performance. The AA1000 framework seeks to help companies engage stakeholders effectively in the development of the indicators, targets and reporting systems they feel are needed to ensure that transparency efforts are effective. There are five modules in AA1000 standards covering: (a) AA1000S assurance standards; (b) governance and risk management; (c) measuring and communicating the quality of stakeholder engagement; (d) integration of accountability processes; and (e) accountability for small and medium organizations (http://www.accountability21.net/default.aspx?id=228).

2   *Fair Labor Association (FLA).* The FLA is a multi-stakeholder initiative, composed of brands and retailers, human rights and labor rights organizations, and universities dedicated to building innovative and sustainable solutions to substandard labor conditions. Factories were recently permitted to participate as well. The FLA conducts independent monitoring and verifies conditions to ensure that the FLA's Workplace Standards are upheld where FLA company products are produced. Through public reporting, the FLA seeks to hold companies accountable and provide consumers and shareholders with credible information to inform buying decisions.

3   *ISO 14001.* ISO 14000 (1999) environmental management standards (EMS) is a series of environmental management standards and guidelines. The ISO 14001 EMS can help organizations minimize how their operations negatively affect the environment by causing adverse changes to air, water or land, and help them to comply with applicable laws, regulations and other environmentally oriented requirements. ISO seeks to establish an organized approach to systematically reduce the impact of the environmental aspects that an organization can control (http://www.iso14000-iso14001-environmental-management.com/). ISO 14020-21 deals with environmental labeling claims.

4   *ISO 18001.* This is an international occupational health and safety management system (OHSMS) that has specifications and guidelines designed to be compatible with ISO 9000 (quality) and ISO 14001 (environment) management system standards.

5 *Organisation for Economic Co-operation and Development (OECD) Guidelines for Multinational Enterprises.* The Guidelines are recommendations addressed by governments to companies operating in or from adhering countries. The guidelines provide voluntary principles and standards for responsible business conduct in a variety of areas including employment and industrial relations, human rights, environment, information disclosure, combating bribery, consumer interests, science and technology, competition and taxation (www.oecd.org/daf/investment/guidelines).

6 *Social Accountability 8000 (SA 8000).* The SA 8000 standard specifies requirements for social accountability to enable a company to develop, maintain and enforce policies and procedures along their global supply chain. The standard is an auditable certification standard based on international workplace norms of ILO conventions, The Universal Declaration of Human Rights and the UN Convention on the Rights of the Child. These standards can be integrated into management systems. Companies that operate production facilities can seek to have individual facilities certified to SA 8000 through audits by one of the accredited certification bodies.

7 *United Nations Global Compact (Global Compact).* The Global Compact is a framework for businesses that are committed to aligning their operations and strategies with ten universally accepted principles in the areas of human rights, labor, the environment and anti-corruption. It is a voluntary initiative with two objectives: (a) to incorporate the ten principles in business activities around the world and (b) to catalyze actions in support of UN goals. Companies who join the Global Compact are required to publish an annual report outlining the ways in which it is supporting the Global Compact and its principles (http://www.unglobalcompact.org/AboutTheGC/index.html).

8 *Worldwide Responsible Accredited Production (WRAP).* WRAP is an independent, non-profit organization dedicated to the certification of lawful, humane and ethical manufacturing throughout the world. WRAP has developed the Apparel Certification Program, designed to independently monitor and certify compliance with the key labor and security standards, ensuring that a given factory produces goods under lawful, humane and ethical conditions. WRAP monitors factories for compliance with detailed practices and procedures implied by adherence to these standards. Historically focused on the apparel industry, WRAP has recently expanded their scope to other sectors (http://www.wrapapparel.org/).

9 *The Global Sullivan Principles (2009).* The principles are intended to be a catalyst and compass for corporate responsibility and accountability (http://www.thesullivanfoundation.org/gsp/default.asp).

11.11.3 Third-party monitoring firms

For companies seeking an independent resource to conduct as assessment of factory conditions, there are a number of accredited organizations:

- Fair Labor Association: http://fairlabor.org/about/monitoring/accredmon;
- Social Accountability International: http://www.sa-intl.org/index.cfm?fuseaction=Page.viewPage&pageID=505;
- WRAP: http://apollo.worlddata.com/wrap/viewMonitorCountryAction.do.

11.12 Naturally colored cotton

Organic production of naturally colored cotton has been tried but there is none currently available as 'certified organic'. Since the late 1980s and early 1990s, there has been a renewed interest in naturally colored cottons, which have existed for over 5000 years (Vreeland, 1993, 1999). The need for higher output cotton production and the availability of inexpensive dyes caused naturally colored cottons to almost disappear about 50 years ago. Yields were low and the fiber was essentially too short and weak to be machine spun. These cotton varieties are spontaneous mutants of plants that normally produce white fiber. Naturally colored cotton exists in various shades of brown and green. Very light blue colored cotton is also available in the germplasm in Uzbekistan. Researchers have tried to develop other colors through conventional breeding, but have not been successful. Some naturally colored cottons have botanically formed material bodies in the lumen (Ryser, 1999) of the fiber (brown, red, mocha, and mauve cottons) that conveys the color, whereas the color of green cotton is due to a lipid biopolymer (suberin) sandwiched between the lamellae of cellulose microfibrils in the secondary wall (Ryser *et al.*, 1983; Schmutz *et al.*, 1993; Ryser, 1999).

The very limited research has led to some improvement in yields, fiber quality, fiber length and strength, and color intensity and variation (Wakelyn and Gordon, 1995; Kimmel and Day, 2001; Öktem *et al.*, 2003). There are claims of the development of a colored cotton equivalent to white cotton in fiber quality, but no colored cotton varieties have been officially approved for commercial cultivation. Progress has been made, but there is still a long way to go before the quality of colored cotton is equivalent to white cotton and the same yields are achieved. Research to improve colored cotton is limited. The availability of suitable germplasm in colored cotton for use in breeding is the biggest hurdle for improvement in colored cotton.

Naturally colored cottons represent a very small niche market. Those available today are usually shorter, weaker and finer than regular Upland cottons, but they can be spun successfully into ring and rotor yarns for many applications (Kimmel and Day, 2001; Öktem *et al.*, 2003). For a limited number of colors, the use of dyes and other chemicals can be completely omitted in textile finishing, which can compensate for the higher raw material price. The color of the manufactured goods can intensify with washing (up to 5–10 washings) (Kang and Epps, 2008); however colors vary somewhat from batch to batch and colors have low light fastness and can start fading in the sun after a few washings (Wakelyn and Gordon, 1995).

The amount of naturally colored cotton available in 2005/2006 in the world was very small, perhaps 10000 US 480lb bale equivalents (about 2180MT). Some naturally colored cotton is presently being grown in China, Peru, India, Brazil and Israel. Colored cotton projects were initiated in many countries including India, Israel and the USA, but the momentum for producing colored organic cotton could not be sustained owing to limited color choice for consumers and lower quality fiber.

In efforts to make naturally colored cotton economically advantageous to produce, organic farming of naturally colored fiber has been tried. There is a market for colored organic cotton but the consumer demand for more color choices is limiting the spread of colored organic cotton production. Some countries – particularly Brazil, China, India and Peru – continue to produce some colored cotton, but not much is certified organic.

11.13 Conclusions

There continues to be worldwide interest in organic cotton as a potentially environmentally friendly way to produce cotton and for economic reasons. Production of organic cotton increased in 2007 to over 0.5% of world cotton production, mainly as a result of increased production in Turkey, India, Syria, China and some African countries. Organic production is not necessarily any more or any less environmentally friendly or sustainable than current conventional cotton production. For consumers of textiles, from a pesticide residue standpoint there is no difference between conventionally grown cotton and organically grown cotton. Growing organic cotton is not suitable for all countries and all farmers. There are limitations to organic cotton production that need to be overcome if organic cotton is to become more than a small niche market. Growing organic cotton is more demanding and more expensive than growing cotton conventionally. Organic production can be a real challenge if pest pressures are high; it offers risks and opportunities. With commitment and experience, it is possible to grow organic cotton successfully and could provide price premiums for growers willing to meet the challenges. *Conventional and organic cotton production can co-exist. Profitability will drive decisions for farmers and throughout the supply chain.*

11.14 Acknowledgements

Dr Keith Menchey, National Cotton Council and Norma Keyes, Cotton Incorporated provided advice and input that was very helpful in preparing this chapter.

11.15 References

AGRONOMY FACTS (1997), *Estimating Manure Application Rates*, Agronomy Facts 55, Pennsylvania State University, College of Agricultural Sciences, Cooperative Extension (http://cropsoil.psu.edu/extension/facts/agfact55.pdf).

ALABAMA COOPERATIVE EXTENSION SYSTEM (2006), *Broiler Litter as a Source of N for Cotton, Timely Information, Agriculture & Natural Resources*, Agronomy Series, Department of Agronomy & Soils, Auburn University, Alabama (http://hubcap.clemson.edu/~blpprt/chick.html).

ANDERSON K (2007), 'Reducing the negative environmental impact in the textile and clothing industries', *[TC]² Bi-Weekly Technology Communicator*, November 7.

ANONYMOUS (2004), 'Kyrgyzstan: organic cotton tested in the South', *Reuters Alert Net*, December 28 (http://www.alertnet.org/).

ANONYMOUS (2008), 'Italian job', *Ecotextile News*, **17**(August/September), 26–27.

ANONYMOUS (2009), 'Organic cotton demand levels off', *Ecotextile News*, **23**(April), 20–22.

ANTHONY WS and MAYFIELD WD (eds) (1994), *Cotton Ginners Handbook*, US Department of Agriculture, Agricultural Handbook 503, Washington, DC.

AOS (AUSTRALIAN ORGANIC STANDARD) (2006), http://www.bfa.com.au/_files/AOS%20 2006%2001.03.2006%20w%20cover.pdf.

ASA (ADVERTISING STANDARDS AUTHORITY) (2008): Anonymous (2008), 'UK bans cotton USA ad over sustainability claim, *Environmental Leader*', March 15; Sweney M (2008), 'US cotton ad ban over green claims', guardian.co.uk, March 12; Anonymous (2008), 'UK bans "sustainable" cotton adverts' *Ecotextile News* 13 (April).

BETTER COTTON INITIATIVE (2008), www.bettercotton.org (see Production Principles & Criteria).

BLUESIGN® (2008) 'bluesign® technologies ag, bluesign® Textile Standard' (http://www.bluesign.com/).

BREMEN COTTON EXCHANGE (2008), 'Analysis of chemical residues on raw cotton' (http://www.baumwollboerse.de/).

BREMEN COTTON REPORT (1993), Special Edition, February 1993, 'Investigation of chemical residues on cotton' (http://www.baumwollboerse.de/cotton/investig.htm).

BUCHANAN R (1995), 'Natural dyes vs. synthetics', *Textile Chemist Colorist*, **27**(8), 11.

BUNDY CS, SMITH PF, RICHMAN DB and STEINER RL (2005), 'Survey of spiders in cotton in New Mexico with seasonal evaluations between Bt and non-Bt varieties', *Journal of Entomological Science*, **40**(4), 355–367.

CHAUDHRY MR (1993), 'Suitable varieties for organic cotton production', Presented at the *International Conference on Organic Cotton*, Cairo, Egypt, September 23–25, 1993. The conference was held under the auspices of the International Federation of Organic Agriculture Movements, Germany, and the Bio Foundation and IMO Institute of Market Ecology, Switzerland.

CHAUDHRY MR (1998), 'Organic cotton production IV', *The ICAC Recorder*, **XVI**(4), December.

CHAUDHRY MR (2003), 'Limitations on organic cotton production', *The ICAC Recorder*, **XXI**(1), March.

COTTON AUSTRALIA (2009), 'BMP Cotton, Best Management Practices' (http://www.cottonaustralia.com/au/ca/).

COTTON INCORPORATED (1996), *Cotton Dyeing and Finishing: Technical Guide*, Cotton Incorporated, Cary, North Carolina.

COTTY PJ, ANTILLA L and WAKELYN PJ (2007), 'Competitive exclusion of aflatoxin producers: farmer driven research and development', Chapter 27, in *Biological Control: A Global Perspective*, Vincent C, Goettel MS and Lazarovitis G (eds), CABI, Wallingford, UK, pp. 241–253.

CROPLIFE INTERNATIONAL (2005), 'Crop protection stewardship activities of the plant science industry: a stocktaking report', p. 15 (http://www.croplife.org/librarypublications.aspx).

CROPNOSIS (2006), *Outlook*, September, Cropnosis Ltd, Edinburgh, UK.

DAAR S (1986), 'Suppressing weeds with alleopathic mulches', *IPM Practitioner*, April, 1–4.

DEFRA (2009), 'Green Claims Code' (http://www.defra.gov.uk/environment/consumerprod/pdf/genericguide.pdf).

DEMERITT L (2006), 'Behind the buzz: what consumers think of organic labeling', *Organic Processing,* **3**(1), 14–17.

DICKERSON WA, BRASHEAR AL, BRUMLEY JT, CARTER FL, GREFENSTELLE WJ and HARRIS FA, eds (2001), *Boll Weevil Eradication in the United States through 1999*, Number Six, The Cotton Foundation Reference Book Series, The Cotton Foundation, Memphis, Tennessee.

DIEHL R (2004), 'Weaving preparation yesterday, today, tomorrow', *International Textile Bulletin*, **3**, 46–48.

DIMITRI C and OBERHOLTZER L (2005), 'Market-led growth vs. government-facilitated growth: development of the US and EU organic agricultural sectors', WRS-05-05, USDA, Economic Research Service, August 2005 (http://www.ers.usda.gov/publications/WRS0505/wrs0505.pdf).

DIMITRI C and OBERHOLTZER L (2006), 'EU and U.S. organic markets face strong demand under different policies', *Amber Waves*, February, USDA, Economic Research Service (http://www.ers.usda.gov/AmberWaves/February06/Features/feature1.htm).

DIVER S (2002), 'Flame weeding for vegetable crops', *ATTRA Current Topics*, June (http://attra.ncat.org/attra-pub/PDF/flameweedveg.pdf).

DOGAN F (2008), 'Logistics, marketing and quality issues affecting the competitiveness of cotton quality', in ICAC 66th Plenary Meeting, October 22, 2007 (http://www.icac.org/meetings/plenary/66_izmir/documents/english/os2/os2_dogan.pdf). Summary of session available at (http://www.icac.org/meetings/plenary/66_izmir/documents/english/minutes/os2_minutes.pdf).

EL-LISSY O and GREFENSETTE W (2002), 'Boll weevil eradication in the U.S. 2001', In *Proceedings of the 2002 Beltwide Cotton Conference*, National Cotton Council, Memphis, Tennessee.

ENVIRONMENTAL JUSTICE FOUNDATION (2007), *The Deadly Chemicals in Cotton*, Environmental Justice Foundation in collaboration with Pesticide Action Network UK, London ISBN 1-904523-10-2.

ETAD (ECOLOGICAL AND TOXICOLOGICAL ASSOCIATION OF DYES AND ORGANIC PIGMENTS MANUFACTURERS) (1997), *Handling Dyes Safely*, ETAD, Washington, DC.

EU DIRECTIVE (2002) Directive 2002/61/EC of the European Parliament and of the Council of 19 July 2002, amending for the 19th time Council Directive 78/769/EEC relating to restrictions on the marketing and use of certain dangerous substances and preparations (azocolourants) (2002), *Official Journal of the European Communities*, **11.9**, L243/15–L243/18.

EU (EUROPEAN UNION) (1991), Regulation 2092/91, Regulations for organic production of agricultural products (http://europa.eu.int/eur-lex/en/consleg/main/1991/en_1991R2092_index.html).

EU (2000), 'The organic logo' (http://europa.eu.int/comm/agriculture/qual/organic/logo/index_en.htm).

EU (2002), 'Eco-label for textiles' (http://eur-lex.europa.eu/LexUriServ/site/en/oj/2002/l_133/l_13320020518en00290041.pdf) (http://eur-lex.europa.eu/LexUriServ/LexUriServ.do?uri=OJ:L:2002:133:0029:0041:EN:PDF).

FARM BILL (1990), Food, Agriculture, Conservation, and Trade Act of 1990 (FACTA), Public Law 101-624, Title XVI, Subtitle A, Section 1603, Organic Foods Production Act of 1990 ('OFPA') as amended (7 US Code 6501 *et seq.*).

FITT GP, WAKELYN PJ, STEWART JM, ROUPAKIAS D, PAGES J, GIBAND M, ZAFAR Y, HAKE K and JAMES C (2004), *Report of the Second Expert Panel on Biotechnology in Cotton,* International Cotton Advisory Committee (ICAC), Washington, DC, November.

FOX SV (1994), 'Organic cotton', in *Proceedings of the International Cotton Conference*, Bremen, pp. 317–319.

FUNK PA (2008), 'Preparing for harvest without chemicals', in *Proceedings of the 2008 Beltwide Cotton Conference*, National Cotton Council, Memphis, Tennessee, CD ROM, pp. 72–77.

FUNK PA, ARMIJO CB, SHOWLER AT, FLECTCHER RS, BRASHERS AD and MCALISTER III DD (2006), 'Cotton harvest preparation using thermal energy', *Transactions of the ASABE*, **49**(3), 617–622.

GARROTT B and PARATTE A (2006, 2007, 2008), Paul Reinhart AG, personal communication. Paul Reinhart AG, cotton merchant that sells/markets organic cotton worldwide (http://reinhart.com (click on organic cotton)).

GLEICH M (2008), 'Cotton made in Africa' (http://www.cotton-made-in-africa.com/Article/en/8).

GLOVER B (1995), 'Are natural colorants good for your health? Are synthetic ones better?', *Textile Chemist Colorist*, **27**(4), 17–20.

GUERENA M and SULLIVAN P (2003), 'Organic cotton production', *ATTRA Current Topics*, July (http://www.attra.ncat.org/attra-pub/PDF/cotton.pdf).

HAE NOW (2008), 'Why choose organic', Organic Cotton Clothing Hae Now Fair Trade (http://www.haenow.com/whyorganic.php).

HAKE SJ, KERBY TA and HAKE KD (eds) (1996), *Cotton Production Manual*, Publication 3352, University of California, Division of Agriculture and Natural Resources, Oakland, California.

HARDY ML (1998), 'Ecolabels and their effect on flame retardants and fire safety', Paper presented at the *Fire Retardant Chemical Association Meeting: Fire Safety and Technology*, Atlanta, Georgia, March 22–25.

HEAD RB and WILLIAMS MR (1996), *Pests, Thresholds and the Cotton Plant*, Publication 1614, Mississippi State University, p. 15.

HOESTRA AY and CHAPAGAIN AK (2007), 'Water footprints of nations: water use by people as a function of their consumption pattern', *Water Resources Management*, **21**, 35–48.

HUBER A (2008), 'Varying fertilizer rates may not affect yields', *Cotton Farming*, April, 16.

ICAC (INTERNATIONAL COTTON ADVISORY COMMITTEE) (1996), *Growing Organic Cotton*, October, ICAC, Washington, DC.

ICAC (1998), 'The cost of producing cotton. 1998', *The ICAC Recorder*, **XVI**(4).

ICAC (2003), 'Irrigation of cotton', *The ICAC Recorder*, **XXI**(4), 4–9.

ICAC (2005), 'COTMAN: a growth monitoring system', *The ICAC Recorder*, **XXIII**(3).

ICAC (2008), 'Expression of Cry 1AC in biotech cotton', *The ICAC Recorder*, **XXVI**(2), 10–14.

IFOAM (INTERNATIONAL FEDERATION OF ORGANIC AGRICULTURE MOVEMENTS) (2008), http://www.ifoam.org/ and http://www.ifoam.org/about_ifoam/standards/norms/ibsrevision/ibsrevision.html.

ISMAL OE, OZGUNEY AF and ARABAI A (2007), 'Oxidative and activator-agent assisted alkaline pectinase preparation of cotton', *AATCC Review*, **7**(4), 34–39.

(ISO) INTERNATIONAL ORGANIZATION FOR STANDARDIZATION (1999), ISO 14021, Environmental labels and declarations – Self-declared environmental claims (Type II environmental labelling) (http://www.iso.org/iso/iso_catalogue/catalogue_tc/catalogue_detail.htm?csnumber=23146).

ISO (2000), ISO 14020, Environmental labels and declarations – General principles (http://www.iso.org/iso/iso_catalogue/catalogue_tc/catalogue_detail.htm?csnumber=34425).

ITMF (INTERNATIONAL TEXTILE MANUFACTURERS FEDERATION) (2007), *Cotton Contamination Surveys* 1999, 2001, 2003, 2005 and 2007, ITMF, Zurich, Switzerland, http://www.itmf.org/cms/pages/publications.php.

JAPANESE AGRICULTURAL STANDARD (2001) (http://www.maff.go.jp/soshiki/syokuhin/hinshitu/organic/eng_yuki_59.pdf).

KANG SY and EPPS HH (2008), 'Effect of scouring on the color of naturally-colored cotton and the mechanism of color change', *AATCC Review*, **8**(7), 38–43.

KASITA I (2008), 'Conventional cotton farmers earn five times more income', *The New Vision*, November 10.

KIMMEL LB and DAY MP (2001), 'New life for an old fiber: attributes and advantages of naturally colored cotton', *AATCC Review*, **1**(10), 32.

KING EG, PHILLIPS JR and COLEMAN RJ (eds) (1996), *Cotton Insects and Mites: Characterization and Management*, The Cotton Foundation Book Series Number 3, The Cotton Foundation, Memphis, Tennessee.

KUEPPER G and GEGNER L (2004), *Organic Crop Production Overview*, ATTRA, National Center for Appropriate Technology (NCAT), Fayetteville, Arizona.

KUSTER B (1994), 'Determination of fibre impurities and contaminants in the cotton finishing process', in *Proceedings of the International Cotton Conference*, Bremen, pp. 159–169.

LACKMAN M and LACKMAN S (2006), 'Certified organic fiber and clothing: fashioning best-practice global standards', *Organic Processing*, **3** October–December, **30–32 33**, 57–62.

LEE JA (1984), 'Cotton as a world crop', Chapter 1, in *Cotton*, KOHEL RJ and LEWIS CF (eds), Agronomy Monograph No. 24, American Society of Agronomy, Crop Sciences Society of America and Soil Science Society of America, Madison, Wisconsin, pp. 1–25.

LE GUILLOU G and SCHARPÉ A (2000), *Organic Farming – Guide to Community Rules*, Directorate General for Agriculture, European Commission, ISBN 92-894-0363-2.

MATTHEWS GI and TUNSTALL J (2006), 'Smallholder cotton production in sub-Saharan Africa: An assessment of the way forward', *International Journal of Pest Management*, **52**(3), 149–153.

MCWILLIAMS DA, WAKELYN PJ and HUGHS SE (2007), 'Considerations of organic cotton production and ginning', in *Proceedings of the 2007 Beltwide Cotton Conference*, National Cotton Council, Memphis, Tennessee, 1957–1973.

MCWORTER CG and ABERNATHY JR (eds) (1992), *Weeds of Cotton: Characterization and Control*, The Cotton Foundation Book Series Number 2, The Cotton Foundation, Memphis, Tennessee.

MINAMI K (2000), 'Nitrous oxide emissions from agricultural fields', In *Trace Gas Emissions and Plants*, Singh SN (ed), Kluwer Academic Publishers, Dordrecht, The Netherland.

MOWBRAY J (2008), 'Light fantastic', *Ecotextile News*, **17**, August/September, 22–24.

MURRAY P and COODY LS (2003), American Organic Standards, Fiber: post harvest handling, processing, record keeping, & labeling, Version 6.14: December 2003, (Approved by OTA Board, January 2004) (*Organic Trade Association's Fiber Processing Standards*, Organic Trade Association, Greenfield, Massachusetts, 2004).

MUWANGA M (2008), 'Uganda: production of organic cotton goods for country', *Opinion*, allafrica.com, 1 April (http://allafrica.com/stories/200804010036.html).

MYERS D (1999), 'The problems with conventional cotton', Chapter 2, in *Organic Cotton – From Field to Final Product*, MYERS D and STOLTON S (eds), Intermediate Technology Publications, Intermediate Technology Development Group, London, pp. 8–20.

MYERS D and STOLTON S (eds) (1999a), *Organic Cotton – From Field to Final Product*, Intermediate Technology Publications, Intermediate Technology Development Group, London.

MYERS D and STOLTON S (1999b), 'Organic cotton'. Interm. Tech. Pub. Ltd, London, UK through The pesticides Trust, 1999, ISBN 1 85339 464 5.

NATIONAL COTTON COUNCIL (2007), 'The first 40 days, the most critical period in cotton production' (www.cottonexperts.com).

NATIONAL COTTON COUNCIL (2008a), 'Boll weevil eradication update, 2004' (http://www.cotton.org/tech/pest/bollweevil/index.cfm).

NATIONAL COTTON COUNCIL (2008b), 'Nematode management practice' (http://www.cotton.org/tech/pest/nematode/practice.cfm?renderforprint=1).

NOP (NATIONAL ORGANIC PROGRAM) (2002), US Department of Agriculture (USDA), Agricultural Marketing Service (http://www.ams.usda.gov/nop/indexie.htm).

NOP (2005), 'Certification of agricultural products that meet NOP standards' (http://www.ams.usda.gov/nop/NOPPolicyMemo08_23_05.pdf).

OEKO-TEX (2008a), Oeko-Tex Standard 100 for textile products (http://www.oko-tex.com/OekoTex100_PUBLIC/index.asp).

OEKO-TEX (2008b) Oeko-Tex Standard 1000 for environmentally friendly textile production (http://ecolabelling.org/ecolabel/oeko-tex-standard-1000/).

OFFICIAL JOURNAL OF THE EUROPEAN UNION (2006), Commission Regulation (EC) No. 780/2006, amending Annex VI to Council Regulation (EEC) No 2092/91 on organic production of agricultural products, 25.5.2006 pp. L137/9–L137/14 (http://eur-lex.europa.eu/LexUriServ/site/en/oj/2006/l_137/l_13720060525en00090014.pdf).

ÖKTEM T, GÜREL A and AKDEMIR H (2003), 'The characteristic attributes and performance of naturally colored cotton', *AATCC Review*, **3**(5), 24–27.

ORGANIC EXCHANGE (2004), 'Organic Exchange guidelines tracking and documenting the purchase, handling and use of organic cotton fiber' (http://www.organicexchange.org/Documents/guidelines0605.doc).

ORGANIC EXCHANGE (2005), Organic Exchange Blended Standard version 2/2005 (http://www.organicexchange.org/, click OE Blended Standard).

ORGANIC EXCHANGE (2007), Organic Exchange 100 Standard (http://www.organicexchange.org/, click OE 100 Standard).

ORGANIC EXCHANGE (2008a), 'Organic cotton: your healthier choice' (http://organicexchange.org/Health/intro.php).

ORGANIC EXCHANGE (2008b), *Organic Farm and Fiber Report 2008*, Organic Exchange, O'Donnell, Texas.

ORGAZ F, MATEOS L and FERRES E (1992), 'Season length and cultivar determine the optimum evapotranspiration deficit in cotton', *Agronomy Journal*, **84**, 700–706.

OTA (ORGANIC TRADE ASSOCIATION) (2005a), 'Global approaches to organic textiles', in *Proceedings of the 4th International Conference on Organic Textiles – INTERCOT*, May 1–3, OTA, Greenfield, Massachusetts.

OTA (2005b), American Organic Standards for Fiber: post-harvest handling, processing, record keeping & labeling (Organic Fiber Processing Standards) (http://www.ota.com/standards.fiberstandards.html).

PATAGONIA (2008), 'Fabric: organic cotton' (http://www.patagonia.com/web/us/patagonia.go?assetid=2077).

PATTERSON P (2008), 'Best practices metrics' (http://www.keystone-group.co.uk/clothing/proceedings/Phil_Patterson.pdf).

PICK S and GIVENS H (2004), *Organic Cotton Survey – 2003 US Organic Cotton Production and the Impact of the National Organic Program on Organic Cotton Framing*, OTA, Greenfield, Massachusetts (http://www.ota.com/2004_cotton_survey.html).

RATTER SG (2004), 'Organic and fair trade cotton in Africa', in *EU Africa Cotton Forum*, July (http://www.cotton-forum.org/docs/presentations/6.3-en.pdf).

RYBICKI E, SWIECH T, LESNIEWSKA E, ALBINSKA J, SZYNKOWSKA MI, PARAYJCZAK and SYPNIEWSKI S (2004), 'Changes in hazardous substances in cotton after mechanical and chemical treatments of textiles', *Fibers and Textiles in Eastern Europe*, **12**(46), 67–73 (http://www.fibtex.lodz.pl/46_18_67.pdf).

RYSER U (1999), 'Cotton fiber initiation and histodifferentiation', Chapter 1, in *Cotton Fibers, Developmental Biology, Quality Improvement, and Textile Processing*, BASRA AS (ed), The Haworth Press, Inc. Binghamton, New York, pp. 21–29.

RYSER U, MEIER H and HOLLOWAY PJ (1983), 'Identification and localization of suberin in the cell walls of green cotton fibres (*Gossypium hirsutium L.*, var. green lint)', *Protoplasma*, **117**, 196–205.

SCHMIDT M and REHM G (2002) *Fertilizing Cropland with Poultry Manure*, University of Minnisota Extension Service (http://www.extension.umn.edu/distribution/cropsystems/DC5881.html).

SCHMUTZ A, JENNY T, AMRHEIN N and RYSER U (1993), 'Caffeic acid and glycerol are constituents of the suberin layers in green cotton fibers', *Planta*, **189**, 453–460.

SECO (2006), 'Organic cotton production and trade promotion project (BioCotton)' (http://www.swisscoop.kg/index.php?navID=22309&langID=1&).

SHELL FOUNDATION (2006), 'Organic cotton: organic cotton is "king"' (http://www.shellfoundation.org/index.php?newsID=295).

SMART© (2008), SMART© Sustainable Textile Standard 2.0 (http://mts.sustainableproducts.com/downloads/sts_v2.doc).

SMITH W F (1996), 'Precision farming overview', In *Proceedings of the Beltwide Cotton Conference*, Vol. 1, pp. 179–180.

SNYDER C and STEWART M (2008), 'Using the most profitable nitrogen rate in your cotton production system', International Plant Nutrition Institute (http://www.ipni.net/ppiweb/ppibase.nsf/$webindex/article=EB4E89D685256CE9002487E4D2FC5106).

STEINER N (1995), 'Evaluation of peracetic acid as an environmentally friendly alternative for hypochorite', *Textile Chemist Colorist*, **27**(8), 29–32.

SULLIVAN P (2001), 'Flame weeding for agronomic crops', *ATTRA Current Topics*, October (http://attra.ncat.org/attra-pub/PDF/flameweed.pdf).

SUSTAINABLE CLOTHING ACTION PLAN (2008), http://www.defra.gov.uk/environment/consumerprod/pdf/sustainable-clothing-action-plan.pdf.

SWEZEY SL (2002), 'Cotton yields, quality, insect abundance, and cost of production of organic cotton in northern San Joaquin Valley, California', in *Proceedings of the 14th IFOAM Organic World Congress*, Canadian Organic Growers, Ottawa, Ontario, Canada.

SWEZEY SL and GOLDMAN PH (1999), 'Organic cotton in California: technical aspects of production', In *Organic Cotton – From Field to Final Product*, MYERS D and STOLTON S (eds), Intermediate Technology Publications, Intermediate Technology Development Group, London, pp. 125–132.

SWEZEY SL, GOLDMAN P, BRYER J, and NIETO D (2007), 'Six-year comparison between organic, IPM and conventional cotton production systems in the Northern San Joaquin Valley, California', *Renewable Agriculture and Food Systems*, **22**(1), 30–40.

TERHAAR A (2008), 'Cotton: focus on sustainability', paper presented at *Cotton Incorporated's 21 Annual Engineered Fiber Selection® System Conference*, Memphis, Tennessee, June 10, Global Lifestyle Monitor V, Synovate (2008).

TRAORÉ D (2005), 'Promotion of organic cotton cultivation in Mali', in *Proceedings of the 4th International Conference on Organic Textiles – INTERCOT*, May 1–3, Organic Trade Association, Greenfield, Massachusetts.

TRIPATHI KS (2005), 'Textile processing of the future: The state of the art', in *Proceedings of the 4th International Conference on Organic Textiles – INTERCOT*, May 1–3, Organic Trade Association, Greenfield, Massachusetts.

TU C, RISTAINO JB and HU SJ (2006), 'Soil microbial biomass and activity in organic tomato farming systems: effects of organic inputs and straw mulching', *Soil Biology and Biochemistry*, **38**(2), 247–255.

UNITED NATIONS (1987), Report of the World Commission on Environment and Development, General Assembly Resolution 42/187, December 11 (http://www.un.org/documents/ga/res/42/ares42-187.htm).

USDA (2008a), *AMS, NOP, Organic Production and Handling Requirements*, 7 CFR Part 205.201-206, USDA, Washington, DC.

USDA (2008b), *National List of Allowed and Prohibited Substances*, 7 CFR Part 205.600-606, USDA, Washington, DC.

USDA (US DEPARTMENT OF AGRICULTURE)-NATIONAL AGRICULTURAL STATISTICS SERVICE (NASS) (2001), *Agricultural Chemical Usage: 2000 Field Crop Summary*, USDA-NASS, Washington, DC.

USDA-NASS (2004), *Farm and Ranch Irrigation Survey*, Vol. 3, Special Studies Part 1AC-02-SS-1, USDA-NASS, Washington, DC.

USDA-NASS (2008), *Agricultural Chemical Usage: 2007 Field Crop Summary*, USDA-NASS, Washington, DC.

US EPA (US ENVIRONMENTAL PROTECTION AGENCY) (1999), 'Assessing health risks from pesticides', US EPA 735-F-99-002 (http://www.epa.gov/cgi-bin/epaprintonly.cgi).

US EPA (2008a), Worker Protection Standard, 40 CFR Part 170.

US EPA (2008b), Federal Insecticide, Fungicide, and Rodenticide Act (FIFRA), 7 US Code 136; Federal, Food, Drug and Cosmetic Act (FFDCA) (1996) as amended by the Food Quality Protection Act (FQPA).

US EPA (2008c), Data Requirements for Registration, 40 CFR Part 158.

US EPA (2008d), Pesticide management and disposal; standards for pesticide containers and containment; Final Rule (40 CFR Parts 9, 156, 165), *Federal Register*, **71**, 47329–47437, August 16, 2006. Also 'Pesticides: regulating pesticides, storage and disposal' (http://www.epa.gov/pesticides/regulating/storage.htm).

US EPA (2008e), 'Sustainability' (http://www.epa.gov/sustainability/basicinfo.htm).

US EPA (2008f), National Emission Standards for Hazardous Air Pollutants (NESHAP): Printing, coating, and dyeing of fabrics and other textiles, 40 CFR Part 63 subpart OOOO.

US EPA (2008g), Final Effluent Guidelines: Rulemaking for the textile mills point source category (http://yosemite.epa.gov/water/owrccatalog.nsf/065ca07e299b464 685256ce50075c11a/c8a94f958ac0137285256b06007239fb?OpenDocument&Car tID=7236-121026).

US FEDERAL TRADE COMMISSION (2008a), 16 CFR Part 260 – Guides for the use of environmental marketing claims (http://www.ftc.gov/bcp/grnrule/guides980427.htm).

US FEDERAL TRADE COMMISSION (2008b), 'Calling it cotton: Labeling and advertising cotton products' (http://www.ftc.gov/bcp/conline/pubs/buspubs/cotton.shtm).

VAN ELZAKKER B and CALDAS T (1999), 'Organic cotton production', in *Organic Cotton – From Field to Final Product*, eds Myers D and Stolton S, Intermediate Technology Publications, Intermediate Technology Development Group, London, pp 21–35.

VREELAND JM, JR (1993), 'Naturally colored and organically grown cottons: anthropological and historical perspectives', in *Proceedings of the Beltwide Cotton Conference*, National Cotton Council, Memphis, Tennessee, pp. 1533–1536.

VREELAND JM, JR (1999), 'The revival of colored cotton', *Scientific American*, **280**, 112.

WAKELYN PJ (1994), 'Cotton: environmental concerns and product safety', in *Proceedings of the 22nd International Cotton Conference*, Bremen, HARIG H and HEAP SA (eds), Faserinstitut Bremen e.V., Bremen, Germany, pp. 287–305.

WAKELYN PJ (1997), 'Cotton yarn manufacturing', in *ILO Encyclopaedia of Occupational Health and Safety*, 4th edn, Chapter 89, 'Textile Goods Industry', IVESTER AL and NEEFUS JD (eds), International Labour Office, Geneva, Switzerland, pp. 89.9–89.11.

WAKELYN PJ, BERTONIERE NR, FRENCH AD, THIBODEAUX DP, TRIPLETT BA, ROUSSELLE MA, GOYNES JR, JR, EDWARDS JV, HUNTER L, MCALISTER DD and GAMBLE GR (2007), *Cotton Fiber Chemistry and Technology*, International Fiber Science and Technology Series, CRC Press (Taylor and Francis Group), Boca Raton, Florida.

WAKELYN PJ and GORDON MB (1995), 'Cotton, naturally', *Textile Horizons*, **15**(1), 36–38.

WAKELYN PJ, MAY OL and MENCHEY EK (2004), 'Cotton and biotechnology', Chapter 57, In *Handbook of Plant Biotechnology*, CHRISTOU P and KLEE H (eds), John Wiley and Sons, Ltd, Chichester, UK, pp. 1117–1131.

WAKELYN PJ, MENCHEY K and JORDAN AG (2000), 'Cotton and environmental issues', in *Cotton – Global Challenges and the Future*, papers presented at a Technical Seminar at the 59th Plenary Meeting of the International Cotton Advisory Committee (ICAC), Cairns, Australia, November 9, pp. 3–11.

WAKELYN PJ, THOMPSON DW, NORMAN BM, NEVIUS CB and FINDLEY DS (2005), 'Why cotton ginning is considered agriculture', *Cotton Gin and Oil Mill Press*, **106**(8), 5–9.

WECKMANN R (2008), 'Analysis of chemical residues on cotton and cotton products', in *Proceedings of the 2008 International Cotton Conference*, Bremen, pp. 211–214.

WILLIAMSON S, FERRIGNO S and VODOUHE SD (2005), 'Needs-based decision-making for cotton problems in Africa: a response to Hillocks', *International Journal of Pest Management*, **51**(4), 219–224.

WILSON A (2008a), 'Labelled with love', *Ecotextile News*, June, 16–19.

WILSON A (2008b), 'Organic aternatives', *Ecotextile News*, May, 25.

WILSON A and MOWBRAY J (2008), *Eco-Textile Labelling*, Ecotextile News and Mowbray Communications Ltd, West Yorkshire, UK.

YAFA S (2005), *Big Cotton: How a Humble Fiber Created Fortunes, Wrecked Civilizations and Put America on the Map*, Viking Penguin, a member of Penguin Group (USA) Inc., London, pp. 290–301.

ZARB J, GHORBANI R, KOCCHEKI A and LEIFERT C (2005), 'The importance of microorganisms in organic production', *Outlooks on Pest Management*, **16**, 52–55.

ZWART SJ and BASTIAANSSEN WGM (2004), 'Review of measured crop water productivity values for irrigated wheat, rice, cotton, and maize', *Agricultural Water Management*, **69**(2), 115–133.

Appendix 11.1 Chemicals allowed and prohibited for use in preparation, dyeing, printing and finishing of organic cotton textiles (Global Organic Textile Standards; Organic Exchange Guidelines; Organic Trade Association)

Synthetic chemicals allowed

- Aluminum silicate (scouring agent, deflocculant, anticoagulant, dispersant)
- Aluminum sulfate (scouring or mordant agent)
- Fatty acids and their esters (softener)
- Hydrogen peroxide (bleaching agent)
- Oxalic acid
- Ozone (bleaching agent)
- Polyethylene (restricted softener)
- Potassium hydroxide (mercerizing, scouring)
- Biodegradable soaps
- Sodium hydroxide (mercerizing, scouring)
- Sodium silicate (bleaching and color brightener)
- Sodium sulfate (bleaching and color brightener)
- Surfactants (that are biodegradable) (scouring agent, emulsifier, wetting agent) (may not be silicon based, contain petroleum solvents or be alkyl phenol ethoxylates).

Non-synthetic chemicals allowed

- Acetic acid
- Chelating agents (stabilizers)
- Citric acid
- Clay-based scours
- Copper, iron, tin (for mordant dyeing)
- Natural dyes (animal and plant)
- Enzymes (non-GMO)
- Flow agents (natural)
- Mined minerals
- Pigment dyes (natural indigo and clays only)
- Potassium acid tartrate
- Sodium carbonate (soda ash) pH adjuster
- Sodium chloride (salt) auxillary
- Tannic acid
- Tartaric acid.

Chemicals prohibited

- Ammonium soaps
- Absorbable halogenated hydrocarbon (AOX)
- Bluing agents (for bleaching and color brightening)
- Chelating agents
- Chlorine compounds (bleaching)
- Dyes non-conformant with criteria
- Formaldehyde
- Synthetic fire retardants
- Functional finishes for anti-crease, antifungal, anti-microbial, anti-pilling, antistatic properties.

Conformant dyes

- Natural dyes (animal and plant)
- Pigment dyes (natural indigo and clays only)
- Most fiber reactive dyes.

Dyes not allowed

- Benzidine and benzidine congener azo dyes; other azo dyes that can undergo reduction decomposition to form carcinogenic aromatic amines (currently this includes the 24 amines classified as substances known to be human carcinogens)
- Dyes should not contain heavy metals (restricted level of residues allowed) except iron, copper, tin (for mordant dyeing)
- Dyes should not contain chelated metals (residues >1 mg metal/kg textile)
- Dyes should not contain AOX and substances that can cause their formation
- Dyes should not contain formaldehyde.

12

The role of nanotechnology in sustainable textiles

S. BLACK, London College of Fashion, UK

Abstract: This chapter gives an overview of current research developments in nanotechnology that have particular relevance to textiles and fashion sectors, and discusses current and future applications. Awareness of such work is increasingly important for design researchers and designers in the product and fashion arenas, in order to facilitate innovative collaborations between design, science and technology for a more sustainably designed future.

Key words: nanotechnology, research developments, commercial scale, applications.

12.1 Key principles of nanotechnology and its use in sustainability

The future holds many possibilities for nanotechnology to provide environmental benefit. Imagine a t-shirt that did not fade in colour after a few washings, but remained as new and could also change its logo, or a skirt that could be re-programmed to become a different colour or pattern for a different occasion. These are future concepts that fashion designers dream of. This chapter gives an overview of current research developments in nanotechnology that have particular relevance to the textiles and fashion sectors, and discusses current and future applications. Awareness of such work is increasingly important for design researchers and designers in the product and fashion arenas, in order to facilitate innovative collaborations between design, science and technology for a more sustainably designed future. Therefore, information has been obtained from symposia and conference presentations where transfer of nanotechnology to business and commercial sectors was a key focus, and from direct dialogue with scientists and technology companies. The last ten years have seen the emergence of new multidisciplinary approaches to textile research: as micro-, nano-, bio- and information technologies and biomaterials have continued to evolve to new stages of maturity, there is an extraordinary array of new possibilities for enhanced functionalities within textiles – from new fibre structures, composite materials and coatings at the nano and micro levels to the visible integration of electronic assemblies into clothing.

Nanotechnology is heralded by many as the next industrial revolution, with unprecedented potential to revolutionise multiple sectors by improving on existing technologies and introducing radical new tools. The development of nanoscience and nanotechnologies across several scientific sectors has taken place only in the last 30 years, as precision tools for engineering at this scale have become available to scientists. Consequently, new applications are emerging from research laboratories around the world (US, Japan, UK and Germany in particular) in key areas such as materials science, precision engineering and electronics, drug delivery for medicine and healthcare, energy production and storage, and water or air purification. While many nanoparticles exist naturally and science at this scale is not entirely new, it is the ability to effectively probe and manipulate matter at the nano scale that has brought the technology to the forefront of current developments. At the level of commercial application, many products incorporating nanotechnologies are already on the market in cosmetics, electronics and textiles sectors, and there are a great deal more in the pipeline which will appear over the next few years. Certain applications of nanotechnology are especially relevant to textiles and, consequently, have important implications for sustainability of textiles.

Nanotechnologies are increasingly seen as a means to address some of the major issues in society: living with environmental change, sustainable energy, global challenges of security, food production and an increasing and progressively ageing population. In Europe, funding for nanoscience and nanotechnology research has increased from €120 million under Framework 4 programmes (1994–1998), to €600 million in 2007 alone under Framework 7, and is set to double by 2013 (Deliyanakis 2008) under its €3500 million programme 'Nanosciences, nanotechnologies, materials and new production technologies', fuelling the pace of research in the field. At the same time, the European Space Agency has announced a new programme to identify the next generation of nanomaterials to be used in future space missions, research that will eventually filter through to society at large. According to the Minister of State for Science and Innovation,[1] the UK Research Councils distributed more than £400 million for projects involving nanotechnology and nanoscience between 1997 and 2007, and further funding of £50 million per annum has been committed. The Engineering and Physical Sciences Research Council has identified a programme of 'Grand Challenges' for nanotechnology research to 2012, the first two challenges being Sustainable Energy and Medicine and Healthcare.

---

[1] Ian Pearson, speaking on 2 October 2008 at the UK government Nano Showcase event organised by the Department for Innovation, Universities and Skills, held at the Department for Business, Enterprise and Regulatory Reform, London.

Nanotechnology represents a fundamentally different platform for technological developments than anything that has preceded it. The term is an umbrella for a vast range of technologies and paradigm-shifting science that take place at the molecular and atomic scales – nanotechnologies are concerned with functionalities at a scale of measurement between 1 and 100 nanometres (nm). A nanometre is one thousand millionth ($1/10^9$ or 10^{-9}) of a metre (called a billionth in the US). If our bodies are measured on the scale of metres, an ant is measured on a scale of millimetres (1000 times smaller) and cells are measured on a scale of micrometres (microns) (1000 times smaller again), then the nanometre measures the scale of viruses and proteins which is 1000 times smaller still. A frequently quoted example of relative measure is that the thickness of a human hair is approximately 80 000 nm. For comparison, atoms measure a tiny fraction (about 0.2) of one nanometre and a water molecule is almost 0.3 nm across.

As with any disruptive technology, debate is ongoing regarding the balance between the risks and benefits of such developments, especially in products such as food, cosmetics and textiles which are consumed, applied directly to the skin, or worn, and therefore come into direct contact with the body. There are hundreds of products already on the market that use different forms of nanotechnology. Policies and codes of conduct are in the process of being developed at international government levels, but critics point out that this is not fast enough to keep up with the pace of commercial applications of new product innovations driven by the all-important first-to-market competitive advantage. A 2006 report by consultants Cientifica analysing market demand for nanotechnology-enabled textiles for the clothing market projected its value at over US$22 billion by 2012 (Cientifica, 2006).

At the nano-scale many properties of materials are changed enormously: they may become extremely strong compared with their weight, or may exhibit counter-intuitive behaviour – for example, normally insulating materials such as nylon (polymer structures) can become conductive to electricity. A key property highly relevant to the textile sector is the high specific surface area of nanomaterials, i.e. the ratio between size or weight and surface area: material structures reveal very large surface areas at the nano-scale relative to their mass. This creates favourable opportunities for engineering of material surfaces at the molecular level, enabling the restructuring and enhancement of the normal surface properties of materials (such as glass, metals and polymers) for specific desired functionalities. Consequently, beneficial effects can be imparted to fibres, fabrics and clothing at a scale that is invisible to the naked eye, and to the touch. The latter aspect is vitally important in the treatment of textile products, particularly where worn on the body.

Nanomaterials may be categorised in relation to the following dimensional typology, according to Dr Alexandre Cuenart of The National Physical Laboratory, Middlesex, UK:

- 0D: nanoparticles – these exist in the form of powders that can be added to liquids homogeneously.
- 1D: nanotubes and nanowires – forms having single dimensional properties that emerge under particular forces, e.g. compression, these are used in composite materials (nanocomposites) and electronics.
- 2D: multilayered structures and thin films – used in electronics and surface coatings.
- 3D: nanostructured materials – used for engineered properties depending on surface patterning to achieve specific functionalities.

For a functional nanotechnology, not all physical dimensions of a material need to be at nano-scale, for example very thin surface coatings have only one dimension at nano-scale; individual nanoparticles have all dimensions at nano-scale.

Much research is ongoing, with many concepts and technologies still at the laboratory stage. Transfer of technology applied in other sectors, such as medical textiles, will eventually impact on consumer textiles. However, some technologies (such as nanoparticles of silver) have already appeared in commercially available products from electronic goods to socks and home textiles (see Section 12.8). At the same time, legislation and regulation is still in development and long-term testing is currently being undertaken to determine the extent of risk, and its appropriate management.

The effectiveness of nano-scale coatings and finishes on fabrics is a key area for sustainability in textiles and fashion, through measurable predicted benefits such as extended product life (for example by reducing the effects of wear from abrasion) and reduction in the need for laundering – the latter being one aspect of textile life cycle in clothing and household textiles that carries great environmental impact via water and energy usage (Allwood *et al.*, 2006; Blackburn and Payne, 2004). Such enhanced performance characteristics can go some way to reversing the consumer trend in recent decades of washing clothes more and more frequently, not to remove dirt, but to refresh them. Anecdotal evidence is appearing that if finishes are applied to reduce odour from bacterial growth, then it may become acceptable for garments such as a shirt or socks to be worn twice and therefore washed half the number of times. A reduction in the need for ironing through effective surface treatments would also lower overall energy consumption (Fisher *et al.*, 2008).

Further areas where nanotechnology can have an impact on textile sustainability and reduce waste include: the improvement of dye uptake through surface fibre effects; prevention of fading by loss of colour through exposure to ultraviolet (UV) light; and coloration without dyes using, for example, gold and silver nanoparticles. In addition, self-repairing textiles would result in textiles being used for longer and discarded less frequently into landfill or incineration cycles. Further attention will be paid to these aspects in the following sections. However, it is important that full life cycle analysis is conducted on any products or composites with regard to environmental impact and potential ecotoxicology of 'free' nanoparticles that may potentially leach out from nano-treated fabrics, or from the manufacturing processes themselves.

12.2   How nanotechnology can be used to reduce environmental impacts

Although the majority of research is still taking place in scientific laboratories, the effects of the funding surge discussed above mean that more applications are emerging into real world-scaled products, and this emergence is set to gain further momentum over the coming years. Consultants Cientifica suggested in their 2006 forecast that high-performance fabrics will impact the sportswear and leisure clothing sectors through higher thermal, comfort and durability properties by 2011–2013; and that by 2022 smart clothing will be available that is responsive to external stimuli, mainly through electronic and sensing textiles (Cientifica, 2006). Early examples of these have been on the market for several years. Smart clothing has been in development since Steve Mann in the mid-1980s first produced wearable computers which are now evolving into wearable electronics (Black, 2007; Tao, 2005). However, the component technologies have developed at different rates, not least the sources of battery power required to operate the systems. The drivers of current research are in the fields of military and medical requirements – for soldier protection and communication, and for monitoring and drug delivery systems for patient care and diagnostics. It appears likely that nano-scale technologies will need to mature further to provide another level of miniaturisation before truly smart and intelligent textiles will be realised, responsive to environmental stimuli and providing mobile information, communication and entertainment. (For a technical discussion on smart textiles and clothing refer to Tao (2001 and 2005) in the Sources of further information and advice section (Section 12.10). Invidious though it is to put specific dates on these developments, it is clear that new products will eventually become available that will change our relationship with textiles and clothing, and thus our behaviour towards product life cycles and the throwaway consumer society.

The next section will consider the key nanotechnologies that are being developed for or transferred to the textile sector, and their use in reducing environmental impact. The main sectors are fabric coatings and finishes, engineered surfaces and fibres, and enhancements to existing processes such as dyeing technologies. In addition, processes such as electrospinning of nano-scale fibres for non-woven fabrics, may eventually transfer from the medical sector to be scaled up for specialist clothing applications.

Aftercare of textiles on both domestic and industrial scales (such as hotel laundry) has already been referred to as a major negative environmental impact in the textile life cycle, therefore the impact of durable surface coatings could be highly significant as they become more widely available and cost effective. Nano surface finishes which render textiles and clothing stain resistant, abrasion resistant, water and oil repellent, self cleaning, antistatic or antibacterial can therefore help prolong the useful life of textile articles and minimise the need for washing or dry cleaning to remove dirt and odours. Certain treatments may also reduce the need for ironing, all with consequent energy and water savings together with reduction in replacement purchases.

Further relevant applications of surface coating technologies are found in the protection of metal surfaces of machinery, including textile production and sewing machines, where friction between surfaces can be significantly reduced by nano-scale coatings, reducing the need for lubrication and its associated costs, which, according to Dr Alexandre Cuenat of the National Physical Laboratory, can account for up to 10% of gross national product (A. Cuenat, personal communication, 7 November 2008).

12.3   Surface coatings and treatment of textile fibres

In the mature textile industry, many fabric finishing technologies already exist that currently use a range of mechanical application processes and micro-scale chemical technologies, to achieve effects such as water-repellent coatings, shrink proofing, antistatic treatments, antibacterial and antiwrinkle effects, flame retardance or insect protection. These textile surface treatments often include use of resins and coating membranes which perform by creating layers of thin films on the surface, normally around 25 microns thick, which fill or cover the interstices of the fabrics to achieve resistance to liquids and so on. Some of these treatments on clothing do not allow moisture from perspiration to escape and the clothing can become uncomfortable. The durability of many treatments is often less than satisfactory, as the effect is literally worn off through abrasion and cleaning. Common treatments for textiles in clothing, such as shower proofing, utilise compounds including fluorocarbons that have negative environmental impact when washed off into the water waste stream during laundering. Furthermore, typical surface treatments applied to furnishing fabrics, such as spraying or padding with resins or silicone-based finishes for water and stain resistance, are unsuitable for clothing products as the fabric handle may be adversely affected. For a full discussion of the environmental impacts of textile finishes, please refer to Slater (2003), *Environmental Impact of Textiles*, and Chapter 6 of the present volume: Key sustainability issues in textile dyeing.

Several types of surface treatments using different nanotechnologies are now available for use on textiles, and these have a number of key advantages; other surface treatments are in development. As the chemical processes used work at the molecular level, nanofinishes do not adversely affect the handle, appearance, flexibility or breathability of the fabric, and in fact are invisible to the naked eye and to the touch. Importantly, the amount of raw materials required to achieve the desired functionality is vastly reduced compared with conventional technologies, reducing the requirement for scarce resources to a minute level, while at the same time greatly increasing functionality. In addition, the chemical bonding achieved at nano-scale is far more durable than conventional finishing processes, reducing the amount of pollution caused when finishing treatments break down. For example, specific nanoparticles (metals or metal oxides such as silver (Ag), titanium dioxide (TiO_2) and zinc oxide (ZnO)) may be embedded into the molecular surface structure, chemically bonding with the fibre molecules to create new, stable and therefore permanent structures that possess antimicrobial (silver) or UV-absorbing properties (TiO_2, ZnO) for example. Nano-scale finishes may also be applied as a protective invisible barrier layer for textile surfaces, and prevent the release of toxic chemicals from excess dyestuffs for instance, or lock in those required in special circumstances such as in protective clothing for hazardous occupations.

12.3.1 Plasma treatments

Nano-structured surfaces can be engineered to create specific properties. Novel plasma coating treatments have now been developed that can restructure the topography of the surface of many fibres and fabrics, to create a required functionalised performance such as: reduced adhesion; sterilisation to prepare one surface to bond with another coating substance; or as pre-treatments for dyeing and printing processes. Plasmas are ionised gases containing charged particles, found in both man-made and natural forms such as neon lights, TV screens or lightning. Owing to the high energy requirements for plasma functionality, research has been carried out, for example under the European Acteco project (EU FP6 Acteco project from 2004 to 2009, see www.acteo.org), to create more ecologically efficient low-pressure and -energy processes for large-scale operations. This has resulted in durable textile coatings which can resist bacterial growth and alter the hydrophobic/hydrophilic and oleophobic/oleophilic properties of textiles (Moore, 2008). These are especially applicable to medical textiles and textile medical implants, which are often constructed from polymer fibres, and for protective clothing and potentially the sports and casual clothing sector. As plasma processes are dry, unlike most wet processing in textiles, atmospheric or low pressure plasma coating holds great potential for resource-efficient and environmentally sound fabric treatments; refer also to Chapter 7 of this book, Environmentally friendly plasma technologies for textiles.

12.3.2 Lotus effect and self-cleaning surfaces

Synthetic polymer fibre surfaces can be engineered during manufacture using well-established techniques, such as embossing, to mimic natural plant surface structures. The well-known 'lotus effect' is based on the surface morphology of the lotus leaf, which was revealed at micro and nano-scale to consist of a mass of closely packed 'points' and troughs, covered in a wax substance, on which drops could not adhere. The lotus effect – a combination of structure and chemistry – was characterised in the mid-1990s by Wilhelm Barthlott and colleagues at the University of Bonn (www.lotus-effect.com; Barthlott and Neinhuis 1997; Neinhuis and Barthlott 1997) (see Fig. 12.1). This was the original inspiration for the technology behind stain- and water-repellent surfaces, and 'self-cleaning' nanotechnologies, which are now routinely applied in painted surfaces for industrial purposes. Self-cleaning glass is currently used in many buildings, based on the photocatalytic effect in which sunlight reacts with a surface coating (containing particles of titanium dioxide) to emit free radicals which combine with the organic molecules (dirt) which in turn are washed off

(a)   (b)

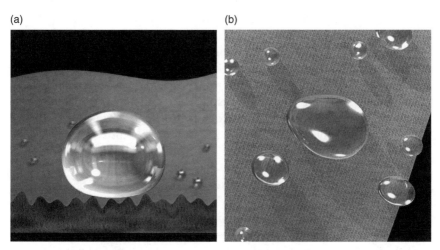

12.1 Lotus effect of water on NanoSphere®-treated textile, (a) schematic of surface morphology, (b) beading of droplets on fabric surface (courtesy of Schoeller AG, Switzerland).

when it rains. A self-cleaning effect on textiles can also be produced as a result of the 'roughening' of fibre surfaces at nanoscale by depositing nanoparticles, preventing liquids and viscous solids (especially food) from adhering, and enabling water droplets to collect loosely attached dirt particles as they roll off the surface. This enables soils to be rinsed off, but does not prevent normal washing action. One research institute, ITV Denkendorf in Germany, has created a self-cleaning quality standard for its fibre and fabric coatings that can be applied to cotton or polymer fibres. For antistatic functionality, especially in automotive applications, a finish has been developed that is claimed to be permanent, withstanding abrasion and remaining unchanged after laundering five times (Scherrieble, 2008).

12.3.3 Sol-gel surface treatments

Much research is underway to transfer the techniques of sol-gel surface chemistry into the textile field. This technology is currently applied to solid hard materials such as glass, metal and wood for protection from corrosion, scratching or UV radiation for example. Impetus here comes from architectural textiles in roofing for example, where resistance to weather conditions is vital; i.e. degradation from UV light, abrasion, and exposure to water, oil-based substances, and microbial attack. Sols are colloidal solutions of small particles such as silicon dioxide (SiO_2), which are transformed through solution and condensation to a nanosol containing nanoparticles. This is then applied through normal finishing techniques such as dipping or padding processes to a textile surface, and cured to remove the solvent and leave the sol as a gel (xerogel) containing a network of metal oxides, in a hybrid organic/inorganic polymer (also termed organically modified ceramics) whose properties can be further modified for predetermined functionality (Textor *et al.*, 2008). Researchers at ITV Denkendorf have applied sol-gel chemistry in a water-based process to automotive textiles combining the functionality of oil and water resistance with antimicrobial effects (Scherrieble, 2008). A sol-gel with supreme insulating functionality has been developed by NASA for space travel, named Aerogel; it is based on silica and has the lowest density of any solid (comprising up to 99.5% air), plus the lowest thermal conductivity (see http://stardust.jpl.nasa.gov/aerogel_factsheet.pdf).

The transfer of sol-gel processes to textiles finishing has been demonstrated as scalable to the commercial realm by Brückman *et al.* (2008) of CHT R Beitlich in Germany, a company which, since 2006, has marketed a binder named iSys MTX which is claimed to be 'the first product adapted specifically for the textile industry on the basis of the Sol-Gel process'. This binder is combined with a second component, iSys AG, containing silver particles that chemically fix on to the sol-gel-treated textile surface to create an antibacterial coating that is said to last up to 60 wash cycles.

According to the company's information, the treatment can be applied to 'fabrics, knitwear and non-wovens, made of cotton, synthetic fibres and their blends'.

12.3.4 Nanoencapsulation

Microencapsulation is already familiar in consumer textile products such as women's tights which have so far been impregnated with a therapeutic cosmetic finish containing aloe vera, or men's socks containing antimicrobial compounds for odour prevention. These effects have the same problems of short life and wash off within a few cycles that were referred to previously. Much research into nanoencapsulation is underway in the medical sector for the purpose of controlled-release drug delivery, however this will inevitably have its impact on more cosmetic and aesthetic areas of textiles. In addition to hygiene and healthcare uses for this technology, clothing with embedded scents or other therapeutic substances could give a competitive advantage to textile and fashion brands. Benefits for sustainability in this arena will no doubt depend on the success and longevity of any technological effect, in order that it is not consigned to the level of a fashion gimmick as seen previously with thermochromic inks used on t-shirts, which partially changed colour in wear.

The future potential for nanoencapsulation to deliver well-being benefits to household and clothing textiles is enormous, and to create multifunctionality where a choice of perfumes, therapies and even responsive warnings against pollution may be inbuilt into clothing.

12.4   Coloration and structural colour

The coloration of textiles through dyeing and printing has been a major factor in the textile industry's environmental impact since its inception (see Chapter 6, Key sustainability issues in textile dyeing). The contribution to be made from nanotechnology towards greater sustainability is significant in a number of aspects. Reducing the high wastage rates from dyeing, which occur because of poor colour accuracy and uneven dyeing results, would have positive impacts on a process that currently utilises large amounts of water and energy, and whose effluents have been responsible for pollution of water courses across the globe, particularly following the invention of synthetic dyes. As discussed in Section 12.3.1, the transfer of plasma technology to textile surfaces enables these surfaces to be better prepared to receive dyes and pigments, therefore improving the uptake of dyestuffs and minimising waste and wash-off. Nanocoatings can also preserve colourfastness by imparting resistance to fading from sunlight (UV light) and to abrasion through washing and wear, enhancing fabric durability and therefore minimising waste.

12.4.1 Colour with metals

One of the exciting potentialities of nanoscience is the discovery of new processes and means to produce desirable effects with minimal environmental impact. However, not all nanotechnologies are new – some are very old, including the use of gold and silver nanoparticles to colour substances, used in seventeenth-century stained glass windows and explained by Michael Faraday in the mid-nineteenth century. Recent research at Victoria University in Wellington New Zealand has applied these techniques to textiles, and demonstrated that colloidal solutions of gold at different particle sizes can produce a good range of colours on cotton and wool fabrics – from purple to lilacs, yellows, browns and greys – which are permanently bonded to the fibres, without conventionally dyeing (Johnston et al., 2006). Although the physics of surface plasmon resonance, which scatters light to produce the colours from nanoparticles, is well understood (particles of size 2–5 nm are red, agglomerates of 50–70 nm are violet, with orange and yellow in between), it is not yet clear how this technology might scale up to the commercial realm. However, the great potential here for sustainability is a one-step process for stable non-fading colours, offsetting the large amounts of water, dyestuffs and energy used in conventional dyeing against the miniscule quantities of metals and far less water used.

12.4.2 Structural colour and nanophotonics

Key nanotechnologies have been developed through biomimicry, and observation of colour in natural species such as butterfly wings and mother of pearl. The colour in butterfly wings relies on the refraction of light against the complex surface nanostructures of the insect wings, reflecting to the eye intense colours in different sections of the light spectrum to appear irridescent blue or green for example (Fig. 12.2). Studies of the Morpho butterfly have resulted in a product that is a nano-scale thin film in 61 layers produced by chemical and fibre company Teijin in Japan; the film reflects four different opalescent colours, and has also been created as a slit-film yarn using only structural coloration. In the UK, other research into nanophotonics is underway to produce structural colour in block copolymers using colloidal crystals (synthetic opals), with potential applications in security, and home and personal care (Jones, 2008; Pursiainen et al., 2005). Nanoscience may yet create the 'invisibility cloak' of popular culture as research develops into optical transformation to produce a 'cloaking' effect by shrinking objects within electromagnetic broadband fields (Li and Pendry, 2008).

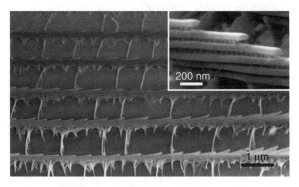

12.2 Morpho butterfly wing structure (courtesy of Zhong Lin Wang, Georgia Tech University).

12.5   Nanofibres

New methods of creating nanofibres and non-woven structures have been developed in research labs and are finding application in medical research to create implants and scaffolds for engineering tissue growth. The electro-spinning process collects extruded polymer nanofibres on to a preform which is built up to the required thickness and shape, suitable for small medical devices. However, centrifugal spinning is a process that is said to show greater potential for being scaled up to industrial levels (Brückmann *et al.*, 2008; Scherrieble, 2008).

Much attention is being paid to the development of carbon nanotubes (CNTs), formed from pure carbon with a hexagonal network tubular structure that renders the material immensely strong. Although these occur naturally in graphite and carbon black, research is underway in many countries to find ways of increasing their production to form, for example, high tensile strength yarns for use in protective body armour. The race is on between researchers around the world to realise commercial quantities of CNTs spun into yarns far stronger than Kevlar for protective body armour, and also for energy storage to power electrical devices, among many other possible applications (Fig. 12.3).

12.6   Electronic textiles

The embedding of metallic nanoparticles into fibres or coating of textile polymer fibres in metals has resulted in new conductive functionalities for yarns and hence for textiles woven or knitted from them. The developments of semiconductors from polymers has also been a major breakthrough, creating new possibilities in electronic or e-textiles – an important new platform for technical textiles for many purposes in protective, automotive

(a)

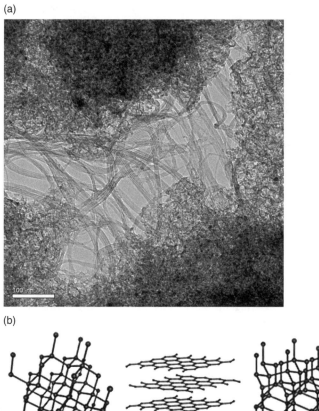

(b)

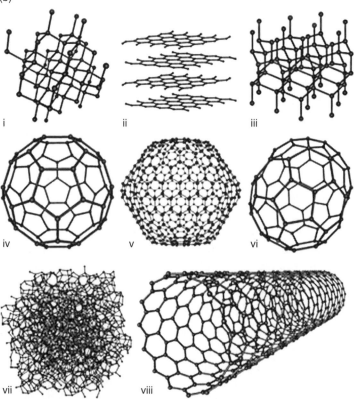

12.3 Carbon nanotubes, (a) micrograph of carbon nanotubes (courtesy of the National Physical Laboratory, Middlesex, UK), (b) structures of eight allotropes of carbon: i, diamond; ii, graphite; viii, carbon nanotube (source: Wikipedia).

and aerospace sectors, and for interior domestic use, performance sports-wear, casual wear and fashion. Already, a number of textile-based systems, such as Soft Switch® and Elek-Tex® are in use in niche consumer outdoor clothing products which allow MP3 players or mobile phones to be operated by textile switching systems on the sleeve or lapel of a garment; see also Section 12.8.3 below (Fig. 12.4).

Yarns containing metals such as steel or silver have for some years been incorporated into knitted and woven products both for antimicrobial function (in wound dressings) and to conduct electricity. X-Static™ is a synthetic polymer yarn covered in a silver layer, used in US army and other military wear for its antimicrobial properties and in other products such as Polartec® PowerDry® fleece gloves for its conductivity. German brand WarmX® offers a range of products that heat targeted sections of

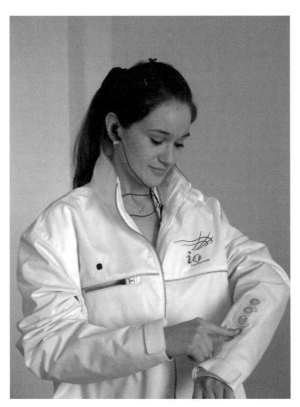

12.4 'Know Where Jacket' prototype by Interactive Wear with MP3, global positioning system (GPS) and Bluetooth capability (courtesy of Interactive Wear AG, Germany).

12.5 WarmX® heated clothing (courtesy of WarmX®, Germany).

their knitted clothing which contain silver yarn, powered by a battery pack (Fig. 12.5).

Particularly in the medical context, the use of soft and familiar textiles is more acceptable to patients than hard components on (or in) the body. Therefore non-invasive textile sensors are being continually developed for healthcare purposes to provide soft interfaces that can detect body vital signs such as heart and respiration rates, and subsequently transmit that data to medical professionals. Examples have been developed in research projects in Europe and the US, such as the MyHeart, Smart Shirt and Vivo Lifeshirt prototype products (produced by MyHeart EU consortium, Sensatex and Georgia Institute of Technology research projects respec-

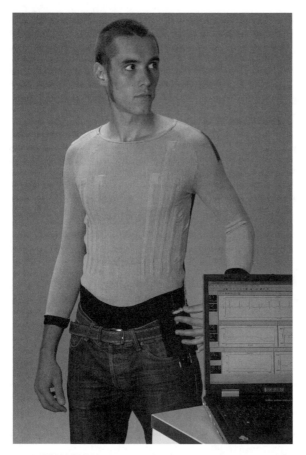

12.6 WEALTHY heart monitoring vest (courtesy of Smartex, Italy).

tively) (Fig. 12.6). Further improvements in textile sensor and biosensor technologies will enable these functions to transfer to mainstream consumer uses.

The discovery in the 1990s of organic light-emitting diodes (OLEDs) via the functionalisation of polymers through nano-scale electroluminescent effects (discovered at Cambridge in 1990 by Richard Friend and colleagues: (Burroughs *et al.*, 1990; Sirringhaus *et al.*, 1998) started a new branch of science named organic (or plastic) electronics. After nearly two decades this research area has begun to bear commercial fruit, for example the UK spin-out company Cambridge Display Technologies was sold to the Japanese company Sumitomo in 2007, and in the same year the UK-based Plastic Logic announced it was to build the first factory to manufacture plastic electronics on a commercial scale (Jones, 2008). A Nokia mobile phone launched in 2008 utilised a screen comprising OLEDs. As the nanotechnology research enabling plastic electronics continues to develop into commercialisation, then the potential for conductive and responsive textile interfaces will provide the long-awaited platform on which flexible textile displays can be achieved for multifunctional purposes.

At the same time, research into power sources and smaller, lighter and more efficient batteries is continuing. Promising nano-scale technologies for harvesting power through kinetic movement or creating flexible arrays of solar cells to harvest radiated energy from the sun are all in development. New research at the Georgia Institute of Technology in the US has attempted to extract energy from fibres with zinc oxide nanowires that convert mechanical energy into electricity using the smallest movements (Kaufmann, 2008), which could realise the dream of being able to power personal mobile devices through human movement – a closed-loop and highly sustainable scenario akin to cycling.

12.7   Risks versus benefits

A key benefit of nano-scale technologies is seen to be the resource-efficient use of both materials and energy. The improved efficiency made possible by nanotechnology, such as miniaturisation and the catalysing of chemical reactions, can have great environmental benefits. The propensity of extremely small nanoparticles to exhibit self-organising behaviour to perform biomimetic processes such as repair can eventually lead to more durable materials and products. However, the dangers of uncontrolled developments by self-reproducing nanostructures can only be estimated by existing risk assessment and prediction tools, and have been the subject of science fiction and other scare scenarios. The 'grey goo' doomsday scenario famously coined by the Prince of Wales is discussed and dismissed as not feasible in The Royal Society and Royal Academy of Engineering Report (2004).

Some nanoparticles, including zinc oxide and titanium dioxide, have been used successfully in products such as sunscreen lotion since the early 1990s, and certain toothpastes incorporate nanomaterials to repair damaged surfaces of tooth enamel. The antibacterial effects of silver were well understood by the ancient Greeks and Romans, and silver is currently employed in a range of medical and consumer products. Whereas silver is non-toxic to humans at normal molecular scale and can be safely worn as jewellery or worn when chemically bonded into a fibre, nanoparticles of silver are toxic to bacteria and micro-organisms at the nano-scale. Silver ions have the effect of destroying the bacteria cell membrane, deactivating the cell metabolism and preventing cell growth. All manufacturers of nano silver-based products, however, claim their products to be totally safe for humans.

The extremely small scale of any nanoparticles and their mobility in potentially moving easily through cell membranes of the body has therefore given rise to serious concerns for human and animal health. Much concern is centred on 'free' nanoparticles and nanofibres (such as carbon nano-tubes) which may escape into the environment during production, handling or disposal, and whose long-term behaviour is unknown. Regulating organ-isations are mindful of previous health risks such as those that became apparent too late with the use of asbestos materials. In the UK, the Natural Environment Research Council set up an environmental Nano Science Initiative in 2006 to fund new research, part of which is a study on the effects of silver surface chemistry on the bactericidal properties of silver nanoparticles. A US study on the safety of nanoparticles of silver was com-pleted in autumn 2008 under the remit of the Project on Emerging Nano-technologies, stating that nano-scale silver will 'challenge regulatory agencies to balance important potential benefits against the possibilities of significant environmental risk' (Luoma, 2008). Further research results are required to reach categorical conclusions on this debate.

In 2004, the Royal Society in collaboration with the Royal Academy of Engineering produced the first key report in the UK on the future prospects and risks of nanotechnologies, entitled *Nanoscience and Nanotechnologies: Opportunities and Uncertainties*. This prompted a number of actions and further reports at government level in the UK, plus a programme of dia-logue and engagement with a range of stakeholders including the lay public (Gavelin *et al.*, 2007). The UK has a major role in a consortium of countries and organisations, including the Organisation for Economic Cooperation and Development (OECD), British Standards Institute, International Organization for Standardization (ISO) and European Committee for Standardisation. These organisations are working towards internationally agreed standards of safety and quality for characterisation, production and manufactured products using nanotechnology. Owing to the enormous pre-dicted economic value of this new industry, risk management and dialogue is seen by governments and stakeholders as key to avoiding the controver-sies associated with other new technologies such as genetic modification of plant crops. In the meantime various countries, including the UK and the US, have implemented voluntary reporting schemes for industry players, such as the Nanoscale Materials Stewardship Programme under the US Environmental Protection Agency's Toxic Substance Control Act, in which companies including DuPont have taken part. In early 2008, the EC set out a code of conduct for researchers in nanoscience and nanotechnology (ec. europa.eu/nanotechnology) and continues to develop and monitor research, including a focus on applications.

The EU legislation REACH (Regulation, Evaluation, Authorisation and Restriction of Chemical substances) introduced in June 2007, seeks to control all manufactured chemicals through compulsory registration; however its volume thresholds are currently too high for nano-scale substances to be covered and discussions are underway to bridge this gap. The regulation of manufactured nano-scale materials has therefore many further crucial stages of legislation to undergo. In tandem with government-level activities, individual research institutes are setting up labelling and certification schemes, for example the respected Hohenstein Institute in Germany introduced a 'nano' textile label in 2005 to indicate textiles that have nanoscale finishes.

12.8   Commercial and consumer applications

As indicated above, hundreds of commercial products that include a nanotechnology component are already on the market across many sectors. Many developments have transferred from military and space research. Significant examples with respect to textiles are discussed in more detail in this section; however this is by no means exhaustive as new products are launched continually, some with more claim to the 'nano' label than others.

12.8.1 Water and stain repellency

In the textile arena, two companies (NanoTex of the US and Schoeller Technologies AG of Switzerland) have taken the lead for commercialising nanotechnology finishes for fabrics used in clothing and interiors, and for technical purposes such as military protection. NanoTex claim to be 'the first to apply nanotechnology to fabrics' and to have set the industry standard. The NanoTex® finishing process was launched in 2000, and the company works with numerous fabric mills and leading fashion brands including Levi's, Eddie Bauer and Gap, as well as supplying the US military. A range of different finishes are produced that resist spills, repel and release stains, neutralise odours, reduce static and wick away moisture (on both synthetics and cotton). Unfortunately, for the sustainability agenda, the company recommends 45 minutes of tumble drying plus ironing to enhance performance after washing. Schoeller has partnered since 2007 with Swiss chemical and dye company Clariant and also works with the research institute ITV Denkendorf in Germany. It developed the NanoSphere® fabric finish, based on the lotus effect, which repels liquids and other viscous food substances. A second generation of NanoSphere® finish has been launched that performs with greater abrasion resistance, withstands 50 wash cycles and, since May 2008, utilises a new formula including C6 fluorochemicals, with no detectable perfluorooctanoic acid

(PFOA) or perfluorooctanesulfonic acid (PFOS). This new finish has been awarded the Bluesign® mark for environmental safety and stewardship. Brands using the finish include Hugo Boss, Henry Lloyd, Polo Ralph Lauren and the Swiss postal and rail services.

In Japan, major chemical and synthetic fibre producers Toray Industries and Kanebo have each developed ultra-fine yarns with moisture-absorbing properties. Kanebo's polyester fibre consists of 20 layers totalling just 50 nm which it is claimed enables the wicking of moisture to a performance 30 times greater than normal polyester. Toray's hygroscopic nylon yarn technology consists of thousands of nano-scale nylon threads bundled together into filaments, which creates the spaces for water to collect, enabling more water absorption than cotton fibres.

12.8.2 Antimicrobial/antibacterial finishes

Several manufacturers produce yarns impregnated with silver, including Silverlon®, SmartSilver® and X-Static®, which has a silver layer permanently bonded to the outside of a nylon fibre core, claiming that performance does not diminish over time. Silver is both thermally conductive and antimicrobial and the yarn is easily incorporated into woven and particularly into knitted structures such as SoleFresh™ socks using silver yarn from JR Nanotech. X-Static® is registered with the US Environmental Protection Agency and several products are approved by the Food and Drugs Agency. At the consumer level, silver yarn has been incorporated into Polartec PowerDry® polyester fleece fabric and army underwear to inhibit the bacterial growth that causes odour; for its conductive properties it is used in WarmX® heated knitwear, and NuMetrex® heart monitoring sportswear. American company ARC Outdoors has launched a branded anti-odour technology E47™ Nano Technology in partnership with SmartSilver® from US-based NanoHorizons, which will be used in its own outdoor clothing and licensed to other manufacturers. Manufacturers of products that contain nano-scale silver claim that it is absolutely harmless to the human body. In Asia, Samsung Electronics have development their Silver Nano Health System™ which is applied to consumer electronic goods such as refrigerators.

With regard to domestic laundry, Arch Chemicals, a US company, has developed Purista® a textile treatment suitable for cotton towels and clothing, which inhibits the formation of bacterial biofilms that can easily remain on fabrics even after washing, especially when low temperatures are used as is now commonplace. The finish is sold to manufacturers and consumers as an aqueous solution, whose active ingredient polyhexamethylene bignanide (PHMB) is a cationic polymer that binds to the cellulose surface for a durable finish. The company commissioned a life cycle study (Blackburn and Payne, 2004) which demonstrated the value of its treatment in reducing the need for washing while effectively preventing the formation of bacterial growth. Here the notion of 'freshness' is the key factor rather than cleanliness, indicated by the marketing slogan 'Stays fresh, wash less'.

12.8.3 Electronic textiles

Issues became apparent with early attempts at the integration of wearable electronic devices into clothing, for example the Levi's ICD jacket in collaboration with Philips electronics in 2001; the issues were mainly related to washability and high cost, but also because people change their clothes more often than their gadgets. The collaboration between Nike and Apple in 2006 (two highly fashion conscious brands) placed sensors in the soles of a pair of running shoes, transmitting data wirelessly to an iPod in the user's pocket to relay speed and performance information and feed back via appropriate music. This created a successful model for commercial application of technology that met consumers' personal and fashion needs at the same time, and has sold extremely well. However, the embedding of electronics into soft textile interfaces is proving a more effective route for integration into clothing.

UK company Peratech invented SoftSwitch technology which is based on the 'quantum tunnelling effect' in which a nanocomposite polymer substance becomes conductive under pressure, deformation or other stimuli. The quantum tunnelling composite is utilised in soft textile sensors which interface with devices that are normally controlled by hard keypads, switches and buttons, providing a relatively low-cost solution for incorporation into clothing. In early 2008, Peratech also acquired Eleksen, owners of another UK-patented interactive textile sensor Elek-Tex®, created from woven fabric less than 1 mm thick, comprising conductive nylon threads oriented in different directions which can sense both pressure and location of the pressure point. Electronic impulses sensed by the fabric field are translated into data by linked software. An iconic product developed by Eleksen is the soft roll-up keyboard which connects to PDA devices and has been integrated into soft, moulded textile keyboards and messenger bags in collaboration with Microsoft.

Other products with electronic functionality are based on the conductive properties of silver. X-Static® yarn is incorporated into Polartec® polyester fleece heated gloves, in addition to applications in electromagnetic shielding fabrics. The NuMetrex® sportswear tops incorporate nylon and stretch Lycra® polyurethane yarn with silver yarn precision knitted to form a close-fitting textile electrode to sense and transmit piezoelectric signals for heart rate monitoring to a receiver module held in a small pocket in the garment. Signals are radioed to a wrist-based digital display for immediate feedback

(a)

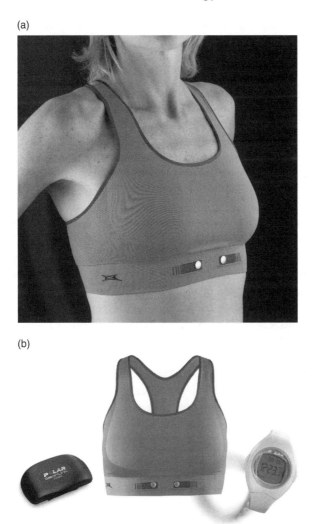

(b)

12.7 NuMetrex® heart monitoring sports top and Polar data system (courtesy of NuMetrex, Delaware, USA).

(Fig. 12.7). This product, developed by Textronics, an offshoot company of Dupont, was launched in 2006 and is achieving commercial success. WarmX® is a range of knitwear products that incorporate silver conductive yarns used to heat a section of the clothing, powered by a battery pack the size of a mobile phone, carried in a pocket in the garment, which can be disconnected for charging and washing. Gradually, as a range of previously disparate technologies and new forms of textile sensors, actuators and other components become available, previous dreams for truly functional and intuitively wearable computing can start to become a reality via the medium of electronic textiles.

12.9   Future trends

The advantages that nanotechnology offers are large specific surface areas which facilitate a high level of functionality, new properties as a result of the extremely small scale and minimal use of resources. These properties are highly compatible with textiles, which already have high specific surface area and are ubiquitous in use in domestic and industrial sectors. The fact that all textile characteristics are retained by nano-scale finishing treatments – handle, flexibility, weight, permeability and texture – adds to their commercial potential. Nano-structured coatings, encapsulating nanoparticles for functionality, are ideally suited to a wide range of textile applications: technical and industrial, architecture and interiors, protective wear, everyday and fashion clothing, and intimate apparel. However, things that people wear and use become fashion, and wearable electronics will not become mainstream until the technology meets genuine needs, becomes invisible and intuitive, and the aesthetics merge with the technology itself. The success of mobile phones and the Apple iPod are testimony to this synergy and cultural integration. As consumers expect and demand more from their clothes and textiles – that they smell pleasant, stay fresh, feel comfortable, stay clean and are simple to care for, while looking great – the opportunity for new experiences through clothes as personal interfaces between private and public domains may be realised through nanotechnology and microtechnology in textiles. In addition, as the sustainability agenda continues to have a major impact on manufacturing and consumer behaviour, these relationships will be forged with new significance.

'Tailor-made' synthetic fibres are increasingly being developed to meet specific technical requirements, for performance in critical medical or safety situations, and these developments will cross over from technical textiles to benefit consumers of everyday textiles to offer well-being, security and entertainment. Research into electronic functionality in fibres and textiles may soon be able to offer colour change effects on a very small scale. Flexible and strong conductive yarns from carbon nanotubes will eventually be realised, creating electronic textiles to enable data transport through clothing for integrated multifunctionality. With the continuing development and miniaturisation of OLEDs and printed electronic circuits, potential is opened up for programmable and downloadable colour changes in clothes. Current developments in plastic electronics, such as colour change markers, are likely to transfer from plastic food packaging on to more flexible polymer substrates suitable for clothing. Delivery of colours and other effects on demand via nano-scale pumps will create a playground of therapies and aesthetic effects. Our clothes will therefore continue to become more therapeutic and responsive to our needs. A personal wish list includes:

- fashion that can delight and endure through variation in surface qualities, density, size and aesthetics;
- fabrics that last longer, stay cleaner and use less energy and resources to make;
- fabrics that are made from biodegradable materials and that decompose benignly;
- multifunctional fabrics that can change in colour and pattern, without using battery power, in response to mood, health, fashion or situation;
- fabrics that can communicate through display;
- fabrics that are truly breathable and can ventilate or insulate in response to conditions;
- fabrics that can renew themselves and recharge their functionalities.

Despite the risks and the high current costs, the potential benefits of nanotechnology research in textiles and beyond to a growing and ageing population are providing results so exciting that they cannot be ignored. They may prove as revolutionary as the industrial revolution of the eighteenth century and the information explosion of the twentieth century, and in doing so become pivotal to a sustainable future.

12.10 Sources of further information and advice

Published information

- *Nanotechnology in consumer products*, October 2006. Available at www.nanoforum.org.
- *Nanoscience and nanotechnology: opportunities and uncertainties*, 2004, Royal Society and Royal Academy of Engineering Report. Available at www.royalsociety.org.
- *Characterising the potential risks posed by engineered nanoparticles*, December 2007, UK HM Government report. Available at www.defra. gov.uk.
- Pearson I (2008) *DIUS statement by the UK Government about nanotechnologies,* February. Runcorn, Cheshire: Department for Innovation, Universities and Skills.
- Black S (2008) *Eco Chic: The Fashion Paradox.* London: Black Dog.
- Fletcher K (2008) *Sustainable Fashion and Textiles, Design Journeys,* London: Earthscan.
- Gavelin K, Wilson R and Doubleday R (2007) *Democratic technologies? The final report of the Nanotechnology Engagement Group.* Involve UK, www.involve.org.uk.

- Jones R (2007) 'Can nanotechnology really be green?', *Nature Nanotechnology*, February, **2** (2), 71–72, DOI 10.1038/nnano, 2007.12.
- Tao X (2001) *Smart Fibres, Fabrics and Clothing*. Cambridge: Woodhead Publishing.
- Tao X (2005) *Wearable Electronics and Photonics*. Cambridge: Woodhead Publishing.

Organisations and websites

Nanotechnology KTN, www.nanotechnologyktn.com
Institute of Nanotechnology, www.nano.org.uk
Technitex, www.technitex.org
Sensors KTN, www.sensorsktn.com
London Centre for Nanotechnology, www.london-nano.org
European Nanotechnology Trade Alliance (ENTA), www.euronano-trade.com
NanoForum, www.nanoforum.org
Nano Cluster Bodensee, Switzerland, www.ncb.ch/en/index
European Commission – Nanotechnology, http://ec.europa.eu/nanotechnology/index-en.html
Safenano information service, www.safenano.org
National Physical Laboratory, www.npl.org.uk

Company websites

NanoTex, www.nano-tex.com
Schoeller, www.schoeller-textiles.com
X-Static, www.noblematerials.com
Morphotex,   www.teijinfiber.com/english/products/specifics/morphotex.html
E47, www.e47nano.com
NuMetrex, www.numetrex.com
WarmX, www.warmx.de
Cientifica, http://cientifica.eu

12.11 References

ALLWOOD JM, LAURSEN SE, DE RODRIGUEZ CM and BOCKEN NMP (2006) *Well Dressed? The Present and Future Sustainability of Clothing and Textiles in the UK*. Cambridge: University of Cambridge Institute for Manufacturing.

BARTHLOTT W and NEINHUIS C (1997) 'Purity of the sacred lotus, or escape from contamination in biological surfaces', *Planta*, **202**, 1–8.

BLACK S (2007) 'Trends in smart medical textiles', in *Smart Textiles for Medicine and Healthcare*. Cambridge: Woodhead Publishing.

BLACKBURN R and PAYNE J (2004) 'Lifecycle analysis of cotton towels: impact of domestic laundering and recommendations for extending periods between washing', *Green Chemistry*, **6**, G59–G61.

BRÜCKMANN R, KOCH M and LUTZ H (N.D.) 'Successful transferring of the nanotechnology Sol-Gel process onto textiles'. CHT R Beitlich GmBH, Tübingen Germany, Company paper, accessed September 2008.

BURROUGHS JH, BRADLEY DDC, BROWN AR, MARKS RN, MACKAY K, FRIEND RH, BURNS PL and HOLMES AAB (1990) 'Light-emitting diodes based on conjugated polymers', *Nature*, **347**, 539.

CIENTIFICA (2006) *Nanotechnologies in the Textile Market*. London: Cientifica, p. 66.

DELIYANAKIS N (2008) 'EU support for nanotechnology', Nanotechnology Unit, Research Directorate-General of the European Commission. Presentation 2nd October 2008 at the UK Government Nano Showcase event, Deptartment for Innovation, Universities and Skills, London. Available at http://www.tuvnel.com/eventarticle.aspx?event_id=168.

FISHER T, COOPER T, WOODWARD S, HILLER A and GOWOREK H (2008) *Public Understanding of Sustainable Clothing: A report to the Department for Environment, Food and Rural Affairs*. London: Defra.

GAVELIN K, WILSON R and DOUBLEDAY R (2007) *Democratic technologies? The final report of the Nanotechnology Engagement Group*. London: Involve.

JOHNSTON JH, RICHARDSON MJ and KELLY FM (2006) 'Gold nanoparticles as colourants and functional entities in high fashion textiles', in *Proceedings of Nanotechnologies and Smart Textiles for Industry and Fashion CD-ROM*, Institute of Nanotechnology, Royal Society London, 11–12 October 2006.

JONES RAL (2008) 'Nanotechnology: Research Councils Approach'. Presentation 2nd October 2008 at the UK Government Nano Showcase event, Department for Innovation, Universities and Skills, London. Available at http://www.tuvnel.com/eventarticle.aspx?event_id=168.

KAUFMANN TC (2008) Reported in *Nano Circle, innovative technologies and trends*, Credit Suisse Bank Investment Services Newsletter, June 2008.

LI J and PENDRY JB (2008) 'Hiding under the carpet: a new strategy for cloaking', *Physical Review Letters*, **101**, 203901.1–4.

LUOMA SN (2008) 'Silver nanotechnologies and the environment: old problems or new challenges?', quoted in *Ecotextile News*, October, 11. Available at www.nanotechproject.org/news/archive/silver, accessed 12 November 2008.

MOORE R (2008) 'Plasma surface functionalisation of textiles', in *Proceedings of Nanotechnologies and Smart Textiles for Industry, Healthcare and Fashion CD-ROM*, Institute of Nanotechnology, Royal Society London, 9 March 2008.

NEINHUIS C and BARTHLOTT W (1997) 'Characterization and distribution of water-repellent, self-cleaning plant surfaces', *Annals of Botany*, **79**, 667–677.

PURSIAINEN OLJ, BAUMBERG JJ, RYAN K, BAUER J, WINKLER H, VIEL B and RUHL T (2005) 'Compact strain-sensitive flexible photonic crystals for sensors', *Applied Physics Letters*, **87**, 101902–4, DOI 10.1063/1.2032590.

ROYAL SOCIETY AND ROYAL ACADEMY OF ENGINEERING (2004) *Nanoscience and nanotechnologies: opportunities and uncertainties*, Report, July, Annex D, p. 109. Available at http://www.raeng.org.uk/policy/reports/pdf/nanotech/annexes.pdf.

SCHERRIEBLE A (2008) 'Textile functionalisation with nano-dimensional material', in *Proceedings of Nanotechnologies and Smart Textiles for Industry, Healthcare and Fashion*, Institute of Nanotechnology, Royal Society London, 9 March 2008.

SIRRINGHAUS H, TESSLER N and FRIEND RH (1998) 'Integrated optoelectronic devices based on conjugated polymers', *Science*, **280**(5370), 1741–1744.

SLATER K (2003) *Environmental Impact of Textiles*. Cambridge: Woodhead Publishing, pp. 69–89.

TAO K (2005) *Wearable Electronics and Photonics*. Cambridge: Woodhead Publishing.

TEXTOR T, SCHRÖTER F, SCHULTZ B and SCHOLLMEYER B (2008) 'Potential application of Sol-gel technique for the functionalisation of textiles, exemplarily shown for architectural textiles', in *Proceedings of Nano Europe 2008*, St Gallen, Switzerland, 16–17 September 2008.

13

The use of recovered plastic bags in nonwoven fabrics

B. R. GEORGE, B. A. HAINES and E. MURPHY,
Philadelphia University, USA

Abstract: Polyethylene plastic bags are widely utilized throughout the world. Recovery of these bags is low, but there are several applications that recovered bags can be used for, such as power generation. These bags may be useful as a binder material in nonwoven fabrics. Various web forming and bonding methods involving shredded plastic bags have been carried out, with positive results: shredded plastic bag pieces can be utilized as a binder material for nonwoven fabrics. Possible end uses include thermal and sound insulation materials, but further research is required to fully understand the behavior of these materials in nonwoven fabrics.

Key words: polyethylene, plastic bags, nonwovens, fabrics.

13.1 Introduction

The practice of using plastic bags instead of paper bags to carry goods from stores has become a worldwide issue. It is estimated that 100 billion to 1 trillion plastic bags are produced worldwide per year.[1,2] It has been reported that China consumes up to 3 billion plastic bags daily.[3] According to the United States Environmental Protection Agency, in 2005 over 3 billion kilograms (kg) of polyethylene plastic bags were generated. Of this amount, only 200 million kilograms of these bags were recovered, about a 6% recovery rate. The majority of the bags generated were discarded.[4] Another source estimates that 380 billion plastic bags are discarded in the United States every year, with a recovery rate of 0.6%.[5] However, it has also been reported that recycling of plastic bags has increased dramatically in the past several years as consumers realize that plastic bags can be recycled.[6] Owing to the perception that plastic bags are harmful to the environment, legislation is being enacted to curtail use of plastic bags. China has forbidden the provision of plastic bags for free, Ireland has a tax on plastic bags, and the city of San Francisco, in the United States, has banned them.[1-3] However, given the lower costs of plastic bag production and distribution compared with those for paper bags, it is unlikely that plastic bags will disappear as a method of handling purchased goods.[1]

There are several alternative uses for recovered polyethylene bags. The easiest method is to incinerate the bags, along with other materials, to generate power such as electricity. This is the least favored method as it does not promote the recycling of the bags. Research has shown that given the proper conditions, carbon oxides and harmful 'polycyclic aromatic hydrocarbons' were produced by combustion of polyethylene.[7] More popular is to convert the recovered bags into new bags or into other products, such as those produced by Trex. Currently, Trex converts approximately half of the recovered plastic bags into composite wood products.[8] Some past research examined the use of recovered polyethylene bags blended with fly ash, a by-product of coal combustion, as a possible building material, with mixed results.[9] It is generally difficult to effectively recycle plastic bags into other products owing to the bags being mixed with other plastics and contaminants such as paper.[10] Although it is possible to separate different plastic types prior to recycling, this is generally time consuming and expensive. However, polymer immiscibility has been an impediment to creating materials from recovered plastics with good mechanical properties. Recent research has focused on producing materials from a mixture of the plastics most commonly recovered – polyethylene, polypropylene, and polyester – with improved mechanical properties. Some research has focused on improving the properties by adding polymers that improve the compatibility of the different plastics to each other, while others have examined the use of liquid carbon dioxide to mill the plastics into a powder form that shows good miscibility between the different polymers.[10–12]

Nonwoven fabrics are fabrics created mostly from fibers, but may also contain yarn and fabric scraps. They offer the advantages of low cost and high productivity in comparison to weaving and knitting. Nonwoven fabric production consists of two stages: (a) web formation, where the fibers are assembled into a fibrous web; and (b) web bonding, where the web is bonded to create a fabric. The dominant method of bonding fibers today is thermal bonding, where heat is utilized to melt fibers together. Of the thermoplastic fibers utilized in the thermal bonding process, the olefins, polypropylene and polyethylene, are preferred, as they have lower melting temperatures and thus lower processing costs in terms of energy, compared with nylon and polyester.

As the cost of petroleum has risen in the past few years, the price of thermoplastic fibers has also risen. Hence, if a steady source of a waste material could be utilized to replace binder fibers it may be possible to maintain or even lower product costs as well as find a second use for the large number of plastic bags that are recycled. Additionally, polymer miscibility or low levels of contamination may not be an impediment to successful product generation. Success in replacing binder fibers with binder material obtained from polyethylene bags might even generate greater demand for these materials, causing an increase in recycling of these materials. A literature search failed to find any previously published work addressing the use of recovered polyethylene bags as a source of binder materials for nonwoven fabrics.

13.2   Experimental approach

Polyethylene shopping bags were gathered by the authors. These bags were cut into flat strips 75 mm long and 5 mm wide with the use of a hand-held rotary cutter. The bag pieces were blended by hand with 6 denier, 76.2 mm Type 610 polyester fibers, made from recycled polyester bottles, supplied by Wellman (Mississippi, USA). The melting temperatures of the bag pieces and the polyester fibers were determined with a Fisher digital melting point analyzer Model 355. Blends of 50/50 and 25/75 bag pieces/polyester, respectively, were created, as was a control of 100% polyester. These fiber combinations were then formed into webs with the use of a 0.91 m wide Rando Webber airlaid unit. All blends were processed twice with the Rando Webber, in order to provide further mixing of the materials. After the second time through the airlaid unit, the webs were rolled up in paper by hand.

Webs of both blends were needlepunched with a 0.30 m wide James Hunter Fiberlocker needlepunch, containing Groz-Beckert $15 \times 16 \times 40 \times 3$ needles, after being placed on spunbond fabrics for support. Initial needlepunching trials were completed at a needlepunch density of 77.5 penetrations per square centimeter. These materials were then thermally bonded with a 200 cm long Tsuji Senki Kogyo through-air oven at 140°C for 3 minutes. Additional needling and thermal bonding trials were performed on the 50/50 blend using spunbond fabrics below and above the airlaid webs during the needling process, with the top spunbond fabric removed after needling. These trials were completed at a needlepunch density of 47 penetrations per square centimeter, and thermal bonding conditions of 150°C for 4 minutes. A final through-air bonding trial was performed on a 25/75 web that was not needled, with conditions of 150°C for 4 minutes.

Both airlaid blends were also thermally bonded with a James Carver Model C hot press with 15 cm × 15 cm heated platens. Two different bonding conditions were utilized: a combination of 180°C and 1.5 metric tons of pressure and a combination of 140°C with a pressure of 0.5 metric tons. The polyester control was bonded at a temperature of 215°C and 0.5 metric tons of pressure.

All three airlaid webs were also bonded with latex. Genflo 3060 latex, supplied by Omnova Solutions (Ohio, USA), diluted to 20% solids, was sprayed on to each side of each web with a hand-held spray bottle. The webs were then dried with the through-air oven at 130°C for 4 minutes.

Additional webs were made via wetlaid web formation. Plastic bag pieces were mixed with 6.4 mm 1.5 denier DuPont Dacron polyester fibers in water with the use of a Hamilton Beach blender. Blends of 50/50 and 25/75 bag pieces/polyester, respectively, were created. The contents were then formed into webs by hand in a 25 cm × 25 cm hand sheet former. The webs were allowed to dry for several days.

The fabrics were characterized via several different methods. The latex-bonded and thermally bonded (hot press) fabrics were allowed to condition under standard conditions, 65% relative humidity and 21.1 °C. These fabrics were then evaluated for thermal transmittance via ASTM D1518 with a Kawabata Evaluation System Thermolabo II. Thickness was measured with a Randall & Stickney thickness tester with a 2.54 cm diameter presser foot. The mass of all test specimens was evaluated with a Sauer scale. Fabric strength was evaluated with an Instron Model 1000 ball burst tester according to ASTM D3787, with a crosshead speed of 300 mm/min.

13.3   Results and discussion

13.3.1 Fabric formation

The polyester fibers utilized in the airlaid process were chosen based on the fact that they were created from recycled polyester bottles, with the idea of creating materials made entirely from recycled products. Airlaid processing of the blends created uniform webs with bag pieces generally integrated throughout the length, across the width, and through the thickness of the webs. Approximately 2 m of each blend were airlaid without any problems. It is conceivable that airlaid web formation of this mixture could be done on a much larger scale. However, wetlaid web formation was not as successful. During several trials, proper mixing of the bag pieces and fibers could not be obtained. Hence, webs created with this method usually were non-uniform, with bag pieces forming clumps in some areas of the web, while other areas were devoid of the bag pieces. This is most likely a result of the larger size of the bag pieces in comparison with the fibers used for these experiments. It may be possible to achieve more uniform results if the bag pieces could be shredded to be a similar size as the fibers. It may be possible to accomplish this via mechanical shredding of the bags, rather than manual shredding. This will be investigated in the future.

Bonding of the webs also had mixed results. Based on past experience with needling airlaid webs, the webs were all placed on a spunbond polyester scrim fabric prior to needlepunching, in order to maintain web integrity during needling. Originally, the webs were needled without a top scrim,

13.1 50/50 bag pieces/polyester fiber fabric formed by airlaid web formation and needlepunching.

which resulted in many bag pieces being punctured by the needles through the thickness and thus remaining on the needles. In addition, many pieces fell out of the web during the needling process. As a result, the tops of the needled fabrics often contained loose bag pieces and minimally entangled pieces, as depicted in Fig. 13.1, all of which were easily removed from the web during handling after needling. Further experiments involved needling with a lightweight polyester spunbond scrim on top in order to minimize needle fouling and removal of the bag pieces during the needling process. However, upon examination of the fabric after the top scrim was removed it was discovered that the bag pieces did not adequately entangle with the polyester fibers to give the fabric sufficient integrity. After needling, several of the fabrics were thermally bonded with a through-air oven, to determine if the entanglement due to needling was sufficient enough to provide better results via thermal bonding, where the bag pieces would act as binder fibers. Although the bag pieces melted during the thermal bonding, they only bonded with other bag pieces, and often shrunk during the bonding process, resulting in fabrics without much integrity or strength. Owing to these experiences, needling and through-air bonding were not explored further, and these fabrics were not evaluated further. However, through-air bonding and needlepunching may be more effective if shorter, lower denier fibers are used, rather than those that were utilized for this study. It is possible that shorter, thinner fibers might allow an increase in fiber and bag piece entanglement which could create stronger fabrics. It is also possible that smaller bag pieces might give greater integration of potential thermal bonding points in the web, thereby increasing the possibility of better fabric integrity.

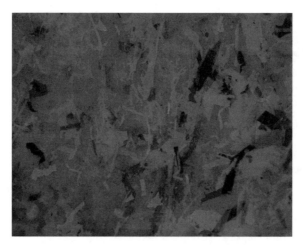

13.2 50/50 bag pieces/polyester airlaid and bonded with a hot press at 140 °C and with a pressure of 0.5 metric tons.

Although through-air bonding was not very effective in providing fabrics with adequate integrity, bonding via heat and pressure was much more successful. The bag pieces again bonded only to themselves, but there was less shrinkage of these pieces in comparison to through-air bonding. Fabrics produced with this method, as depicted in Fig. 13.2, had adequate stability so that the bag pieces were not easily removed. However, in all samples produced, there were some areas of the fabrics that did not have enough bag pieces to hold the polyester fibers adequately, and as a result these materials had loose fibers on their surfaces. However, this might be solved if narrower bag strips were utilized, as there would be more bag pieces blended throughout the fabric, thereby resulting in more bonding. The use of shorter, lower denier fibers may also rectify this problem.

Given the difficulties with other methods of bonding, latex bonding was also evaluated as a method of creating fabrics with adequate strength for different applications. The latex was sprayed on and then dried in a through-air oven, preserving the thickness of the original web. However, pad application of the latex should also be possible and may be preferable, depending on the desired end use of the fabric. Latex bonding of these webs was probably the most successful of all methods utilized, as there were no problems with poor bonding between the bag pieces and the polyester, nor were there unbonded areas resulting in loose fibers on the surface of the fabric. There were some problems with lack of latex penetration through the fabric, but these could be remedied with the use of a vacuum system to pull the latex into the interior of the fabric, or by passing the web through a pad after latex application.

13.3.2 Fabric evaluation

The latex-bonded, and thermally and pressure-bonded fabrics were evaluated to determine their properties. Basis weight, thickness, bursting strength, and thermal conductivity were measured. However, owing to the length of time required to cut the bags into small pieces, and difficulties with several of the methods of bonding, only small amounts of bonded fabric could be generated. As a result, only three specimens of each fabric were evaluated in each test. Additionally, owing to the manual nature of rolling up the airlaid webs and bonding them, the fabric specimens exhibited variation during evaluation. Various average properties are listed in Table 13.1.

The polyester fabric was thermally bonded at 215 °C as greater temperatures resulted in severe melting of the fabric during the bonding process, owing to thermal inconsistencies across the platens of the hot press. Because of the limited amount of polyester web produced and the difficulties encountered in thermally bonding it, there was not enough to thermally bond at a pressure of 1.5 metric tons, corresponding to the blends bonded at 180 °C. The increased bonding temperature and pressure were utilized in an attempt to decrease the amount of loose polyester fibers in the fabrics. Although this was not specifically evaluated, visual examination seems to suggest that this was accomplished.

The 25/75 bag/polyester fabrics exhibited the greatest basis weights and thicknesses, indicating that this web had a greater basis weight. All of the airlaid webs were processed on the same day, on the same machine with the same settings, thus the cause of this increased basis weight is not clear, but it may be a result of manual wind-up of the web. The thermal conductivity values do not depict any clear trends. However, this is not necessarily negative, as it seems to indicate that there are no major differences between polyester and the bag pieces in terms of fabric insulating capabilities. Thus it may be possible to produce insulation materials containing bag pieces that do not suffer loss of insulating performance when compared with pure materials.

The normalized bursting strength, the bursting strength values divided by the basis weight, do depict some trends in the thermally bonded fabrics. As the percentage of bag pieces increases the bursting strength also increases. This is most likely a result of the greater amount of bond points within the fabric, as it was discovered that the bag pieces only bond to themselves, and not the polyester fibers. In addition, it is likely that with the increase in bond points, the ball of the ball burst tester encountered bond points as it passed through the fabric during testing, thereby resulting in greater strength values.

Based on the mechanical evaluation of the fabrics, it is clear that fabrics that incorporate bag pieces can be utilized in the creation of nonwoven

Table 13.1 Average properties for fabrics made from various mixtures of bag piece/polyester fibers

Property (averages)	100% fiber, hot press, 215°C	25/75 bag fiber, hot press, 140°C	50/50 bag fiber, hot press, 140°C	25/75 bag fiber, hot press, 180°C	50/50 bag fiber, hot press, 180°C	100 fiber, latex bonded	25/75 bag fiber, latex bonded	50/50 bag fiber, latex bonded
Basis weight (g/m^2)	177	260	164	253	118	196	381	213
Thickness (mm)	0.74	5.2	1.6	0.99	0.40	20	30	9.8
Thermal conductivity (W/m K)	0.007	0.011	0.012	0.013	0.008	0.063	0.090	0.043
Bursting strength (kg)	3.0	3.6	4.3	15	19	25	31	22
Normalized bursting strength ($kg/(g/m^2)$)	0.017	0.014	0.025	0.075	0.42	0.13	0.080	0.10

fabrics. These fabrics appear to have similar performance values to those consisting of pure fibers. However, if thermal bonding is utilized then it is possible that the fabrics will suffer a decrease in strength compared with fabrics containing only fibers. Nonetheless, there are several possible applications where fabric strength is not a determining factor in the choice of a fabric for a given end use. Such materials include sound absorption materials for automotive and building applications, thermal insulation for buildings, carpet underlayment, and re-usable weed-control mats.

13.4   Conclusions

It is possible to create nonwoven fabrics that contain shredded bags, and to utilize these shredded bag pieces as a binder material in thermal bonding. Thus far, airlaid web formation has proven to be the most effective method of producing the web precursors of the fabric. However, it may be possible to use the wetlaid process effectively if the bag pieces are shredded into smaller pieces. Needlepunching proved to be ineffective in entangling the bag pieces with the fibers utilized in this study. However, changes in the needling parameters might yield different results. Through-air bonding did not provide effective bonding of the fabrics, as the bag pieces tended to shrink and bond only to themselves. However, thermal bonding utilizing pressure, such as calendering, has proven to be successful. Latex bonding of these materials is also an effective way of producing fabrics. No major decreases in properties have been noted from the evaluation of the properties of these materials. However, further evaluation is required to fully understand the role that shredded bag pieces may have as binder materials in nonwovens. As such, further work will focus on comparing the shredded bag pieces directly to polyethylene fibers, commonly utilized as binder fibers, as well as to evaluating different blend ratios of these materials. However, as produced, these materials might be used for sound or thermal insulation or in other applications where high strength is not of vital importance.

Incorporation of these bags into nonwoven fabrics could provide the nonwovens industry with an alternative to more expensive binder materials or an alternative to other fibers, if these bags are utilized as filler materials. By incorporating recovered plastic bags into nonwoven fabrics, the nonwovens industry might be able to increase recovery of these materials through increased demand for them, which would allow the industry to increase its environmentally friendly image.

13.5   Acknowledgements

The authors wish to thank Bob Averell and Wellman Incorporated for their contribution of polyester fibers.

13.6   References

1   RUBIN, I., 'Bag Ban is Misguided Solution,' *Plastics News*, November 26, 2007: 6.

2   CONWAY, C., 'Taking Aim at All Those Plastic Bags,' *The New York Times*, April 1, 2007: 2.

3   'China Wants to End Dependence on Three Billion Plastic Bags a Day,' *National Post*, January 10, 2008: A3.

4   UNITED STATES ENVIRONMENTAL PROTECTION AGENCY, OFFICE OF SOLID WASTE, 'Municipal Solid Waste in the United States 2005 Facts and Figures,' October, 2006: http://www.epa.gov/msw/pubs/mswchar05.pdf.

5   BELL, A., '13 Easy Ways to go Green,' *USA Weekend*, April 20–22, 2007: 8.

6   TRUINI, J., 'Plastic Bag, Film Recycling Hits All Time High,' *Waste News*, March 31, 2008: 3.

7   FONT, R., ARACIL, I., FULLANA, A., and CONESA, J., 'Semivolatile and Volatile Compounds in Combustion of Polyethylene,' *Chemosphere*, **57**(7), November, 2004: 615–627.

8   KANG, C., 'For Trex, Selling Composite Has Been Tough,' *The Washington Post*, November 26, 2007: D07.

9   ALKAN, C., ARSLAN, M., CICI, M., KAYA, M., and AKSOY, M., 'A Study on the Production of a New Material From Fly Ash and Polyethylene,' *Resources, Conservation, and Recycling*, **13**, 1995: 147–154.

10   CAVALIERI, F., and PADELLA, F., 'Development of Composite Materials by Mechanochemical Treatment of Post-Consumer Plastic Waste,' *Waste Management*, **22**, 2002: 913–916.

11   FORTELNÝ, I., MICHÁLKOVÁ, D., and KRULIŠ, Z., 'An Efficient Method of Material Recycling of Municipal Plastic Waste,' *Polymer Degradation and Stability*, **85**, 2004: 975–979.

12   BERTIN, S., and ROBIN, J.J., 'Study and Characterization of Virgin and Recycled LDPE/PP Blends,' *European Polymer Journal*, **38**, 2002: 2255–2264.

14

Environmentally friendly flame-retardant textiles

S. NAZARÉ, University of Bolton, UK

Abstract: Flame-retardant textile products go through many processing steps that contribute to their overall chemical footprints and sometimes the final chemical residues remaining in the finished product. This chapter first discusses ecotoxicological issues of flame retardants and the particular risk of flame-retardant textiles to human health. Legislative and regulatory drivers for minimising environmental as well as human health implications are also discussed. Strategies for the development of sustainable environmentally friendly flame retardants are also reviewed briefly. Finally, important governmental and non-governmental organisations that are directly associated with sustainability, renewability and recyclability of flame-retardant chemicals are listed.

Key words: environmentally friendly flame retardants, health and environmental issues, risk assessment, nanocomposites, plasma treatments, formaldehyde-free treatment.

14.1 Introduction

Millions of people are exposed to flame-retardant consumer products everyday in the home, the workplace and public buildings. Flame-retardant textiles can be found in various uses including apparels, industrial work-wear, upholstered furniture, furnishing fabrics in mass transport vehicles and military applications. These flame-retardant textile products go through many processing steps that contribute to their overall chemical footprints and sometimes the final chemical residues remaining in the finished product. Many stages of fibre production and subsequent processing – and especially dyeing, printing and finishing processes – all involve the use of chemicals and consequently lead to direct release of chemicals to the environment. Finished products also have the potential for long-term releases of chemicals from laundering and final disposal. The majority of flame-retardant textiles are non-durably and durably finished cottons which find applications in both domestic and contract sectors; on the other hand, inherently flame-retardant fibres are used in high-performance applications where their increased cost is justified.[1]

Methods of rendering textiles flame retardant are well established and have been described elaborately elsewhere.[2-4] Flame retardation of textiles mainly exploits two flame-retardant mechanisms, viz. condensed phase and gas phase mechanisms. The condensed phase mechanism generally operates in phosphorus- and/or nitrogen-containing flame retardants to alter the pyrolysis process and is mainly employed for char-forming polymers, e.g. cellulose and wool. The gas phase mechanism, on the other hand, alters flame chemistry and hence flame retardants functioning by this mechanism are applicable to all types of fibres. Halogen-containing compounds in particular function via gas phase flame inhibition and are used to dilute the flame and/or increase fuel ignition temperature.[2]

Depending on the nature of the flame retardant and the end use of the textile product, the concentration of flame retardant in a textile can vary from 3 to 30 wt%. Levels of flame retardant loading also depend upon the area density and structure of the fabric.[2] In response to the introduction of stringent flammability standards, especially in the UK, Europe and North America, the demand for flame-retardant textiles is expected to increase. World demand for flame-retardant chemicals has been predicted to grow 4.7% per year to 2.2 million metric tonnes in 2011.[5] Despite claims regarding the potentially damaging environmental and health effects of halogenated flame retardants,[6-11] demand for brominated flame retardants is expected to rise owing to their low price and superior performance. Most bromine used in the world goes into flame retardants.[12] It is important to note here that only 10–20 wt% of the total world production of halogenated flame retardants, especially decabromodiphenyl ether (decaBDE), is used in textile applications.

There are obvious benefits in using flame retardants, as many human lives and property are saved from fire, and it is the risk–benefit balance regarding their use that requires continual attention. While the use of flame retardants can reduce the risk of fire in some situations, the application of flame retardants to textiles, for example, itself is likely to present a risk to human health or the environment. At present, knowledge of long-term effects resulting from exposure to flame retardants and their breakdown products is limited. Nonetheless, most of the related studies concluded that the use of these flame retardants presents no risk to human health or to the environment.[13-17] In case of fire and depending on conditions of burning and the type of flame-retardant chemistry, highly toxic substances can be formed and released in the atmosphere. Most people are reported to die in fires due to inhalation of toxic gases including carbon monoxide.[18]

Several national regulatory bodies have implemented regulations on specific substances associated with flame-retardant applications. Brominated flame retardants, in general, are perceived to be widespread environmental contaminants as they are persistent, bioaccumulative and form dioxins and furans on combustion. Very recently, some states in the USA have banned the use of all brominated and chlorinated flame retardants in mattresses as well as in furniture.[19] UK law[20] enforces the use of materials with low flammability or the use of flame retardants in certain high risk areas such as domestic furnishings, whereas Scandinavian countries as well as Germany are more cautious of possible risks to human health and the environment from the use of flame retardants. These countries do not collect fire death statistics as comprehensively as the UK, and so fire risks are not as fully identified. This debate over fire safety and the use of flame retardants continues in spite of the clearance of certain commonly used flame retardants like decaBDE following environmental risk analysis.

Research on so-called 'environmentally friendly' flame retardants has now been ongoing for more than a decade and a variety of non-halogenated chemical alternatives are readily and commercially available although more often than not their performance is marginal when assessed in terms of their ability to give consistent passes to defined testing regimes. This chapter highlights the risk of flame-retardant textiles to human health as well as the ecosystem in general. Recent developments in halogen-free flame retardants for various textile applications, including environmentally friendly methods of creating flame retardant textiles and various strategies for replacing halogenated flame retardant-systems, are discussed briefly. The developments in eco-labelling of flame-retardant textiles are also described.

14.2   Key issues of flame retardants

A flame retardant is a chemical or mixture of chemicals that can be added to, or incorporated into, a material to prevent its ignition by a small ignition source. A large variety of compounds, from inorganic to complex organic molecules, are used as flame retardants, synergists and smoke suppressants. In case of ignition, the flame retardant should be able to slow down the rate of burning. However, a flame retardant can only inhibit, not prevent a material from burning in small-scale fires. Furthermore, a flame-retardant treatment does not necessarily improve the thermal stability of a textile material. In fact, the decomposition temperature of a flame-retardant textile material is lowered compared with that of untreated fabric.[4]

Flame retardants used in textiles generally include the following four types of chemical compounds.[21,22]

- Inorganic minerals such as aluminium hydroxide, calcium carbonate and boron compounds exhibit their flame-retardant activity through physical effects, i.e. by insulating the fibre from the applied heat and oxygen. For example, boric acid and its hydrated salts undergo strong endothermic thermal decomposition and form a foamed glassy surface on the fibre. These retardants are often used as coatings on textiles, especially in carpet-backings, and their efficiency depends on the chemical composition of the fibre. Large amounts of filler loading (50–60% w/w) are required to achieve acceptable levels of flame retardancy.[23]

- Inorganic and organic phosphorus-containing compounds alter the pyrolysis process to promote char formation and decrease volatile species. These are primary flame retardants based on the condensed phase mechanism of flame retardancy. Add-on levels of about 15 wt% are usually suggested for desired flame-retardant effects.
- Halogen-containing compounds interfere with the free radical reactions occurring in the flame. These gas phase retardants can be used for all types of fibres since the reactions in the flame are similar for all polymer types. Since the mode of operation is through modification of flame chemistry, it is necessary for the flame retardant compound to volatilise at the right time and in the right part of the flame to interact with pyrolysis products and inhibit oxidation reactions.[3] Halogenated flame retardants are very effective and hence add-on levels of 10–15 wt% on solids are sufficient to achieve flame-retardant effects.
- Antimony oxides and nitrogen compounds have very small flame-retarding effects by themselves; however, they may act as synergistic retardancy enhancers for primary flame retardants. Phosphorus–nitrogen synergy is well documented[24–26] and the antimony–halogen system is a classic example of the synergistic flame-retardant phenomenon. With such synergisms, both the gas phase and condensed phase mechanisms are believed to contribute to the overall flame-retardant effect.[26]

Inorganic compounds – including aluminium trihydroxide, ammonium polyphosphates and antimony trioxide – represent the greatest volume of production (50%), followed by halogenated compounds (25%) and organo-phosphorus compounds (20%). Nitrogen-containing melamine compounds amount to 5% by volume.[27]

Durability of flame retardants is an important consideration in textile applications because of laundering requirements. Flame retardants are often classified according to the durability of the flame-retardant textile.

- *Non-durable flame retardants*. These are generally based on water-soluble salts such as ammonium phosphates, ammonium sulphate, boric acid and their combinations. These systems are applied to mattresses, draperies or theatre curtains which are only rarely, if ever, washed. Non-durable treatments can be applied using simple pad/dry techniques and the flame retardant is only physically bonded to the fibre surface. The process is relatively cheap and can be easily accommodated within the routine textile finishing sequence.
- *Semi-durable flame retardants*. In addition to the flame retardant, a semi-durable flame-retardant formulation typically contains an acrylic binder and a melamine-formaldehyde resin.[23] A curing stage at higher temperature (130–160 °C) is often introduced to allow the finish to interact with the fibre, thereby imparting some degree of resistance to

water soak and gentle laundering treatments – such as dry cleaning. Semi-durable flame retardants are used for tents, carpets and curtains.

• *Durable flame retardants.* These are reactive types of flame retardants that react with the fibre or form cross-linked structures on the fibre. The flame retardant may be incorporated into the polymer chain by co-polymerisation. As a result, the flame-retardant fibre/fabric is durable to washing, leaching, infant sucking and the flame-retardant is less likely to be absorbed by the skin. With increasing public interest and awareness in environmental and toxicological matters, the incorporation of flame retardants as part of the polymer backbone is becoming more popular. Durable flame-retardant systems include reactive organophosphorus and halogenated compounds.

A number of factors govern the selection of the type of flame retardant to be used in a specific application. These include the flammability of the polymer matrix, processing and performance requirements, chemical properties and possible hazards to human and environmental health. Performance and selection criteria for flame retardants to be used in textile applications include their ability to:

• achieve low or high levels of test performance on flame retardancy depending on the purpose;
• not affect the processing characteristics of the polymer;
• cause only minor changes in the properties of fibres by their addition, e.g. properties such as strength, rigidity, colour, gloss or heat resistance;
• possess acceptable levels of durability and stability;
• have an acceptable cost.

For most of the common textile materials there are a number of limitations on the use of flame retardants. Firstly, a flame-retardant system is often specific for a particular type of textile. Secondly, the amount of flame retardant used to impart flame retardancy is not always directly proportional to the level of flame retardancy achieved on the fabric. Often it is observed that lower amounts of flame retardants can actually enhance the rate of burning with respect to an untreated fabric.[28] This is attributed to the fact that small add-on of the retardant provides the fuel for the flame at lower temperatures. In the presence of sufficient amounts of flame retardant, the decomposition reactions are directed more towards dehydration mechanisms thereby inhibiting emission of volatile species.

14.2.1 Health aspects of flame retardants in textiles

Every chemical can have toxicological consequences in some situation. The toxicological hazard of any flame-retarding chemical is unavoidable. This section will concentrate on the particular risk of flame-retardant textiles to human health. Possible toxicological hazards of the flame-retardant chemical itself range from local irritation of the skin to neurotoxicity and carcinogenicity. Sensitisation of the skin and the respiratory organs can occur at low concentrations of flame retardants in fire-resistant textiles in contact with human skin. Acute or chronic toxicity usually occurs as a result of non-durable, leachable flame retardants and also due to dusting-off of a flame retardant. Direct exposure of flame retardants in textiles to humans is possible via three main routes: dermal exposure, inhalation and oral exposure. Various factors affecting each of these exposure routes are discussed below.[27]

Dermal exposure

This form of exposure occurs when wearing clothes or by contact with bed linen, carpets or furnishing fabrics, etc. For the flame retardant to penetrate into the skin, the chemical must leach from the fibres, which essentially depends on the chemistry of the flame retardant, the type of fibre, the impregnation process, the extraction medium and other conditions such as time and temperature. Leaching of the flame retardant may occur as a result of contact with sweat, blood or urine, and the use of skin creams and body lotions. The potential for extraction also depends on the durability or the inclusion of the flame retardant in the polymer matrix. For example, non-durable salts are more likely to be prone to leaching than reactive flame retardants which form a constituent of the polymer. Inclusion in the polymer matrix of the fibre can impede the diffusion and consequent volatility of a flame retardant.

Estimated human exposure (EHE) of the skin to the flame retardant depends on various parameters including the concentration of the flame retardants in the textile, the area of skin in contact, the duration or frequency of the contact with skin and the loss of flame retardants with time by volatilisation, sublimation, dusting off, washing out, etc. For example, EHE for a nightdress made from a lightweight fabric with an area density of $100\,g/m^2$, impregnated with 20% flame retardants having a leachability of 0.01% in 24 hours can be calculated assuming a contact area of $1.5\,m^2$ and a contact duration of 8 hours/day for 7 days a week. For a body weight of 70 kg and assuming 100% absorption through skin, the EHE for an individual can be estimated. Thus, EHE can be estimated for each individual situation, however the risk evaluation can not be performed for individual flame-retardant chemicals but has to be performed for the flame-retardant-textile system and the specific situation. However, EHE can be generalised for flame-retardant textiles by defining a 'standard end product' with 'standard end use'.

Inhalation

Inhalation of the flame retardant or its degradation products can occur if the flame-retardant chemicals volatilise or release toxic gases during the fire. Important factors for estimation of human exposure via inhalation are the molecular weight of the flame retardant chemical, the vapour pressure of the chemical and the room ventilation – in order to calculate saturation concentration in a room. It is also well known that dioxins produced from low-temperature burning of halogenated flame retardants cause acute and chronic toxicity even at very low concentrations. Toxic products of combustion is a complex field and no internationally accepted standard method for determining the toxicity of combustion gases exists.

Oral exposure

The only realistic way of oral entry of flame retardant is by sucking on impregnated flame-retardant textiles. Exposure may also occur indirectly via the environment. Flame retardants may be released to water or air during manufacturing, processing or disposal. They could be inhaled or may find their way to drinking water or food and could be ingested by humans.

14.2.2 Environmental aspects of flame retardants in textiles

As mentioned earlier, flame-retardant species may be released to the environment during production, processing, use and disposal of flame-retardant textiles. Production and handling of the flame-retardant chemical itself could release toxic dust or vapour. During use of flame-retardant fabrics, flame-retardant formulation components could be released by wear and tear, by volatilisation, during laundering and by leaching under the action of rain, oil, solvents, etc. Moreover, since the textiles are not labelled as containing hazardous substances, most of the textiles are disposed of as municipal waste and not as hazardous waste.[29] Disposal of textiles by land filling or incineration can lead to leaching in the dump or emissions of combustion gases respectively. Furthermore, flame retardants and related fumes or toxic gases may also be released to the environment during accidental combustion of an impregnated fabric. Usual toxic combustion gases may range from carbon monoxide to ammonia as well as pyrolysis products and related oxidised species.

The environmental evaluation of flame retardants requires a basic understanding of the ecotoxicological hazard and environmental fate and behaviour of the flame-retardant chemical. Flame retardants may not produce local toxic effects at the point where they enter the environment, yet the flame-retardant chemicals may be transported by water or air, depending on the water solubility and vapour pressure. Furthermore, the environmental fate of a chemical depends on the biological or abiotic degradations, bioaccumulation and adsorption properties. Factors affecting environmental exposures include:

- the volume of flame-retardant application;
- physical–chemical properties of the flame retardant;
- amount released during different phases of the life cycle;
- emission control measures;
- biotic and abiotic degradation;
- persistence and bioaccumulation.

Very little is known about the ecotoxicological properties of flame retardants. Short-term and chronic effects may occur as with other chemicals in the various compartments of the ecosystem. An Austrian study[30] in 1999 suggested determination of basic information on acute toxicity in fish, daphnia and rats, and growth inhibition of algae, etc. for each flame-retarding chemical. Other studies[30,31] suggest determination of a few physical–chemical properties such as molecular weight, melting temperature, vapour pressure, water solubility, n-octanol/water partition coefficient to understand the environmental fate and behaviour of a flame-retardant chemical. Bioaccumulation and adsorption in the soil are usually estimated from these physical–chemical properties, i.e. mainly from the n-octanol/ water partition co-efficient of the chemical.[32]

Owing to their high molecular weight and their low water solubility, most brominated flame retardants are not bioaccumulative, in other words they do not stay or build up in human bodies. A few of the many halogenated flame retardants detected in the environment include decabromodiphenyl oxide (DBDPO) (or deca-BDE), tetrabromobisphenol A (TBBPA), hexabromocyclododecane (HBCD), and various tetra-, penta-, and hexa-BDEs. The amounts and locations of accumulation vary and depend on physico-chemical properties of the molecule and the product end use.[12]

14.3   Legislative and regulatory drives for minimising environmental implications

In the UK and the EU, there are several legislations relevant to the textiles industry, particularly in relation to the use of certain chemicals and substances in production. Concerns surrounding toxic combustion products from halogenated flame retardants being released into the environment have resulted in regulatory restrictions in Europe, UK and in the USA. Other concerns include both the processing and recycling of flame-retardant-treated materials. Research[31–33] suggests that the toxic effects from long-term exposure as a result of the persistence and bioaccumulation of flame retardants are of greater concern than the toxicity effects of such substances themselves. Countries such as North America, China, Japan and Korea have initiated studies to provide information on increasing concentrations of polybrominated diphenylethers (PBDEs) in a wide range of environmental samples.[7-11] For a comprehensive review of various government-initiated testing programmes of brominated flame-retardant chemicals as environmental contaminants, readers are directed elsewhere.[10,13,14] As a consequence of results from these studies, various European and US governments began to consider the need to restrict the production, use and disposal of brominated flame retardants. Some countries include restrictions on the use of compounds because of potential toxic effects in humans. For example, detection of brominated flame retardants in sediments and fish downstream of textile and plastic industries has led to the prohibition of use of the so-called 'most harmful' brominated flame retardants in Sweden. Production of polybrominated biphenyls (PBBs) was banned in 1976 in the USA and subsequently in Europe in 1980 following an accident that happened in Michigan, USA. PBB (sold under the trade name FireMaster), was accidentally mixed with animal feedstuff and entered the food chain through milk and other dairy products, beef products and contaminated swine, sheep, chickens and eggs.[6] As a result of this incident, production of PBB was prohibited in the USA. Since 1983, Switzerland and Austria have also prohibited use of PBB in textiles, particularly those that come in contact with skin.[14]

The UK Environmental Protection Act was introduced in 1990 and was fully enforced by late 1995. By this time, the European Commission was implementing several environmental legislations specifically targeting reduction in effluents and waste from the textile flame-retardant finishing industry.[1] Directive 76/769/EEC[34] and its subsequent amendments restrict the use of hazardous substances. European Council Directives 79/663/EEC, 83/264/EEC and 2003/11/EC amending or supplementing the Annex to 76/769/EEC, restrict the use of tris (2,3 dibromopropyl) phosphate, tris-(aziridinyl)-phosphinoxide, PBBs, pentabromodiphenyl ether (pentaBDE) and octabromodiphenyl ether (octaBDE) in fireproofing garments. This, together with other legislation on the use of azocolourants, lead and cadmium limits the potentially hazardous substances that may be found in textile products.[35] European legislation on labelling of textile products is discussed in Section 14.3.2.

On 1st of June 2007, the EU introduced regulation on Registration, Evaluation, Authorisation and Restriction of Chemical substances (REACH). The primary aim of REACH is to protect human health and the environment through early identification of risks posed by chemicals. Implementation of the REACH legislation forces industry to assess and manage the risks posed by the chemicals that they produce and also to provide appropriate safety information to their customers. Industries that manufacture or import more than 1 tonne of a chemical substance per year were required to register the substance in a central database administered by the new EU Chemicals Agency in Finland by 31st December 2008. REACH thus aims to streamline and improve the former legislative framework for chemicals within the EU.

14.3.1 Risk assessment, management and reduction

Risk assessment is the process of quantitatively determining the likelihood of adverse effects resulting from exposures to flame-retardant chemicals. The risk assessment of a flame retardant or a chemical in general must comprise the toxicity data to evaluate risk to human health and environment, release scenarios and estimations of release rates. As mentioned earlier, it is extremely difficult to determine or estimate the release of flame-retardant chemical in a given textile at each stage of its life cycle. However, for performing risk assessment of a specific textile–flame retardant system, it is suggested that the whole life cycle of the product, including the evidence of the necessity to lower the flammability of a product and the efficiency of the flame-retardant chemical, should be taken into consideration.[36]

The quantitative environmental risk evaluation is based on the comparison of the predicted environmental concentrations (PEC) and the predicted 'no effect' concentration (PNEC). The minimum data required for evaluation of the ecotoxicological hazard of a flame-retardant chemical include:

- information on production processes including geographical distribution;
- amount of flame retardant used per year for various applications;
- means of disposal of flame retardant;
- information on applications which may depend on whether or not (a) non-durable flame retardant textiles are present or (b) durable flame-retardant textiles are present;
- information on processing in terms of (a) the stage at which the flame retardant is introduced and (b) the procedures used.

Traditionally, government authorities have focused their research activities on establishing fundamental assessment aspects including occurrence, persistence, toxicology and possible routes of exposure.[14]

In 1980, the World Health Organisation (WHO) established the International Programme on Chemical Safety (IPCS) to establish the scientific basis for safe use of chemicals in general, and to strengthen national capabilities and capacities for chemical safety. The IPCS assessed the health and environmental effects of existing chemicals so that preventive actions could be taken against adverse health and environmental impacts.

The Organisation for Economic Co-operation and Development (OECD) established a risk reduction programme in early 1990 to reduce the risk from chemicals and investigate selected brominated flame retardants.[37] The Brominated Flame Retardant Industry Panel (BFRIP), the Chemical Manufacturers Assosiation (CMA) and the European Flame Retardant Industry Panel (EBFRIP) participated in the Risk Reduction Programme through business and industry advisory committees.[14] Three main technical PBDE formulations, the penta-, the octa- and deca-congeners were particularly addressed for their environmental life cycle and risk reduction measures.[37] During 1995–1996, international flame-retardant producers agreed voluntary commitment to OECD. The Voluntary Industry Commitment (VIC) provides assurance to industry and customers regarding the future of their product. The key points of the flame-retardant industry's voluntary commitment to OECD are that the global brominated flame-retardant manufacturers will:[37]

- inform and educate their customers on the proper handling, use, recycling and disposal of products;
- commit not to produce or import/export PBBs, with the exception of decabromodiphenyl; this is a worldwide commitment achieved without the need for numerous international regulations;
- commit to minimising levels of release of pentaBDE during manufacturing;
- commit to an average purity of 97% or greater for decaBDE and to minimise levels of hexa- and lower brominated diphenylether congeners;
- not manufacture the non-commercial polybromodiphenylether or oxide (PBDPO) congeners as individual flame retardants except when present as part of the commercial deca-, octa- and penta-products.

The commitment successfully addressed the concerns expressed by the OECD member states regarding PBB and PBDPO without having to ban, phase-out or limit the use or substitution of PBDPO.

In parallel, the European Commission sponsored a separate study on the determination of toxicity and ecotoxicity of flame retardants used in the industries associated with upholstered furniture and related articles.[15] Twenty-two chemicals were subjected to detailed analysis and it was concluded that these chemicals do not pose risk for consumer health through their use, nor risk to the environment through end-of-life product disposal. Toxicological and environmental effects of penta-, octa- and deca-BDE were also reviewed in 1995 under the EU 'Existing Chemical's Regulation' on evaluation and control of existing chemicals. After 10 years of extensive studies, the final risk assessment[16] was completed in 2004 and confirmed in

2005. The report concludes that the use of these flame retardants presents no risk to human health or to the environment. However, the report recommends that further studies should be continued for environmental monitoring.

In 1999, the UK Department of Trade and Industry (DTI) Consumer Safety Unit commissioned the report 'Risks and benefits in the use of flame retardants in consumer products'.[17] The report reviews data on the use, utility and toxicity of flame retardants used in the UK. The study concludes that the benefits of many flame retardants in reducing the risk from fire outweighed the risk to human health. The report also emphasised the fire death incidence resulting from residential fires and that flame retardants play a significant role in reducing the number of fire deaths. The risk to the environment or to humans was found to be highly uncertain and no chronic effects were observed in the environment or on human health. The study, however, did not include toxicological analysis of flame-retardant compounds and combustion products.

In 2000, perhaps the most comprehensive risk assessment of flame retardants was carried out by the National Academy of Sciences of the USA. Published under the title of '*Toxicological Risks of Selected Flame-Retardant Chemicals*', the study reviews potential non-cancer and cancer effects of 16 selected flame retardants used in furniture fabrics.[33] The selection of flame retardants was based on:

(a) the existence of scientific evidence that the flame retardant presented a hazard to human health and/or the environment;
(b) the possible use, persistence, accumulation or degradation of the flame retardant showing that there may be significant human or environmental exposure;
(c) the size and nature of populations at risk (both human and other species) and risks for the environment;
(d) international concern, i.e. the flame retardant was of major interest to several countries;
(e) adequate data on the hazards being available.

Out of the 16 chemicals assessed, the study concluded that the following 8 flame retardants were considered to be safe, even under worst-case exposure assumptions.

1 Hexabromocyclododecane.
2 Decabromodiphenyl oxide.
3 Alumina trihydrate.
4 Magnesium hydroxide.
5 Zinc borate.
6 Ammonium phosphates.

7   Phosphonic acid ester.
8   Tetrakis hydroxymethyl phosphonium salts.

The same study recommended further exposure studies on the following 8 flame retardants.

1   Antimony trioxide.
2   Antimony pentoxide and sodium antimonates.
3   Calcium and zinc molybdates.
4   Organic phosphonates (dimethyl hydrogen phosphate).
5   Tris(monochloropropyl)phosphates.
6   Tris(1,3-dichrolopropyl-2) phosphate.
7   Aromatic phosphate plasticisers (tricresyl phosphate).
8   Chlorinated paraffins.

In order to address issues related to some of the flame retardants mentioned above, the US Consumer Product Safety Commission (CPSC) started a series of research and development investigations. The safety of phosphorus-based flame retardants has been assessed in Switzerland as well as in Germany. The Swiss Federal Health Office (BAG), analysed concentrations of ten phosphorus-based flame retardants in indoor air and concluded that the risk is very low and that no additional measures are required to minimise the risk of phosphorus compounds.[38] In the German study,[39] eight organophosphate esters widely used in automobile interiors were evaluated for their exposure risk.[39] The study concluded that even under extreme experimental conditions, the recommended maximum exposure levels were not reached in automobile interiors.

In order to fully describe the environmental impact of flame-retardant textiles, the entire process chain needs to be described including the raw material handling, transport, energy and electric power consumption, actual flame-retarding process, waste handling and disposal of the product. Life cycle assessment (LCA) is a versatile tool to investigate the environmental impact of a product, a process or an activity by identifying and quantifying energy and material flow for the system. The Fire-LCA, developed by the SP Technical Research Institute of Sweden, is a realistic LCA method that incorporates fires as one possible end-of-life scenario. The Fire-LCA tool, however, does not provide information on the effect of the toxicity of chemicals used in the product, the number of lives saved, the cost associated with different cases or the societal effect of manufacturing practice.[36]

The risk management of a flame-retardant-textile system is based on a comparison of the benefits and the risks of a product. For example, there could be a risk to health from the introduction of a flame retardant in textile products such as children's nightwear, bedding and mattresses – where flame retardation is not a significant benefit as the risk of fire is low. In contrast, the benefits of impregnating flame retardants into bed linen and mattresses in hotels and high-risk areas such as prison cells and hospitals may be considered to outweigh the risks to health and the environment, by preventing ignition from smouldering cigarettes or intentionally started fires.

Flame-retardant exposure risk can be reduced or controlled by different ways throughout the life cycle of the flame-retardant product. During the manufacture of the flame-retardant chemical, its incorporation into the textiles and waste disposal, the exposure of the flame-retardant chemical can be controlled by careful monitoring of losses to the environment. To reduce both the human and environmental exposure risk during use of flame-retardant textiles, intrinsically safe formulations should be used. In cases where use of environmentally safe flame-retardant formulations is not possible, care must be taken that the flame retardant is chemically incorporated into the fabric either by chemical grafting or by cross-linking.[40] Strategies for the development of 'environmentally friendly' flame retardants are discussed in Section 14.5.

14.3.2 Developments in eco-labelling and flame-retardant materials

Eco-labels provide consumers with guidance in choosing those products that are considered to be least hazardous to the environment. They support the development of ecofriendly products by prohibiting the use of undesirable chemicals. The fire safety of the product is excluded from eco-label criteria on the grounds that the eco-label criteria address environmental issues only. The first eco-label was introduced in 1989 by the Nordic Council of Ministers in the form of the Nordic Swan.[14] In Sweden, Denmark, Norway, Iceland and Finland the textile products labelled with the Nordic Swan did not contain brominated flame retardants. In 1994, the Danish government introduced the Blue Angel label which restricted use of flame retardants with the potential to form dioxins and furans. The EU flower mark is now used within the eco-labelling scheme in the EU. The general Eco-label criteria adopted within the EU exclude flame retardants subject to regulatory hazard labelling phrases (so called R-phrases); however, the use of decaBDE is not restricted under these Eco-labels. The European Flame Retardant Association (EFRA) did not support the definition of Eco-label criteria using R-phrases since these criteria do not take into consideration the exposure of consumers or the environment to flame retardants, and therefore do not consider the risk to health and the environment.[35] The EFRA, however, recommended inclusion of fire safety in Eco-label criteria for two main reasons. Firstly, the consumers expect labelled products to offer best performance and safety and secondly, fires have significant impacts on the environment.[41]

The revised EU criteria for eco-labelled products, particularly textile products where the prohibition of a flame-retardant chemical could make it hazardous in terms of flammability, excludes many flame retardants, especially those that are chemically bound to the textile substrate. Flame retardants, although classified as dangerous to the environment according to Directive 67/548/EEC, are excluded from EU Eco-label criteria for mattresses and certain high-risk textile products. Many flame retardants – including melamine, nitrogen-based flame retardants, certain phosphorus-based and brominated flame retardants – are not subject to any regulatory hazard labelling phrases.

Given the global nature of the trade, the World Summit on Sustainable Development (WSSD) approved the Globally Harmonised System of Classification and Labelling of Chemicals (GHS) in 2002. The system aims to harmonise the human health and environment information provided by the manufacturers worldwide.

14.4   Desirable properties of an ideal flame-retardant chemical used in textile applications

For a flame retardant to be commercially successful, the flame-retardant textile product must meet the required flammability standards, retain aesthetic and physiological properties of textile materials and should be durable to repeated home launderings, tumble drying and dry cleaning processes. Durability of the flame retardant is also an important criterion such that the more durable the flame retardant, the lower is the potential for human exposure and environmental release.[42] In addition to this, it is very important that the flame-retarding chemical is relatively cheap and its method of application is simple, cheap and compatible with the use of normal textile machinery. The flame-retardant chemical must not be biologically harmful, i.e. the chemical must not be harmful if swallowed, inhaled or absorbed through the skin.

For flame retardants to be environmentally friendly, they should not be persistent, bioaccumulative or toxic to humans, other animals and the ecosystem in general. Persistency does not necessarily mean that the chemical is detrimental to human health or that it is toxic to the environment. The important requirement is that the chemical must readily break down in the environment. Environmental transformations of persistent organic pollutants can be subdivided into three processes: biotransformation, abiotic oxidation and hydrolysis, and photolysis. The relative importance of these processes depends on the rates at which they occur under natural environmental conditions. These rates are, in turn, dependent on the chemical structure and properties of the substance and its distribution in the various compartments of the environment.[32]

14.5   Strategies for development of 'environmentally friendly' flame retardants

The substitution of current commercially acceptable but environmentally suspect flame retardants in textiles is a complicated process. It requires an understanding of the effects of the substitution not only on the flame resistance, but also on the thermal and mechanical stability of the textile material and processing parameters. Substitution strategies include direct replacement of brominated flame retardants, use of inherently flame-resistant fibres as replacements or as underlying barrier layers and complete redesign of the product.[42]

Finding a replacement for bromine in flame retardants is a real challenge because bromine-based formulations have the advantage of working across a wide spectrum of polymer and textile products. As pointed by Horrocks *et al.*,[43] the challenge of replacing halogen flame retardants in textiles is particularly great because of the need to address both aesthetic and durability issues. Some of the pioneering work in developing sustainable flame retardants for textiles has been carried out by Horrocks and co-workers. These include introduction of substantive intumescence (reference 44 and references therein), formaldehyde-free treatments,[45] antimony–halogen-free formulations,[46,47] development of phosphorus flame-retardant strategies[48,49] and inclusion of nanoparticulate additives in backcoating formulations[43] as well as within the fibre matrix.[50] Several authors have reviewed attempts to replace brominated flame retardants[43–50] and develop inherently flame-resistant fibres[23,51,52] and significant aspects are presented below.

Inherently fire-resistant fibres are often used where high durability is required. Some synthetic fibres – such as aramid, novoloid and melamine formaldehyde – require no additional flame retardants as their base polymer is inherently flame retardant. Traditionally these are used in applications such as firefighter's clothing, protective industrial workwear and barriers. However, conventional synthetic fibres can be manufactured to be inherently fire resistant by the addition of a flame retardant during the spinning process. Most of the flame retardants for synthetic fibres were developed in 1980 and no significant developments have been seen in the area of inherently flame-resistant fibres since then. Common examples include the use of phosphorus-based additives in polyamide, polypropylene and polyester fibres. For good flame-retardant effects, high concentrations of phosphorous-based additives are required and so few, if any, acceptable commercial examples exist for polyester (cyclic phosphonates, e.g. Amgarad CU (Rhodia)) and none exist for polyamides. Most successful flame retardants for polyester include phosphorus-containing comonomer. Modified polyester thus formed does not promote char but the phosphorus compound is considered to act in gas phase.[53] In applications where fire safety is a requirement for polyamide textiles, polyamide fabric can be made flame retardant with a thiourea formaldehyde resin finish. The topic of flame retardancy of polyamides has been thoroughly reviewed by Weil and Levichik.[54] Polypropylene is the most common polylefin used in textiles. Different flame retardants used for polypropylene fibres have been comprehensively reviewed by Zhang and Horrocks.[24]

For more information on inherently fire-resistant fibres, their commercial availability and flame-retardant properties, readers are referred to articles by Bajaj,[51] Weil and Levichik,[23] Horrocks *et al.*,[1,2] Bourbigot and Flambard[52] and Bourbigot.[53] Within the scope of this chapter, however, it is worth noting that synthetic fibres pose a particular challenge of non-biodegradability and hence environmentally acceptable disposal.[1]

Complete redesign of the product using non-woven barrier technology and surface modification are becoming very popular, particularly in the development of environmentally friendly flame-retardant products. Barrier fabrics play an important role in fire protection of foam filling in furniture and mattresses. The fabric structure has a significant effect on its burning behaviour. Heavier and tighter woven fabrics will burn slowly, whereas lightweight fabrics with open structures will burn quickly. Another approach to the development of flame-retardant products is to blend inherently fire-resistant fibres with cotton or polyester to produce fabrics with the required fire performance, durability, comfort and affordable cost. Kozlowski and coworkers[55] have developed flexible fire barriers based on natural non-woven textiles for use in furnishing and bedding. Their studies suggest that blended wool/hemp fabrics exhibit moderate flame-retardant properties without the addition of chemical treatments and the fact that they are difficult to ignite makes them ideal for barrier fabrics.

Fibre blends, particularly combinations of natural and high-performance flame-retardant fibres, are seen as a promising approach to produce hybrid yarns with low flammability, good handle and enhanced mechanical properties, along with good asthetics. Hybrid core yarns are cost-effective and hence could be manufactured on an industrial scale in the near future.

Nanocomposites

Compared with conventional flame retardants, nanocomposites offer significant advantages in the area of flame retardancy. Only very low concentrations of silicate are necessary in nanocomposites, resulting in commercial advantages such as low density, lower cost and ease of preparation. Moreover, these materials are seen as an environmentally friendly alternative to some types of fire retardants, as they contain no halogens, phosphates or aromatics other than those that may be present in the polymer matrix; and they do not produce the increases in carbon monoxide and soot levels during combustion that are associated with conventional flame retardants.[56] While traditional fillers very often severely degrade the physical properties of the polymer or discolour it, an important feature of nanocomposites is the simultaneous improvement in many physical properties, without any change in the polymer colour.

One promising approach is to achieve flame retardant effects from nano-structures within the fibre bulk. Dispersion of ~2–5 wt% nanoclay in the polymer matrix significantly improves the mechanical, thermal, barrier and flame-retardant properties of the base polymer.[57] In bulk polymers, the layered silicates are known to form a protective barrier when exposed to heat.[58] This protective barrier slows fuel pyrolysis and also reduces the flame temperature.[59] However, this mechanism of forming a thermal protective barrier does not seem to be efficient in textile materials such as fibres, films and fabrics. The surface of the fibres will be fuel-rich, and their high surface area could lead to enhanced oxygen contact and turbulent combustion.[60] Horrocks and Kandola[61] have aptly pointed to the physical and thermally thin structure of textile materials as the key to this inefficiency. For such thermally thin materials, they suggested that the 'shield-forming' mechanism could be too slow for any effective improvement in fire performance and that the presence of nanoclay alone in fibres, films and textiles may not be sufficient to impart flame retardancy at higher heat fluxes. Moreover, nanocomposites by themselves only lower the heat release rate but do not cause self-extinguishment in polymers generally. In an attempt to achieve regulatory fire safety performances, Wilkie and Morgan[62] suggested use of polymer nanocomposites in combination with a conventional flame retardant. The idea is to replace a major proportion of conventional flame retardant by a very small fraction of nanoclay to obtain same level of flame-retardant performance. This approach of minimising flame-retardant additive concentrations has significance, especially in synthetic fibres where levels in excess of 10% w/w usually reduce the ease of polymer extrusion and subsequent processing, as well as adversely affecting their normal desirable properties.[63]

Nanofillers such as montmorillonite clays have been known for decades but only recently used to promote nanocomposite structures in polymers. There are two possible ways for such materials to be used in textile applications: (a) melt spinning of nanocomposite polymer filaments, allowing them to be woven or knitted into fabric forms, and (b) using nanocomposite formulations for coating fabrics. However, the processing of polymer–clay nanocomposites is challenging in terms of maintaining nano-dispersion and the influence of nanoparticles on rheological properties can be crucial in fibre extrusion. Very few studies[50,60,64–69] are reported in the literature wherein the polymer–clay nanocomposite fibres are extruded and flammability properties tested on the fabrics woven/knitted from such composite fibres. Devaux *et al.*[70] and Horrocks *et al.*[43] have reported on the inclusion of nanocomposite coatings.

14.5.1 Environmentally friendly flame-retardant treatments

Formaldehyde-free treatments

Most durable flame-retardant treatments for cellulosics are based on formaldehyde (e.g. Proban®/Pyrovatex®). Formaldehyde is used in various steps of producing flame-retardant textiles. Owing to its volatility, formaldehyde can be transferred readily from treated garments to the skin. Formaldehyde is listed as a dangerous substance, in Annex 1 of Council Directive 67/548/EEC. Here, formaldehyde is classified as: toxic, R23/24/25 (by inhalation, in contact with skin and if swallowed); corrosive, R34 (causes burns); Carc3, R40 (possible risk of irreversible effects); and R43 (may cause sensitisation by skin contact).[35] Given the negative impact of formaldehyde on human health, it has been a primary focus of the cotton apparel and textile finishing industries to create equivalent non-formaldehyde technologies. Because of its widespread use in major durable finishes such as Pyrovatex® and Proban®, most of the research efforts has been concentrating on the development of new formaldehyde-free cross-linking agents for cellulose-containing materials, and mainly cotton.[71–74] Phosphorus- and nitrogen-containing synergistic flame retardants are often suggested for cotton; however, use of formaldehyde for their respective methylolation is an essential feature for their subsequent reactivity with anhydroglucopyranose-OH groups.[44] Moreover, to achieve acceptable levels of multiple laundering durability, Wu and Wang[72] have suggested application of methylolated resin species like dimethylol dihydroxyethylene urea (DMDHEU) or methylated formaldehyde-urea. Thus, the quest for a truly formaldehyde-free, durable and effective flame retardant for cellulosics in particular still remains a challenge. Horrocks *et al.*[44] have suggested use of low-formaldehyde resins in combination with organophosphorus compounds as successful retardants for cotton. Pioneering work in this area includes use of char-forming polycarboxylated species like butyl tetracarboxylic acid (BTCA) along with other functional species to interact with cellulose. However, since the BTCA–cellulose ester links are susceptible to hydrolysis, durability to domestic laundering is limited. More recent work has combined BTCA with phosphorylated species such as the hydroxyalkyl organophosphorus oligomer to enhance both flame retardancy and durability.

Encapsulation of flame retardant

In the early 1960s, microcapsules of flame retardants were first developed which could resist rainwater while providing fire resistance for forests. Microencapsulation is a technique in which microscopic amounts of solid or liquid can be encapsulated by a film-forming material to form tiny particles. The encapsulated solid or liquid core is isolated from the environment, thereby completely preserving the properties of the core. Under appropriate conditions, the shell is completely destroyed and the core material is released to react. Generally, the polymers for encapsulation of flame retardants are char-forming or form an intrinsic intumescent formulation in combination with encapsulated flame retardants. The microencapsulation process consists of three main steps:

(a) emulsification of a mixture containing cross-linking agent, catalyst and water-soluble core material;
(b) formation of microcapsules by dispersing the emulsion into hydrophilic polymer; followed by
(c) filtration, washing and drying of microcapsules.

The capsule wall grows by diffusion of molecules between the inner phase and outer phase through the intermediate phase. The resultant microcapsule can be applied to the fabric together with a suitable binder. The size of the microcapsule is dependent on various processing factors such as the concentration of hydrophilic polymer and cross-linking agent in aqueous solution. The speed of diffusion is also very important in the microencapsulation process. Flame-retardant finishes could be microencapsulated to improve the durability of the treatment to leaching, domestic washing and dry cleaning. Among the very few studies in this area, are the attempts to microencapsulate ammonium phosphate with polyurethane and polyurea shells to make an intrinsic intumescent system compatible with normal polyurethane (PU) coatings for textiles[75] and microencapsulation of water-soluble dimethyl phosphorate (DMMP) using the acetal product of polyvinyl alcohol (PVA) and glutaraldehyde (GA) as shell material.[76] Addition of encapsulated flame retardant to the polymer melt prior to fibre spinning is also a possibility for producing a durable system.

Sol-gel processing

A new environmentally friendly, halogen-free flame-retardant finishing process has been developed by Cireli *et al.*[77] They treated cotton fabrics with phosphorus-doped silicon-based solution using a sol-gel process. The sol-gel technique is claimed to impart durable flame retardancy to cotton fabrics without requiring after-treatment with formaldehyde. Moreover, significantly lower flame-retardant concentrations are required for sol-gel finishing processes compared with those required for conventional flame retarding processes where flame-retardant concentrations of 300–500 g/l are required. This essentially reduces the environmental loading.

Radiation-curable coatings

Radiation-induced grafting is a widely known technique and finds application mainly in the modification of surfaces of materials used in electronics, ion-exchange membranes and battery separators, etc. However, this technology has also found applications in protective coatings used in wound dressings, as well as in developing heat-resistant fabrics. Solvent-free coatings on textile materials can be accomplished by low-energy electron beams. The growing environmental concerns in the textile coating industry have led to the development of solvent-free formulations that can be thoroughly cured within a fraction of a second upon ultraviolet (UV) light exposure at ambient temperature, to generate polymer materials without emission of volatile organic compounds. Radiation-curable flame-retardant coatings offer several advantages – such as high cure speed, low energy consumption (since most radiation curing occurs at room temperature or at temperatures up to about 60 °C) and being environmentally friendly since radiation-curable formulations are solvent-free.[78,79] Conventional radiation-curable coatings for most applications are organic materials and are thus flammable. However, for flame-retardant properties, flame-retardant monomers and/or oligomers are often used. Additive-type flame retardants are not very common in radiation-curable coatings for several reasons. Firstly, the high concentration of additives and opacity of the formulation leads to difficult curing and even degradation of cured properties. Secondly, use of inorganic (insoluble) additives leads to viscosity increase which is not compatible with the radiation-curing process and finally, the curing speed is low for coating formulations containing inorganic particles. Several reactive-type flame-retardant monomers and oligomers used in radiation-curable coating formulations are reported in the literature.[80–86]

Many processes for the radiation treatment of natural polymers, though known for a long time, have not yet been commercialised, either because of the high cost of irradiation (high dose) or because of the reluctance on the part of the industry to adapt to the radiation technology.[87]

Plasma treatments

Traditional finishing methods usually involve high energy consumption, large amounts of chemical substances, frequent use of toxic organic solvents and the production of liquid and gaseous effluents that require expensive purification treatments before being released to the environment. Cold plasma treatments are seen as environmentally friendly processes for textiles characterised by low consumption of chemicals and energy. With plasma processing, expected results are obtained just by using air, nitrogen, oxygen or other inert gases. Moreover, atmospheric pressure cold plasma processing is extremely versatile and operates at room temperature, thus limiting the amount of energy necessary for heating water or for inducing chemical reactions.[88] However, it is worth noting here that achieving high levels of surface depositions using atmospheric pressure plasma could be a challenge in itself.[89]

One of the pioneering works on flame-retardant treatment of fabrics using cold plasma was reported in the 1980s by Simionescu et al.,[90] where surface grafting of rayon fabrics was carried out with phosphorus-containing polymers and improvements in fire-retardant properties were observed. The more recent studies of low pressure argon plasma graft polymerisation by Tsafack and Levalois-Grützmacher[91] report the successful grafting of phosphorus-containing acrylate monomers (diethyl(acryloyloxyethyl) phosphate (DEAEP), diethyl-2-(methacryloyloxyethyl)phosphate (DEMEP), diethyl(acryloyloxymethyl) phosphonate (DEAMP) and dimethyl(acryloyloxymethyl)phosphonate (DMAMP)) on to polyacrylonitrile fabrics. The surface-coated fabrics thus obtained have covalently bonded polymer on the surface of the fabric and the flame-retardant treatment is claimed to be resistant to washing at higher temperatures. In a more recent work by Vannier et al.[92] at Lille using plasma-induced graft polymerisation (PIGP), polyethylene methacrylate phosphate was grafted on to the cotton fabric. An increase in limiting oxygen index (LOI) values from 21 vol.% for the pure cotton to 32 vol.% for the flame-retardant cotton was reported. Plasma-treated flame-retardant fabric exhibited better flame-retardant performance compared with untreated cotton.

14.6   Future trends

Flame-retardant textiles are under close scrutiny, particularly because of the proximity of the end product to the consumer whether as a clothing item or as furnishing fabrics. Use of traditional halogenated flame retardants in textile materials is becoming increasingly subject to restrictions and some may be completely banned in the near future. PentaBDE and octaBDE have already been banned in the EU and the USA and their production has been voluntarily stopped since 2004. DecaBDE, one of the most widely used brominated flame retardants in textile applications passed the EU chemical risk assessment. However, since it is used with antimony trioxide (Sb_2O_3) as synergist, decaBDE is still under scrutiny until toxicity assessments of antimony trioxide and antimony pentoxide indicate no toxicity concerns. Some states in the USA and Sweden have already banned the use of decaBDE in electronic devices as well as in textiles. Despite there being no environmental and human health hazard, all the flame retardants produced in volumes greater than 1000 tonnes are now considered as a high-volume substance under REACH and will be assessed for REACH registration by December 2010.

Currently, many hazardous substances are used despite the existence of safer alternatives, simply because there is no legislative or economic requirement to systematically substitute them with safer alternatives. The Principle of Substitution states that hazardous chemicals should be systematically substituted by less hazardous alternatives or preferably alternatives for which no hazards can be identified. While the health and environmental impacts of using hazardous substances are difficult to quantify, more and more companies, non-governmental organisations and decision makers are advocating the solution of precaution.

14.7   Sources of further information and advice

To counteract global warming, numerous government and non-governmental organisations are now active in the areas of sustainability, renewability and recyclability of resources. Important organisations that are directly associated with, and are monitoring, flame-retardant chemicals are listed below.

- *The Alliance for Consumer Fire Safety in Europe (ACFSE).* A non-profit organisation founded in 1998. The organisation educates consumers about the potential hazards of fire in the home, and raises awareness of fire safety in consumer products among manufacturers and decision makers. It also advocates effective standards of fire safety in consumer services such as transportation, travel and accommodation. The ACFSE also monitors the causes and effects of domestic fires, and maintains compatible fire statistics across Europe.
- *Voluntary Emissions Control Action Programme (VECAP).* This programme was initiated in 2004 by the European industry sector comprising flame-retardant manufacturers and their user chains in the plastics and textile sectors. This innovative programme aims to increase awareness and understanding of chemicals management throughout the supply chain.
- *European Flame Retardants Association (EFRA).* This organisation represents the flame-retardant industry within the European Economic Area (EEA) and its main objectives are to promote fire safety, to coordinate studies and research relating to flame-retardant chemicals and to liaise with the institutions of the EU and, with other relevant authorities

and sister associations. The Association is also instrumental in initiation of environmental and toxicological programmes, and liaising with testing institutes and laboratories. The findings of research programmes and views of the Association are then conveyed to consumer authorities and other interest groups.

- *European Commission Scientific Committee on Toxicity, Ecotoxicity and the Environment (CSTEE)*. The CSTEE was created by the European Commission to address scientific and technical questions relating to examination of the toxicity and ecotoxicity of chemical, biochemical and biological compounds whose use may have harmful consequences for human health and the environment.
- *The US National Academy of Sciences (NAS)*. A private, non-profit, self-perpetuating society of distinguished scholars engaged in scientific and engineering research, dedicated to the furtherance of science and technology and to their use for the general welfare.
- *Swiss Federal Health Ministry (BAG)*. The main aim of the organisation is to promote and maintain the good health of all people living in Switzerland. One of the main objectives is assessment and regulatory checks on chemicals and toxic products.
- *Greenpeace*. An independent global campaigning organisation that acts to change attitudes and behaviour, to protect and conserve the environment and to promote a toxic-free future.

14.8   References

1 HORROCKS A.R., HALL M.E. and ROBERTS D., 'Environmental consequences of using flame retardant textiles – A simple life cycle analytical model', *Fire Materials*, **21**: 229–234 (1997).

2 HORROCKS A.R., 'Textiles', Chapter 4 in *Flame Retardant Materials*, HORROCKS A.R. and PRICE D. (eds), Woodhead Publishing, Cambridge, pp. 128–181 (2001).

3 GORDON P.G., 'Flame retardants and textile materials', *Fire Safety Journal*, **4**: 109–123 (1981).

4 SCHINDLER W.D. and HAUSER P.J., 'Flame retardant finishes', Chapter 8 in *Chemical Finishing of Textiles*, Woodhead Publishing, Cambridge, pp. 98–116 (2004).

5 *World Market for Flame Retardants*, Published by Freedonia Group, USA (2007).

6 MICHIGAN DEPARTMENT OF COMMUNITY HEALTH (MDCH), Division of Environmental and Occupational Epidemiology (DEOE), 1-800-MI-TOXIC (1-800-648-6942), http://www.michigan.gov/documents/mdch_PBB_FAQ_92051_7.pdf.

7 YOGUI G.T. and SERICANO J.L., 'Polybrominated diphenyl ether flame retardants in the U.S. marine environment: A review', *Environment International*, **35**(3): 655–666 (2009).

8 ZHAO G., ZHOU H., WANG D., ZHA J., XU Y., RAO K., MA M., HUANG S. and WANG Z., 'PBBs, PBDEs, and PCBs in foods collected from e-waste disassembly sites and

daily intake by local residents', *Science of The Total Environment*, **407**(8): 2565–2575 (2009).

9 LAW R.J., 'Tetrabromobisphenol A: Investigating the worst-case scenario', *Marine Pollution Bulletin*, **58**(4): 459–460 (2009), doi: 10.1016/j.marpolbul. 2009.02.023.

10 FREDERIKSEN M., VORKAMP K., THOMSEN M. and KNUDSEN L.E., 'Human internal and external exposure to PBDEs – A review of levels and sources', *International Journal of Hygiene and Environmental Health*, **212**(2): 109–134 (2009).

11 XU J., GAO Z., XIAN Q., YU H. and FENG J., 'Levels and distribution of polybrominated diphenyl ethers (PBDEs) in the freshwater environment surrounding a PBDE manufacturing plant in China', *Environmental Pollution*, **157**(6): 1911–1916 (2009), doi: 10.1016/j.envpol.2009.01.030.

12 BROMINE SCIENCE AND ENVIRONMENTAL FORUM, www.bsef.com.

13 HARDY M.L., 'Regulatory status and environmental properties of brominated flame retardants undergoing risk assessment in the EU: DBDPO, OBDPO, PeBDPO and HBCD', *Polymer Degradation and Stability*, **64**: 545–556 (1999).

14 KEMMLEIN S., HERZKE D. and LAW R., 'Brominated flame retardants in the European chemicals policy of REACH-Regulations and determination in material', *Journal of Chromatography A*, **1216**(3): 320–333 (2009).

15 BENETTI *et al.*, EU Study contract no ETD/91/88-5300/MI/44, Final report, December 1992.

16 EUROPEAN COMMISSION JOINT RESEARCH CENTRE, http://ecb.jrc.it/.

17 STEVENS G., 'Risks and benefits in the use of flame retardants in consume products', A report for the UK Department of Trade and Industry, January 1999, DTI References URN 98/1028, http://www.dti.gov.uk/homesafetynetwork/bs_rfret.htm.

18 PURSER D., 'Toxicity of fire retardants in relation to life safety and environmental hazards', Chapter 3 in *Fire Retardant Materials*, HORROCKS A.R. and PRICE D. (eds), Woodhead Publishing, Cambridge, pp. 69–127 (2001).

19 WAKELYN P.J., 'Environmentally flame resistant textiles', Chapter 8 in *Advances in Fire Retardant Materials*, HORROCKS A.R. and PRICE D. (eds), Woodhead Publishing, Cambridge, pp. 188–212 (2008).

20 CONSUMER PROTECTION ACT (1987), The Furniture and Furnishing (Fire) (Safety) Regulations, 1988, SI 1324, HMSO, London (1988).

21 LEWIN M. and SELLO S.B., *Handbook of Fiber Science and Technology*: Volume 2: *Chemical Processing of Fibers and Fabrics: Functional Finishes*; Dekker, New York (1984).

22 ZAIKOV G.E. and LOMAKIN S.M., 'Ecological issue of polymer flame retardancy', *Journal of Applied Polymer Science*, **86**: 2449–2462 (2002).

23 WEIL E.D. and LEVCHIK S.V., 'Flame retardants in commercial use or development for textiles', *Journal of Fire Sciences*, **26**: 243–281 (2008).

24 ZHANG S. and HORROCKS A.R., 'A review of flame retardant polypropylene fibres', *Progress in Polymer Science*, **28**(11): 1517–1538 (2003).

25 MENEZES E. and PARANJAPE M., 'Flame retardants in textiles', *Colourage*, **51**: 19–26 (2004).

26 LEWIN M. and WEIL E.D., 'Mechanisms and modes of action in flame retardancy of polymers', Chapter 2 in *Fire Retardant Materials*, HORROCKS A.R. and PRICE D. (eds), Woodhead Publishing, Cambridge, pp. 31–68 (2001).

27 HOFER H., 'Health aspects of flame retardants in textiles', Report for Austrian Standards Institute Consumer Council (October 1998).

28 KHATTAB M.A., KANDIL S.H., GAD A.M., EL-LATIF M. and MORSI S.E., 'Effect of condensed-phase and gas-phase flame retardants on the ignition behaviour of cotton fabrics', *Fire Materials*, **16**: 23–28 (1992).

29 Recycling and disposal: End of life of products containing flame retardants, www.ceifc-efra.eu.

30 HOFER H., 'Environmental aspects of flame retardants in textiles', Report for Austrian Standards Institute Consumer Council (April 1999).

31 Report on Environmental Health Criteria-192, WHO, International Programme on Chemical Safety, Switzerland.

32 Review of selected organic persistent chemicals, http://www.who.int/entity/ipcs/assessment/en/pcs_95_39_2004_05_13.pdf.

33 Subcommittee on Flame-Retardant Chemicals, Committee on Toxicology, Board on Environmental Studies and Toxicology, National Research Council, *Toxicological Risks of Selected Flame Retardant Chemicals*, National Academies Press, Washington DC (2000).

34 Council Directive 76/769/EEC of 27 July 1976 on the approximation of the laws, regulations and administrative provisions of the Member States relating to restrictions on the marketing and use of certain dangerous substances and preparations.

35 European Commission Green Public Procurement (GPP) Training Toolkit-Module 3: Purchasing Recommendations. http://ec.europa.eu/environment/gpp/pdf/toolkit/textiles_GPP_background_report.pdf

36 SIMONSON M. and ANDERSSON P., 'Life cycle assessment of consumer products with a focus on fire performance', Chapter 13 in *Advances in Fire Retardant Materials*, HORROCKS A.R. and PRICE D. (eds), Woodhead Publishing, Cambridge, pp. 331–359 (2008).

37 OECD. Risk Reduction Programme no. 3, Selected brominated flame retardants. Environmenta Directorate, Paris (1994).

38 http://www.cefic-efra.com/pdf/0303/BAG_%20report.pdf.

39 Healthy Buildings 2003. 7th International Conference, http://hb2003.nus.edu.sg/.

40 EMSLEY A.M. and STEVENS G.C., 'The risk and benefits of flame retardants in consumer products', Chapter 14 in *Advances in Fire Retardant Materials*, HORROCKS A.R. and PRICE D. (eds), Woodhead Publishing, Cambridge, pp. 364–397 (2008).

41 EUROPEAN FLAME RETARDANTS ASSOCIATION, www.cefic-efra.org.

42 Decabromodiphenylether: An Investigation of Non-halogen Substitutes in Electronic Enclosure and Textile Applications, Publication of the Lowell Centre for Sustainable Production, MA, USA (2005).

43 HORROCKS A.R., DAVIES P.J., ALDERSON A. and KANDOLA B.K., 'The challenge of replacing halogen flame retardants in textile applications: Phosphorus mobility in back-coating formulations', in Proceedings of 10th European Meeting of Fire Retardant Polymers, FRMP'05, Berlin, 6–9th September 2005, *Advances in the Flame Retardancy of Polymeric Materials: Current perspectives presented at FRPM'05*, SCHARTEL B. (ed.), Norderstedt, Germany, pp. 141–158 (2007).

44 HORROCKS A.R., KANDOLA B.K., DAVIES P.J., ZHANG S. and PADBURY S.A., 'Developments in flame retardant textiles – a review', *Polymer Degradation and Stability*, **88**: 3–12 (2005).

45 HORROCKS A.R. and ROBERTS D. Minimization of formaldehyde emission, in *Proceedings of the Conference Ecotextile '98: Sustainable Development*, Bolton, UK, Woodhead Publishing, Cambridge (1999).

46 HORROCKS A.R., WANG M.Y., HALL M.E., SUNMONU F. and PEARSON J.S., 'Flame retardant textile back-coatings. Part 2. Effectiveness of phosphorus-containing flame retardants in textile back-coating formulations', *Polymer International*, **49**: 1079–1091 (2000).

47 HORROCKS A.R., WANG M.Y., HALL M.E., SUNMONU F. and PEARSON J.S., 'Flame retardant textile back-coatings. Part 1: Antimony-halogen system interactions and the effect of replacement by phosphorus-containing agents', *Journal of Fire Sciences*, **18**: 265–294 (2000).

48 HORROCKS A.R., DAVIES P., ALDERSON A. and KANDOLA B.K., 'The potential for volatile phosphorus-containing flame retardants in textile back-coatings', *Journal of Fire Sciences*, **25**(6): 523–540 (2007).

49 DAVIES P.J., HORROCKS A.R. and ALDERSON A., 'The sensitisation of thermal decomposition of APP by selected metal ions and their potential for improved cotton fabric flame retardancy', *Polymer Degradation and Stability*, **88**: 114–122 (2005).

50 HORROCKS A.R., KANDOLA B.K., SMART G., ZHANG S. and HULL T.R., 'Polypropylene fibers containing dispersed clays having improved fire performance. I. Effect of nanoclays on processing parameters and fiber properties', *Journal of Applied Polymer Science*, **106**(3): 1707–1717 (2007).

51 BAJAJ P., 'Heat and flame protection', in *Handbook of Technical Textiles*, HORROCKS A.R. and ANAND S.C. (eds), Woodhead Publishing, Cambridge, pp. 223–263 (2000).

52 BOURBIGOT S. and FLAMBARD X., 'Heat resistance and flammability of high performance fibres: a review', *Fire Materials*, **26**: 155–168 (2002).

53 BOURBIGOT S., 'Flame retardancy of textiles: new approaches', Chapter 2 in *Advances in Fire Retardant Materials*, HORROCKS A.R. and PRICE D. (eds), Woodhead Publishing, Cambridge, pp. 9–37 (2008).

54 WEIL E.D. and LEVICHIK S.V., 'Current practice and recent commercial developments in flame retardancy of polyamides', *Journal of Fire Sciences*, **22**(3): 251–264 (2004).

55 KOZLOWSKI R., MIELENIAK B., MUZYCZEK M. and KUBACKI A., 'Flexible fire barriers based on natural nonwoven textiles', *Fire Materials*, **26**: 243–246 (2002).

56 PAVLIDOU S. and PAPASPYRIDES C.D., 'A review on polymer–layered silicate nanocomposites', *Progress in Polymer Science*, **33**(12): 1119–1198 (2008).

57 KANDOLA B., 'Nanocomposites', Chapter 6 in HORROCKS A.R. and PRICE D. (eds), *Fire Retardant Materials*, Woodhead Publishing, Cambridge, pp. 204–220 (2001).

58 GILMAN J.W., 'Flammability and thermal stability studies of polymer layered-silicate (clay) nanocomposites', *Applied Clay Science*, **15**: 31–49 (1997).

59 BETTS K.S., 'New thinking on flame retardants', *Environmental Health Perspectives*, **116**(5): 211–213 (2008).

60 SHANMUGANATHAN K., DEODHAR S., DEMBESY N.A., FAN Q. and PATRA P.K., 'Condensed-phase flame retardation in nylon 6-layered silicate nanocomposites: films, fibres and fabrics', *Polymer Engineering and Science*, **43**(4): 662–675 (2008).

61 HORROCKS A.R. and KANDOLA B.K., 'Potential applications of nanocomposites for flame retardancy', Chapter 11 in *Flame Retardant Polymer Nanocomposites*, MORGAN A.B. and WILKIE C.A. (eds), Wiley-VCH, Verlag GmbH & Co, KGaA, Weinheim, Germany, pp. 325–354 (2007).

62 WILKIE C.A. and MORGAN A.B., Current developments in nanocomposites as novel flame retardants, Chapter 5 in *Advances in Fire Retardant Materials*, HORROCKS A.R. and PRICE D. (eds), Woodhead Publishing, Cambridge, pp. 95–123 (2008).

63 HORROCKS A.R., 'Applications for nanocomposite-based flame retardant systems', Chapter 6 in *Advances in Fire Retardant Materials*, HORROCKS A.R. and PRICE D. (eds), Woodhead Publishing, Cambridge, pp. 124–158 (2008).

64 SOLARSKI S., MAHJOUBI F., FERREIRA M., DEVAUX E., BACHELET P., BOURBIGOT S., DELOBEL R., MURARIU M., DA SILVA FERREIRA A., ALEXANDRE M., DEGÉE P. and DUBOIS P., '(Plasticized) polylactide/clay nanocomposite textile: thermal, mechanical, shrinkage and fire properties', *Journal of Materials Science*, **42**(13): 5105–5117 (2007).

65 WANG D.-Y., WANG Y.-Z., WANG J.-S., CHEN D.-Q., ZHOU Q., YANG B. and LI W.-Y., 'Thermal oxidative degradation behaviours of flame-retardant copolyesters containing phosphorous linked pendent group/montmorillonite nanocomposites', *Polymer Degradation and Stability*, **87**: 171–176 (2005).

66 BOURBIGOT S., LE BRAS M., FLAMBARD X., ROCHERY M., DEVAUX E. and LICHTENHAN J., *Fire Retardancy of Polymers: New Applications of Mineral Fillers*, LE BRAS M., BOURBIGOT S., DUQUESNE S., JAMA C. and WILKIE C.A. (eds), Royal Society of Chemistry, London, pp. 189–201 (2005).

67 HORROCKS A.R., HICKS J., DAVIES P.J., ALDERSON A. and TAYLOR J., 'Synergistic flame retardant copolymeric polyacrylonitrile fibres containing dispersed phyllosilicate clays and ammonium polyophosphate', Chapter 20 in *Fire Retardancy of Polymers: New Strategies and Mechanisms*, HULL T.R. and KANDOLA B.K. (eds), RSC Publication, Cambridge, pp. 307–329 (2009).

68 HORROCKS A.R., KANDOLA B.K., SMART G., ZHANG S. and HULL T.R., 'Polypropylene fibers containing dispersed clays having improved fire performance. I. Effect of nanoclays on processing parameters and fiber properties', *Journal of Applied Polymer Science*, **106**(3): 1707–1717 (2007).

69 SMART G., KANDOLA B.K., HORROCKS A.R., NAZARÉ S. and MARNEY D., 'Polypropylene fibers containing dispersed clays having improved fire performance. Part II: characterization of fibers and fabrics from PP-nanoclay blends', *Polymers for Advanced Technologies*, **19**(6): 658–670 (2008).

70 DEVAUX E., ROCHERY M. and BOURBIGOT S., 'Polyurethane/clay and polyurethane/POSS nanocomposites as flame retarded coating for polyester and cotton fabrics', *Fire Materials*, **26**: 149–154 (2002).

71 YANG H. and YANG C.Q., 'Nonformaldehyde flame retardant finishing of the Nomex/cotton blend fabric using a hydroxyl-functional organophosphorus oligomer', *Journal of Fire Sciences*, **25**: 425–446 (2007).

72 WU W. and YANG C.Q., 'Comparision of DMDHEU and melamine-formaldehyde as the binding agents for a hydroxyl-functional organophosphorus flame retarding agent on cotton', *Journal of Fire Sciences*, **22**: 125–142 (2004).

73 YANG C.Q. and WU W., 'Combination of a hydroxyl-functional organophosphorus oligomer and a multifunctional carboxylic acid as a flame retardant finishing

system for cotton: Part I. The chemical reactions', *Fire Materials*, **27**: 223–237 (2003).

74 YANG C.Q. and WU W., 'Combination of a hydroxyl-functional organophosphorus oligomer and a multifunctional carboxylic acid as a flame retardant finishing system for cotton: Part II. Formation of calcium salt during laundering', *Fire Materials*, **27**: 239–251 (2003).

75 GIRAUD S., BOURBIGOT S., ROCHERY M., VROMAN I., TIGHZERT L., DELOBEL R. and POUTCH F., 'Flame retarded polyurea with microencapsulated ammonium phosphate for textile coating', *Polymer Degradation and Stability*, **88**: 106–113 (2005).

76 LIN M., YANG Y., XI P. and CHEN S., 'Mircroencapsulation of water-soluble flame retardant containing oprganophosphorus and its application on fabric', *Journal of Applied Polymer Science*, **102**: 4915–4920 (2006).

77 CIRELI A., ONAR N., EBEOGLUGIL F., KAYATEKIN I., KUTLU B., CULHA O. and CELIK E., 'Development of flame retardancy properties of new halogen-free phosphorus doped SiO_2 thin films on fabrics', *Journal of Applied Polymer Sciences*, **105**: 3747–3756 (2007).

78 CLELAND M.R., PARKS L.A. and CHENG S., 'Applications for radiation processing of materials', *Nuclear Instruments and Methods in Physics Research Section B: Beam Interactions with Materials and Atoms*, **208**: 66–73 (2003).

79 RANDOUX T.H., VANOVERVELT J.-CL., VAN DEN BERGEN H. and CAMINO G., 'Halogen-free flame retardant radiation curable coatings', *Progress in Organic Coatings*, **45**: 281–289 (2002).

80 CHEN X., HU Y., JIAO C. and SONG L., 'Preparation and thermal properties of a novel flame-retardant coating', *Polymer Degradation and Stability*, **92**(6): 1141–1150 (2007).

81 LIANG H., SHI W. and GONG M., 'Expansion behaviour and thermal degradation of tri(acryloyloxyethyl) phosphate/methacrylated phenolic melamine intumescent flame retardant system', *Polymer Degradation and Stability*, **90**(1): 1–8 (2005).

82 CHEN-YANG Y.W., CHUANG J.R., YANG Y.C., LI C.Y. and CHIU Y.S., 'New UV-curable cyclotriphosphazenes as fire-retardant coating materials for wood', *Journal of Applied Polymer Science*, **69**(1): 115–122 (1997).

83 ZHU S.W. and SHI W.F., Combustion behaviour and thermal properties of UV cured methacrylated phosphate/epoxy acrylate blends', *Polymer Degradation and Stability*, **81**(2): 233–237 (2003).

84 AVCI D. and ALBAYRAK A.Z., 'Synthesis and copolymerization of new phosphorus-containing acrylates', *Journal of Polymer Science and Polymer Chemistry*, **41**(14): 2207–2217 (2003).

85 LIANG H.B. and SHI W.F., 'Thermal behaviour and degradation mechanism of phosphate di/triacrylate used for UV curable flame-retardant coatings', *Polymer Degradation and Stability*, **84**(3): 525–532 (2004).

86 REDDY P.R.S., AGATHIAN G. and KUMAR A., 'Ionisation radiation graft polymerized and modified flame retardant cotton fabric', *Radiation Physics and Chemistry*, **72**: 511–516 (2005).

87 CHMIELEWSKI A.G., HAJI-SAEID M. and AHMED S., 'Progress in radiation processing of polymers', *Nuclear Instruments and Methods in Physics Research Section B: Beam Interactions with Materials and Atoms*, **236**(1–4): 44–54 (2005).

88 MARCANDALLI B. and RICCARDI C., 'Plasma treatments of fibres and textiles', Chapter 11 in *Plasma Technologies for Textiles*, SHISHOO R. (ed.), Woodhead Publishing, Cambridge, pp. 282–301 (2007).

89 HORROCKS A.R., 'Flame retardant/resistant textile coatings and laminates', Chapter 7 in *Advances in Fire Retardant Materials*, HORROCKS A.R. and PRICE D. (eds), Woodhead Publishing, Cambridge, pp. 159–188 (2008).

90 SIMIONESCU C.I., DÉNES F., MACOVEANU M.M., CAZACU G., TOTOLIN M., PRICE S. and BALAUR D., 'Grafting of rayon fabrics with phosphorus containing polymers in cold plasma in order to obtain flame-retardant materials', *Cellulose Chemistry and Technology*, **14**: 869–883 (1980).

91 TSAFACK M.J. and LEVALOIS-GRÜTZMACHER J., 'Plasma-induced graft-polymerisation of flame retardant monomers onto PAN fabrics', *Surface and Coatings Technology*, **200**: 3503–3501 (2006).

92 VANNIER A., DUQUESNE S., BOURBIGOT S., DELOBEL R., MAGNIEZ C. and VOUTERS M., in Proceedings of International Conference on *Textile Coating and Laminating*, Barcelona (Spain), November 2006, pp. 8–29.

15
Systems change for sustainability in textiles

K. FLETCHER, London College of Fashion, UK

Abstract: This chapter explores systems thinking in the context of enhancing the sustainability of the textile sector. It describes ways in which the sector could advance on a path of deep change towards sustainability and investigates a series of 'leverage' or intervention points in the textile industrial system where sustainability goals can be most effectively pursued.

Key words: sustainable textiles, systems thinking, holism, design, change.

15.1 The blind men and the elephant

There is a Buddhist parable about six blind men who were gathered together to examine an elephant. When all six have felt the elephant, a wise man says to each, 'Well, blind man, have you seen the elephant? Tell me, what sort of thing is it?' They variously assert that the elephant is like a serpent (trunk), a great wall (body), a tree trunk (leg), a leathery bird with huge wings (ears), a rope (tail) or a brush (tip of the tail). Knowing only the parts, and blind to the whole, the men come to blows over their very different and inaccurate conclusions about the animal.

Sustainability in textiles is just as multi-dimensional. While many have described parts of it, few have captured its totality. A glance at the pages of any textile industry journal quickly confirms this distortion. What we see is a growing body of sustainability work that focuses on optimising parts of the textile production chain, that is on improving discrete processes, separate lifecycle phases or aspects of the supply chain. These improvements save valuable resources, minimise pollution effects and improve the conditions of workers. Yet – and this is the important point here – they tell us very little about the sustainability of the textile industry as a whole. In focusing exclusively on the sustainability of the trunk or the tail of the textile sector elephant, we relate in an unknown way to the sustainability of the entire industry 'animal'. We simply don't know whether our actions are helpful or harmful to the whole. This Chapter describes what this author believes are the limits of our current approaches to developing more sustainable textile practices and explores some emerging ways of thinking, tools and opportunities for bringing fundamental change to the way the textile sector works.

15.2   From a narrow to a holistic view of sustainability in the textile sector

As mentioned briefly above, attempts to approach sustainability through an assortment of partial perspectives tends to give a distorted – or even false – picture of the ecological, economic and social 'health' of the textile industrial system. For example, depending on which part of the textile industry you examine or which indicator you measure, things can be shown to be getting better or worse. On the one hand, the sector can be shown never to have been more sustainable or environmental policies more all embracing. New innovations in dyeing technology and techniques, for instance, are cutting chemical usage rates, energy consumption and pollution levels. Likewise supply chain transparency initiatives are opening up production routes to unprecedented scrutiny and potential improvement. Major high street retailers have also set challenging targets for sustainable sourcing of lower impact fibres and have set new standards in ethical trading.

On the other hand, when different areas of sustainability in textiles are examined, serious problems come to light. Consumption rates are increasing – in the UK over the last 4 years, every person has bought on average one-third more textiles and clothing (Allwood *et al.*, 2006, p. 12), intensifying pressure on resources. Pressure on individual consumers is also escalating: garments (the end product for many textiles) are shopped for addictively – frequently with money consumers do not have, trapping people with record levels of credit card debt (Gilmore, 2007). The societal drive to constantly reformulate identity in the light of changing fashion trends helps feed psychological insecurity and rising levels of mental illness, while fashion imagery is linked to serious medical conditions such as anorexia (Thomsen *et al.*, 2002), with recent statistics revealing that it is now reaching record levels in young men as well as women (Macfarlane, 2008). At the same time, working conditions in factories are being forced ever lower in what is known as a 'race to the bottom', as manufacturers compete on price for a place in the supply chain of big brands (FEI, undated).

Yet, while rarely acknowledged as such, issues such as consumption, fashion, globalisation and mental and physical health (all of which reflect the cultural worldviews and societal norms that influence the textile industry) are as much a part of the sustainability debate in textiles as technical and commercial details of dye technologies, supply chain initiatives and alternative fibre sourcing. And if we truly wish to build towards a more sustainable textile industry, then it is imperative that we start a dialogue within the sector around some of the broadest socio-cultural, political and structural questions of our times. If we don't, we risk dealing in a piecemeal fashion with the symptoms of the environmental and social crises, while never dealing with the root causes.

Sustainability issues in textiles require us to take account of influences emerging from outside the boundaries of the conventional textile industry. These 'external' influences – everything from farming practices to international energy policy, and consumption patterns to the level of ecological literacy in society – have great influence over the whole sector's sustainability. This helps explain why, for example, it sometimes proves difficult (even with large amounts of money, time and resources) to 'solve' environmental problems that arise in a particular part of the supply chain: their root cause lies elsewhere. Put simply, ecological and social issues are systems extending beyond the boundaries of individual companies and industries. Thus in order to develop a more sustainable textile industry we need to engage with these issues *on their level* and connect with other industries, disciplines, communities and international groups beyond the boundaries of our own. Change at the level of the industrial system does not replace changes made at the level of fibre, dye or polymer; rather it works to direct and shape them. It provides us with guidance to form our combined actions and a checking mechanism to confirm that our day-to-day individual commercial choices lead us ultimately in the direction we want to head.

15.3   Ways of thinking

As yet we have precious little experience of working for holistic, systems level change to promote sustainability in textiles. This is likely to be for a host of reasons, including among others: the highly fragmented structure of the textile processing chain, typically involving a large number of small and medium-sized companies who tend to work on bringing change to processes that chiefly bring benefit to themselves; and the dominance of binary and linear ways of thinking associated with Western industrial capitalism. Susan Kaiser (2008, p. 144) describes how binary (i.e. either/or) thinking puts production and consumption in opposition, whereas the two functions used to be closely connected in homes or villages prior to the industrial revolution; industrialisation reframed them as the inverse of each other. The legacy of this disconnection between production and consumption is still evident in sustainability discussions today where there is on-going polarisation between demand- and supply-side sustainability initiatives. Western industrialisation has also tended to encourage a one-dimensional or linear view on how resources flow through the supply chain, even though a great deal of lip service is paid to non-linear ideas like life-cycle thinking. For the educationalist Stephen Sterling (2001, p. 16), it is these non-holistic ways of thinking that blind us to the connective and dynamic reality of sustainability challenges. Our fundamental problem, he states, is one of 'inadequate perception'.

Changing perceptions and developing new learning are pivotal to making sustainability happen; though the role that key institutions like companies and governments have played in this to date is unclear. It appears, for example, that many policy decisions perpetuate old ways of doing things and the commercial status quo; albeit a status quo that has, in recent years, strived to use water and energy more efficiently. Take the European Commission's Integrated Pollution Prevention and Control regime for example, which sets out 'best available techniques' for a wide range of textile processes, but makes little play of interconnected lifecycle innovation or whole systems improvement. This type of regulation has the effect of cosseting business from the (inevitable) knowledge that in order to achieve sustainability, consumption must change. Successive governments are not alone, however, in failing to address consumption as one of the root causes of environmental and social problems. According to industrial ecologist John Ehrenfeld (2004, p. 4), 'virtually all suggestions by the powerful institutions of the modern world for solving the sustainability challenge are based on quick technological fixes'. But Ehrenfeld suggests that there are no quick fixes in dealing with our 'addiction' to consumption. He goes on, 'Achieving positive results requires drastic action. We need to shift from our reductionist, problem-solving mode to one that is driven by a vision of a sustainable future we all share. We need to reflect carefully on our current state of affairs and replace ineffective ways of thinking and acting.'

15.4   Recognising the limits of eco-efficiency

John Ehrenfeld, like many other sustainability advocates, articulates a concern about the limited effectiveness of one of the most popular industrial responses to sustainability: eco-efficiency. Eco-efficiency (i.e. delivering more value to the consumer at a lower environmental cost) has traditionally involved 'greening' existing products through a process of making small incremental changes. Companies commonly favour it because it requires easy-to-achieve change, the benefits of which are typically felt quickly and fit in with business and economic cycles. The main criticism made of eco-efficiency is that its strategies fail to moderate fundamentally inefficient industrial systems, because they focus on optimising one small part of the system, rather than the whole. Furthermore, they are limited by the efficiency potential of the very products or processes they are applied to and will therefore always struggle to bring the magnitude of change required by sustainability. The authors of *Cradle to Cradle*, William McDonough and Michael Braungart (2002, p. 62), pull no punches in their assessment of eco-efficiency.

Eco-efficiency is an outwardly admirable, even noble concept, but it is not a strategy for success over the long term, because it does not reach deep enough. It works within the same system that caused the problems in the first place, merely slowing it down with moral proscriptions and punitive measures. It presents little more than an illusion of change.

Illusion or no, the fact remains that this is where most change tends to start. It is also where the vast majority of sustainability actions in the textile sector have been concentrated to date, largely because it tends to be in the direct influence of most companies. The challenge for advocates of a more holistic approach to sustainability in textiles is to transform this focus on eco-efficiency into the start of a process of questioning, thinking and improving to drive fundamentally deeper and longer-term change. The task is to create a more sustainable future that is not just an extension of how we do things today but instead flows from ways of thinking and acting that are different to those we have used in the past.

15.5   Making a transition

It has been said that the challenge of sustainability is to design and develop ways in which we can live better while consuming much less. Small signs of a promising response to this challenge are emerging. In the United Kingdom, for example, there is currently a groundswell of grassroots activity around these ideas, catalysed by the Transition Towns movement. Transition Towns involve communities in a collective process of imagining and creating a future for themselves that addresses the twin challenges of diminishing oil and gas supplies and climate change, while building the kind of community residents want to be part of. Similar principles of developing a planned response to the inevitable changes implied by sustainability can also be seen in some textile businesses. Interface, for example, the modular flooring company and long-time environmental pioneer has a company commitment to produce zero impact on the environment by 2020. The use of petrochemicals is of particular concern to Interface as the bulk of its business is based on nylon. Yet in the context of the challenge of climate change, Interface is assessing its product portfolio so that profit is made in ways other than just selling oil-based carpet tiles to customers. Its decade-old Evergreen Lease™ carpet hiring service makes money by serving more clients with fewer tiles. It has also developed product lines such as Just™, a Fair Trade, hand-woven carpet tile made using natural materials sourced local to the artisan communities who produce them. Yet with the exception of a small (but growing) band of innovators, the textile industry seems very ill prepared for deep transition to sustainability and the prospect of global warming and so-called 'peak oil'. Perhaps the most likely reason for this is that deep sustainability questions take us out of our comfort zone and we are, of course, more at home in familiar terrain. Stepping too far outside this terrain feels risky and, particularly in difficult economic times, we tend to want to limit rather than expand our exposure to risk.

One route to expanding the limits of our sustainability comfort zone lies in initiating a dialogue about the kind of industry that individuals and companies want to be part of. In asking: what changes need to be imagined in industry so that the development of a more sustainable society becomes as ingrained into the textile industrial system as the delivery of fibres and fabrics? Developing this more 'relational' industrial system (i.e. one whose success is more dependent on its healthy relationships with environments, workers, etc.) requires us to make a very important distinction between first-order change (and learning) and second-order change (and learning). The difference between the two is that first-order change takes place within accepted boundaries and leaves basic values unchanged and unexamined (with obvious parallels with eco-efficiency); whereas second-order change involves examining – and even transforming – these basic values. Second-order change transforms not only what we learn, but also how we learn. It precipitates a 'double learning' process (Sterling, 2001, p. 15) and is part of what Einstein referred to as a shift in consciousness, essential because (in his words) 'we cannot solve our problems with the same thinking we used when we created them'. When applied to the textile industry as a whole we begin to see that in order to make the change towards sustainability in companies, product lines, the lives of workers, consumers etc., the industrial system itself needs to engage in deep change. We have to begin a process of scrutinising the goals and rules of the textile industry. We have to ask the questions that few people are prepared to pose, or even fewer prepared to answer: who benefits from the current set-up? Who does it serve?

15.6   Places to intervene in a system

According to Donella Meadows (1997), a visionary systems theorist, asking such questions about the goals of an industry has the potential to bring big change. Meadows has set out a list of systems intervention or 'leverage points' where change could be made in complex systems (like the textile industry). Her list is structured such that the most common strategies for change, which also often turn out to be the least effective, are first. Each builds on the previous in terms of both influence and significance to culminate in complete system realignment. Donella Meadows' list helps explain why certain types of change bring short-term benefits and why others have deeper, broader and longer-lasting effects. It also builds confidence and shows us how our actions (including small actions) can affect the bigger system. Meadows sets out nine 'places to intervene in a system', which are a set of tactics, of varying effectiveness, for changing current practices. Each of these is explored further below (for a fuller investigation of these ideas, see *Sustainable Fashion and Textiles: Design Journeys* (Fletcher, 2008)). From least to the most effective, these leverage points are:

9   Numbers (subsidies, taxes, standards).
8   Material stocks and flows.
7   Regulating negative feedback loops.
6   Driving positive feedback loops.
5   Information flows.
4   The rules of the system (incentives, punishment, constraints).
3   The power of self-organisation.
2   The goals of the system.
1   The mindset or paradigm out of which the goals, rules, feedback structure arise.

15.6.1 Numbers and standards (point 9)

For Donella Meadows, a focus on numbers tallies with the eco-efficiency strategies described above and tends to bring the most limited change because it involves only minor adjustments to a product or process. Adjusting numbers may change efficiency ratings, but because these improved ratings are being applied to the same fibres, processed with the same machinery, sold by the same retailers as before, the system does not change much. So for example, if we filter wastewater, it cuts down on river pollution downstream from the dye house, but it does not show us alternative ways to colour fabric. Likewise, if we substitute recycled for virgin packaging, it positively reduces the amount of waste being sent to landfill, but it does not tackle problems linked to over-consumption of resources more generally.

Yet even though the benefits of focusing on numbers are only ever going to be small, this is where most change tends to start. Manufacturers, retailers or designers can, for example, fine-tune an inefficient process far quicker and more easily than they can redesign the overall system. Yet Meadows' point is that it is important not to focus on making changes only at this level because while the benefits tend to be felt quite quickly and are particularly important for the company whose process is being improved, they will not on their own completely transform the sector into a more sustainable one.

15.6.2 Material stocks and flows (point 8)

Substituting one material for another and introducing innovative products can have an enormous effect on how a system operates, particularly in a manufacturing-led industry like textiles. A key way to support more sustainable materials flows in our industry is to build large, stable material stocks of, say, low impact fibres. This will help ensure that the lower impact alternatives are widely available and readily taken up. To increase the stocks of these fibres, suppliers need to be drawn into the market. To make this happen, there needs to be strong, dependable demand. Guaranteed markets for fibres like organic cotton protect farmers from price fluctuations in commodity markets; and for recycled fibres like polyester or wool, encourages research and development in the building of successful new markets for second-hand materials.

15.6.3 Regulating negative feedback loops (point 7)

Negative feedback loops help maintain systems within safe limits. They focus on keeping undesirable factors under control by looking at the output of the system and reducing it (hence the term 'negative') to keep the system in check. In the case of the textile sector these impact-reducing loops are provided by legislation or pressure from consumers and non-governmental organisations. Rafts of new European legislation – including IPPC, REACH and producer responsibility 'take back' legislation which require companies to take their products back from consumers at the end of their lives – are all negative feedback loops and act to reduce the impact of industry. Yet to continue to be effective and keep impacts in check, the strength of codes and regulations will have to increase (with ever more punitive legislation) unless we address issues to do with consumption and relationships between brands and suppliers in other ways.

15.6.4 Driving positive feedback loops (point 6)

Positive feedback loops are powerful, self-reinforcing and can drive growth with important effects on the bigger system. There is, for example, a loop of positive feedback around encouraging the growth of sustainable textiles. The more consumer interest there is in environmental and social issues, the more products will be offered by companies keen to have a share of the market. This increased choice and raised profile leads to more consumer interest in the issues and greater demand for sustainable products. And so the loop goes on, positively reinforcing this market. Another positive loop could form around consumer education, where industry leads knowledge building around real sustainability issues (rather than marketing sound bites) which then empowers consumers to act as part of an integrated solution and drive real advancements and not just the consumption of 'green' products.

15.6.5 Information flows (point 5)

Adding or changing the flows of information between companies in a supply chain or between retailers, designers and consumers can bring big change for relatively little effort – compared say to changing a company's equipment or factory set-up. The power of information in bringing more sustainable change is recognised for example by the European Commission in its IPPC recommendations, where exchanging information on the type and load of chemicals used by upstream partners in the supply chain can help create 'a chain of environmental responsibility for textiles'. Other examples include corporate social responsibility (CSR) programmes that use information strategically to bring change inside corporations.

15.6.6 The rules of the system (point 4)

The system rules define the scope and boundaries of the fashion and textile industry and set out who benefits. They can for example set out new relationships with workers, a different approach to resource use or describe a shorter supply chain. They can establish businesses in resource-efficient clusters or cascades where waste from one becomes the raw material for another.

15.6.7 The power of self-organisation (point 3)

Self-organisation is a process in which a system's internal organisation increases in effectiveness without being guided or managed by an outside source. It happens without any controlling 'brain' and instead involves co-operative working between each part of the system, adapting as needed to help with the functioning of the whole. Providing the people in the system are armed with information, knowledge and choice, then it can lead to positive and substantial change. To promote self-organisation we have to promote diversity and build the biggest stock possible of sustainability-related ideas, materials, behaviours and culture from which to seed the building of new, or more effective versions of existing, systems.

15.6.8 The goals of the system (point 2)

The goals of the system profoundly influence its dynamics, impacts and products. The textile industry's whole-system goal is to make profit through selling fibres, fabrics and garments. These profits help businesses grow and control more of the market, which then reduces their exposure to financial risk. Changing these goals to promote sustainability and to balance the making of profit with social and environmental quality would lead to big change (impacting on all of the other leverage points on this list). What would happen if the textile system goal was to decouple economic growth from consumption?

15.6.9 The mindset or paradigm out of which the system arises (point 1)

Paradigms or the accepted models of how ideas relate to one another are the sources of systems. If we influence things at the level of a paradigm, then a system can be totally transformed. Paradigms affect ideas and thoughts and are information-led. Generally speaking, we resist changes to our paradigms more than any other type of change, although as Meadows states: 'there's nothing physical or expensive or even slow about paradigm change. In a single individual it can happen in a millisecond. All it takes is a click in the mind, a new way of seeing' (Meadows, 1997).

15.7   Working at the level of rules, goals and paradigms

Meadows' list builds a rationale for deep change based both on eco-efficiency and on developing new rules, goals and paradigms in which the textile sector could operate. In identifying the places where a system can be leveraged to bring change, we begin the process of transforming it. We begin the long-overdue conversation about what the industry is for, who it serves and who benefits. We begin to scrutinise the profit motive. We start to ask questions such as: What do workers want from the textile industry? And more broadly: What do citizens want from it? We start to work with goals such as zero water use, zero energy use and zero waste. We begin to ask about the ways in which fabrics can be designed to regenerate environments as they are produced. We start to think about garments designed to build resilience in local communities. We ask about fashion and how it can be designed and produced to make us flourish as human beings. We think about which business models foster sustainability values and about the experience of doing business as a holistic, interconnected textile enterprise.

Questions like these present the biggest on-going sustainability challenge in textiles. Starting the process of exploring them provides us with an opportunity for big change. Yet contrary to common expectations, big change does not just flow from high-level international meetings of company directors, for (as Meadows suggests) single, small actions can have big effects. Development practitioner Nabeel Hamdi (2004, p. xix) reinforces this view and suggests that, 'going to scale' with many initiatives requires, paradoxically, that we start small: 'in order to do something big ... one starts with something small and one starts with where it counts'.

In order to act 'where it counts' and engage industry fully in the transition towards sustainability we need to both critique much current thinking and practice, and also envision and design credible alternatives. This chapter has included very little about the role of design in developing a more sustainable future for textiles in an explicit sense. Yet implicitly, design is the glue that binds together the arguments built here. For design is about change. This broad view of design has been described by the author John Thackera (2005, p. 1) who cites scientist Herb Simon: 'Everyone designs who devises courses of action aimed at changing existing situations into preferred ones'. It includes the work of community groups, big companies, individual consumers as well as professional designers – but what they have in common is the sort of thinking characterised by design, that is: working between or across disciplines or functions; creating things that did not exist before; and making metaphorical leaps that allow us to transcend the limits of our current way of doing things.

For the designer, environmental pioneer and geodesic dome architect Buckminster Fuller, his life's work involved making metaphorical leaps at the right place in the system. He compared himself and his practice to a 'trim tab' – a small surface connected to a boat's rudder. Fuller referred to the function of the trim tab in nautical and aeronautical design to demonstrate how small amounts of energy and resources applied at precisely the right time and place can produce maximum advantageous change. When a large ship, such as a tanker, moves through the ocean, it has great momentum. Turning the rudder to change the direction of the ship requires great effort. However by turning the trim tab – the trailing edge of the main rudder – it creates a small amount of turbulence, allowing the main rudder to turn with less effort, thus slowly pulling the whole ship around.

If we extend Fuller's metaphor to the textile sector, and attempt to 'pull our whole industry around', we need to begin to identify the textile sector's trim tab(s). Yet for this to happen we first have to begin a process of seeing and understanding the sector as a complex system intimately connected with other technical, structural, social, cultural and economic systems all situated within the context of sustainability. We have (to go back to the parable in the opening lines of this chapter) to see the elephant not as a series of separate parts but as an interconnected and dynamic whole. The good news is that when we do this, the small changes we make often become synergistic and their combined effect is transformative and far greater than the individual actions.

15.8   References

ALLWOOD, J.M., LAURSEN, S.E., MALVIDO DE RODRIGUEZ, C. and BOCKEN, N.M.P. 2006. *Well Dressed?* Cambridge: University of Cambridge Institute of Manufacturing.

EHRENFELD, J.R. 2004. Searching for Sustainability: No Quick Fix. *Reflections* **5**(8): 4.

FEI (FASHIONING AN ETHICAL INDUSTRY). Undated. *The structure of the fashion industry.* Bristol: FEI Factsheet 2. Available from: http://fashioninganethicalindustry. org/resources/factsheets/industry/factsheet2/ (accessed 3 August 2008).

FLETCHER, K. 2008. *Sustainable Fashion and Textiles: Design Journeys.* London: Earthscan.

GILMORE, G. 2007. Debt worries hit record levels as festive buyers face payback time. *The Times,* 27 December 2007. Available from http://www.timesonline.co.uk/tol/ money/borrowing/article3097640.ece (accessed 3 August 2008).

HAMDI, N. 2004. *Small Change.* London: Earthscan.

KAISER, S. 2008. Mixing metaphors in the fiber, textiles and apparel complex: moving toward a more sustainable fashion. In J. HETHORN and C. ULASEWICZ (eds), *Sustainable Fashion: Why Now?* New York: Fairchild Books.

MACFARLANE, J. 2008. Now men fall prey to anorexia as they seek a body like Beckham. *The Mail,* 2 August 2008. Available from http://www.dailymail.co.uk/ news/article-1041004/Now-men-fall-prey-anorexia-seek-body-like-Beckham. html (accessed 3 August 2008).

MCDONOUGH, W. and BRAUNGART, M. 2002. *Cradle to Cradle,* New York: North Point Press.

MEADOWS, D.H. 1997. Places to intervene in a system. *Whole Earth,* Winter, 91.

STERLING, S. 2001. *Sustainable Education: Revisioning Learning and Change.* Totnes: Green Books.

THACKERA, J. 2005. *In the Bubble: Designing in a Complex World.* Boston: MIT Press.

THOMSEN, S.R. WEBER, M.M. and BROWN, L.B. 2002. The relationship between reading beauty and fashion magazines and the use of pathogenic dieting methods among adolescent females. *Adolescence,* Spring. Available from http://findarticles.com/p/ articles/mi_m2248/is_145_37/ai_86056749?tag=artBody;col1 (accessed 3 August 2008).

Index